Differential Quadrature and Its Application in Engineering

Springer-Verlag London Ltd.

Chang Shu

Differential Quadrature and Its Application in Engineering

Springer

Chang Shu, PhD
Department of Mechanical & Production Engineering,
National University of Singapore, 10 Kent Ridge Crescent, Singapore 119260

ISBN 978-1-4471-1132-0

British Library Cataloguing in Publication Data
Shu, Chang
Differential quadrature and its application in engineering
1.Differential equations - Numerical solutions 2.Numerical integration
I.Title
515.3'5
ISBN 978-1-4471-1132-0

Library of Congress Cataloging-in-Publication Data
Shu, Chang, 1962-
Differential quadrature and its application in engineering/Chang Shu
p. cm.
ISBN 978-1-4471-1132-0 ISBN 978-1-4471-0407-0 (eBook)
DOI 10.1007/978-1-4471-0407-0
1. Differential equations--Numerical solutions. 2. Numerical integration. 3. Engineering mathematics. I. Title.
QA372.S454 2000
515'.353--dc21 99-049542

Originally published by Springer-Verlag London Berlin Heidelberg in 2000
Softcover reprint of the hardcover 1st edition 2000

Typesetting: Camera ready by author

69/3830-543210 Printed on acid-free paper SPIN 10742078

Preface

The differential quadrature (DQ) method is a numerical technique for solving differential equations. It was first developed by the late Richard Bellman and his associates in the early 1970s. The DQ method, akin to the conventional integral quadrature method, approximates the derivative of a function at any location by a linear summation of all the functional values along a mesh line. The key procedure in the DQ application lies in the determination of the weighting coefficients. Initially, Bellman and his associates proposed two methods to compute the weighting coefficients for the first order derivative. The first method is based on an ill-conditioned algebraic equation system. The second method uses a simple algebraic formulation, but the coordinates of the grid points are fixed by the roots of the shifted Legendre polynomial. In earlier applications of the DQ method, Bellman's first method was usually used because it allows the use of an arbitrary grid point distribution. However, since the algebraic equation system of this method is ill-conditioned, the number of the grid points usually used is less than 13. This drawback limits the application of the DQ method.

The DQ method and its applications were rapidly developed after the late 1980s, thanks to the innovative work in the computation of the weighting coefficients by other researchers and the author. As a result, the DQ method has emerged as a powerful numerical discretization tool in the past decade. As compared to the conventional low order finite difference and finite element methods, the DQ method can obtain very accurate numerical results using a considerably smaller number of grid points and hence requiring relatively little computational effort. So far, the DQ method has been efficiently employed in a variety of problems in engineering and physical sciences. In 1996, Bert and Malik presented a comprehensive review of the chronological development and the application of the DQ method up to that year. Currently, there is no book which systematically describes both the theoretical analysis and the application of the DQ method. It is strongly felt that the time has come for a reference book on the DQ method and its application since there are many achievements in these two aspects over the past decade. This motivated the author to write the present book.

The author first obtained his interest in the DQ method in 1988 when he happened to read a book entitled "*Methods in Approximation: Techniques for Mathematical Modelling*". This book was written by Bellman and Roth in 1986. There are many innovative ideas contained in this book. The author was particularly impressed by its Chapter 6, which introduces the DQ method. Through some sample problems, it shows that very accurate numerical results can be obtained by using merely a few grid points. Encouraged by the results in the book, the author started to study the DQ method systematically. At first, he tried to apply the DQ method to simulate the two-dimensional driven cavity flow by solving the incompressible Navier-Stokes equations. He applied Bellman's first method to compute the weighting coefficients and ambitiously used 21 grid points in each direction. He, however, failed to get any converged solution. Later, he realized that the failure is actually due to the use of a large number of grid points for an ill-conditioned equation system. He also noted that Bellman and his associates actually proposed two methods to compute the weighting coefficients. Two questions came to his mind. First, why should there be two methods to compute the weighting coefficients? Second, how can we be sure that the weighting coefficients computed using these two methods are the same? After careful consideration for about one month, he finally realized that these two questions can be easily answered by employing linear vector space analysis and polynomial approximation. Subsequently, he presented a simple algebraic formulation to compute the weighting coefficients of the first order derivative without any restriction on the choice of the grid points, and a recurrence relationship to compute the weighting coefficients of the second and higher order derivatives. Using the same analyses, the author also derived some simple algebraic formulations to compute the DQ weighting coefficients when a function is approximated by a truncated Fourier series. In the past decade, the author and his colleagues have successfully applied the DQ method with explicit computation of the weighting coefficients to the simulation of many incompressible viscous flows, free vibration analysis of beams, plates and shells, as well as the waveguide analysis. These applications include the use of the DQ method in a single domain and multi-domains, in the Cartesian coordinate system and the curvilinear coordinate system, as well as the parallel implementation. Apart from the work on the explicit computation of the weighting coefficients and the application in various areas, the author also made significant contributions in the theoretical analyses such as the error estimates, relationship between the DQ method and conventional discretization techniques, effect of the grid point distribution on the accuracy of the DQ results and the stability condition. In this book, the author wishes to share with the reader his understanding and experiences in the development and applications of the DQ method.

The outline of the book is as follows. Chapter 1 presents the fundamentals of the DQ method, that is, linear vector space analysis and function approximation. In this chapter, some properties of a linear vector space are reviewed, and polynomial approximation and Fourier series expansion are introduced. Chapter 2 describes in detail the explicit computation of the DQ weighting coefficients when a function is approximated by a high order polynomial. The error estimates, relationship between the DQ method and the spectral collocation method as well as the highest order finite difference scheme are discussed in this chapter. The differential cubature (DC) method for multi-dimensional problems is also introduced in this chapter. In addition, this chapter shows that the DQ formulations can be easily used to generate any low order finite difference scheme. The explicit computation of the DQ weighting coefficients for the first, second and third order derivatives is shown in Chapter 3 for the case in which a function is approximated by a truncated Fourier series. Chapter 4 discusses the properties of the DQ weighting coefficient matrices such as the rank, structure and effect of the grid point distribution on eigenvalues of the DQ discretization matrix. Chapter 5 shows how to efficiently solve the DQ resultant ordinary differential equations and the DQ algebraic equations as well as the DQ eigenvalue problems. The application of the DQ method to solve incompressible Navier-Stokes equations is given in Chapter 6. Both the version of the vorticity-stream function formulation and the version of the primitive variable form are solved using different approaches. Chapter 7 shows the application of the DQ method in vibration analysis of beams, plates and shells. The focus in this chapter is on the implementation of the boundary conditions. The miscellaneous applications of the DQ method in other areas are described in Chapter 8. Chapter 9 discusses the application of the DQ method for complex problems. The multi-domain DQ method and coordinate transformation technique are introduced in this chapter. This chapter also gives the complete governing equations in the curvilinear coordinate system for flow simulation and vibration analysis of irregular plates. Chapter 10 presents the generalized integral quadrature (GIQ) method and its application to solve boundary layer equations. The GIQ method was developed by the author, which is also based on the analyses of linear vector space and polynomial approximation. It is suitable for the solution of integral and differential-integral equations. The author is well aware of the difficulties which are often encountered in implementing computer programs from their listings. Hence, three Fortran programs are attached in the appendix. Appendix A shows a complete Fortran program for the simulation of natural convection in a square cavity by the DQ method. Appendix B lists a complete Fortran program for vibration analysis of a rectangular plate. A complete Fortran program for analysis of a *L*-shaped waveguide is given in Appendix C. This program is based on the multi-domain DQ method.

This book is written primarily for scientists and engineers who are interested in applying the DQ method to real problems. It is hoped that the book is suitable for graduate-level study, conveying to students that the DQ method is a robust technique for solving practical problems.

The author would like to devote this book to his wife, Wang Hang, for her infinite patience and encouragement, and also to his son, Xin Zhe, to ease Daddy's guilt from spending too little play and homework time. Thanks are also due to the many graduate students who took the time reading and checking the draft of the manuscript and proposed significant improvements. Special thanks go to Mr Zhu Yongda, who carefully checked the manuscript several times and helped to draft Chapter 8. I would also like to thank Mr Teo Chiang Juay for his suggestions in improving the quality of the book.

C. Shu

Singapore
October, 1999

Acknowledgements

The author would like to sincerely acknowledge the use of figures and tables reproduced from the following sources.

Bert CW, Malik M, Differential quadrature method in computational mechanics, Applied Mechanics Review, Vol 49(1), pp 1-28, Copyright (1996), with permission from American Society of Mechanical Engineers.

Civan F, Rapid and accurate solution of reactor models by the quadrature method, Computers in Chemical Engineering, Vol 18(10), pp 1005-1009, Copyright (1994), with permission from Elsevier Science.

Malik M, Bert CW, Differential quadrature solutions for steady-state incompressible and compressible lubrication problems, Journal of Tribology – Transactions of the ASME, Vol 116, pp 296-302, Copyright (1994), with permission from American Society of Mechanical Engineers.

Malik M, Civan F, A comparative-study of differential quadrature and cubature methods vis-à-vis some conventional techniques in context of convection-diffusion-reaction problems, Chemical Engineering Science, Vol 50 (3), pp 531-547, Copyright (1995), with permission from Elsevier Science.

Shu C, Free vibration analysis of composite laminated conical shells, Journal of Sound and Vibration, Vol 194(4), pp 587-604, Copyright (1996), with permission from Academic Press.

Shu C, Application of differential quadrature method to simulate natural convection in a concentric annulus, International Journal for Numerical Methods in Fluids, Vol 30, pp 977-993, Copyright (1999), with permission from John Wiley & Sons, Ltd.

Shu C, Chew YT, Fourier expansion-based differential quadrature and its application to Helmholtz eigenvalue problems, Communications in Numerical Methods in Engineering, Vol 13, pp 643-653, Copyright (1997), with permission from John Wiley & Sons, Ltd.

Shu C, Chew YT, Khoo BC, Yeo KS, Application of GDQ scheme to simulate incompressible viscous flows around complex geometries, Mechanics Research Communications, Vol 22(4), pp 319-325, Copyright (1995), with permission from Elsevier Science.

Shu C, Chew YT, Khoo BC, Yeo KS, Solutions of three-dimensional boundary layer equations by global methods of generalized differential-integral quadrature, International Journal of Numerical Methods for Heat & Fluid Flow, Vol 6(2), pp 61-75, Copyright (1996), with permission from MCB University Press Limited.

Shu C, Chew YT, Richards BE, Generalized differential and integral quadrature and their application to solve boundary-layer equations, International Journal for Numerical Methods in Fluids, Vol 21, pp 723-733, Copyright (1995), with permission from John Wiley & Sons, Ltd.

Shu C, Du H, Implementation of clamped and simply supported boundary conditions in the GDQ free vibration analysis of beams and plates, International Journal of Solids and Structures, Vol 34(7), pp 819-835, Copyright (1997), with permission from Elsevier Science.

Shu C, Du H, A generalized approach for implementing general boundary conditions in the GDQ free vibration analysis of plates, International Journal of Solids and Structures, Vol 34(7), pp 837-846, Copyright (1997), with permission from Elsevier Science.

Shu C, Khoo BC, Chew YT, Yeo KS, Solutions of three-dimensional boundary layer equations by global methods of generalized differential-integral quadrature, Computer Methods in Applied Mechanics and Engineering, Vol 135(3-4), pp 229-241, Copyright (1996), with permission from Elsevier Science.

Shu C, Richards BE, Application of generalized differential quadrature to solve 2-dimensional incompressible viscous Navier-Stokes equations, International Journal for Numerical Methods in Fluids, Vol 15(7), pp 791-798, Copyright (1992), with permission from John Wiley & Sons, Ltd.

Shu C, Xue H, Solution of Helmholtz equation by differential quadrature method, Computer Methods in Applied Mechanics and Engineering, Vol 175(1-2), pp 203-212, Copyright (1999), with permission from Elsevier Science.

Striz AG, Chen WL, Bert CW, Free vibration of plates by the high accuracy quadrature element method, Journal of Sound and Vibration, Vol 202(5), pp 689-702, Copyright (1997), with permission from Academic Press.

Table of Contents

1 Mathematical Fundamentals of Differential Quadrature Method: Linear Vector Space Analysis and Function Approximation 1

1.1 Introduction 1

1.2 Derivative Approximation by Differential Quadrature (DQ) Method 3

1.2.1 Integral Quadrature 4

1.2.2 Differential Quadrature 5

1.3 Analysis of A Linear Vector Space 6

1.3.1 Definition of A Linear Vector Space 6

1.3.2 Properties of A Linear Vector Space 8

1.4 Solution of Partial Differential Equations (PDEs) and Function Approximation 11

1.4.1 Two Basic Types of Solution for PDEs 11

1.4.2 High Order Polynomial Approximation 13

1.4.3 Fourier Series Expansion 18

1.4.3.1 General Function 18

1.4.3.2 Even Function 21

1.4.3.3 Odd Function 23

2 Polynomial-based Differential Quadrature (PDQ) 25

2.1 Introduction 25

2.2 Computation of Weighting Coefficients for the First Order Derivative 26

2.2.1 Bellman's Approaches 26

2.2.2 Quan and Chang's Approach 28

2.2.3 Shu's General Approach 29

2.3 Computation of Weighting Coefficients for the Second and Higher Order Derivatives 32

2.3.1 Weighting Coefficients of the Second Order Derivative 32

2.3.2 Shu's Recurrence Formulation for Higher Order Derivatives 34

2.3.3 Matrix Multiplication Approach 36

2.4 Error Analysis 38

2.4.1 The Function Approximation 38

2.4.2 The Derivative Approximation 40

2.5 Relationship Between PDQ and Other Approaches 44
2.5.1 Relationship Between PDQ and Finite Difference Scheme 44
2.5.1.1 Generation of Finite Difference Scheme 44
2.5.1.2 Relationship Between PDQ and Highest Order Finite Difference Scheme 48
2.5.2 Relationship Between PDQ and Chebyshev Collocation Method 52
2.6 Extension to the Multi-dimensional Case 55
2.6.1 Direct Extension for Regular Domain 55
2.6.2 Differential Cubature Method 60
2.7 Specific Results for Typical Grid Point Distributions 62
2.7.1 Uniform Grid 62
2.7.2 Chebyshev-Gauss-Lobatto Grid 63
2.7.3 Coordinates of Grid Points Chosen as the Roots of Chebyshev Polynomial 64
2.8 Generation of Low Order Finite Difference Schemes by PDQ 65

3 Fourier Expansion-based Differential Quadrature (FDQ) 69
3.1 Introduction 69
3.2 Cosine Expansion-based Differential Quadrature (CDQ) for Even Functions 70
3.3 Sine Expansion-based Differential Quadrature (SDQ) for Odd Functions 81
3.4 Fourier Expansion-based Differential Quadrature (FDQ) for Any General Function 86
3.5 Some Properties of Fourier Expansion-based Differential Quadrature 91

4 Some Properties of DQ Weighting Coefficient Matrices 95
4.1 Introduction 95
4.2 Determinant and Rank of DQ Weighting Coefficient Matrices 96
4.2.1 Definition and Properties of Determinant and Rank 96
4.2.2 Determinant and Rank of DQ Weighting Coefficient Matrices 98
4.3 Structures and Properties of DQ Weighting Coefficient Matrices 100
4.3.1 Definition of Centrosymmetric and Skew Centrosymmetric Matrices 101
4.3.2 Properties of Centrosymmetric and Skew Centrosymmetric Matrices 102
4.3.2.1 Properties of Centrosymmetric Matrices 102

4.3.2.2 Properties of Skew Centrosymmetric Matrices 105
4.3.3 Structures of DQ Weighting Coefficient Matrices 107
4.3.3.1 Structure of First Order DQ Weighting Coefficient Matrix 107
4.3.3.2 Structures of Higher Order DQ Weighting Coefficient Matrices 109
4.4 Effect of Grid Point Distribution on Eigenvalues of DQ Discretization Matrices 110
4.4.1 Stability of Ordinary Differential Equations 111
4.4.2 Eigenvalues of Some Specific DQ Discretization Matrices 112
4.4.2.1 The Convection Operator 112
4.4.2.2 The Diffusion Operator 117
4.4.2.3 The Convection-Diffusion Operator 119
4.5 Effect of Grid Point Distribution on Magnitude of DQ Weighting Coefficients 120

5 Solution Techniques for DQ Resultant Equations 123
5.1 Introduction 123
5.2 Solution Techniques for DQ Ordinary Differential Equations 124
5.3 Solution Techniques for DQ Algebraic Equations 128
5.3.1 Direct Methods 130
5.3.2 Iterative Methods 134
5.3.2.1 Iterative Methods for Conventional System 134
5.3.2.2 Iterative Methods for Lyapunov System 137
5.4 Implementation of Boundary Conditions 140
5.5 Sample Applications of DQ Method 143
5.5.1 Burgers Equation 143
5.5.2 Two-dimensional Poisson Equation 145
5.5.3 Helmholtz Eigenvalue Problem 148

6 Application of Differential Quadrature Method to Solve Incompressible Navier-Stokes Equations 153
6.1 Introduction 153
6.2 Governing Equations 154
6.2.1 Dimensional Form 154
6.2.2 Non-dimensional Form 157
6.2.3 Vorticity-Stream Function Formulation 159
6.3 Solution of Vorticity-Stream Function Formulation 160
6.3.1 Discretization of Governing Equations 160
6.3.2 Implementation of Boundary Conditions 161

6.3.2.1 Implementation of Boundary Condition for Vorticity 162
6.3.2.2 Implementation of Boundary Condition for Stream Function 162
6.3.2.3 Implementation of Boundary Condition for Temperature 167
6.3.3 Solution Procedures 168
6.3.4 Some Numerical Examples 170
6.3.4.1 The Flow Past A Circular Cylinder 170
6.3.4.2 The Natural Convection in A Concentric Annulus 172
6.4 Solution of Incompressible Navier-Stokes Equations in Primitive Variable Form 175
6.4.1 Introduction 175
6.4.2 Pressure Correction Method 176
6.4.3 Two Approaches to Specify Boundary Condition for p' and to Enforce Continuity Condition on the Boundary 178
6.4.3.1 Approach 1 178
6.4.3.2 Approach 2 180
6.4.4 Computational Sequence 181
6.4.5 Sample Application and Comments on the Two Approaches 182
6.4.5.1 Importance of Enforcing Continuity Condition on the Boundary 182
6.4.5.2 Comments on Performance of Two Approaches 184

7 Application of Differential Quadrature Method to Structural and Vibration Analysis 186
7.1 Introduction 186
7.2 Differential Quadrature Analysis of Beams 188
7.2.1 Governing Equations and Boundary Conditions 188
7.2.2 Numerical Discretization 189
7.2.3 Implementation of Boundary Conditions 190
7.2.3.1 The δ-technique 190
7.2.3.2 Modification of Weighting Coefficient Matrices 191
7.2.3.3 Direct Substitution of Boundary Conditions into Discrete Governing Equations 194
7.2.4 Numerical Example: Free Vibration Analysis of A Uniform Beam 196
7.3 Differential Quadrature Analysis of Thin Plates 197
7.3.1 Governing Equations and Boundary Conditions 197
7.3.2 Numerical Discretization 199

7.3.3 Implementation of Boundary Conditions 200
7.3.3.1 The δ-technique 200
7.3.3.2 Modification of Weighting Coefficient Matrices 201
7.3.3.3 Direct Substitution of Boundary Conditions into Discrete Governing Equations 202
7.3.3.4 General Approach 205
7.3.4 Numerical Example: Free Vibration Analysis of Square Plates 207
7.4 Differential Quadrature Analysis of Shells 209
7.4.1 Governing Equations and Boundary Conditions 209
7.4.2 Numerical Discretization 218
7.4.3 Implementation of Boundary Conditions 219
7.4.4 Numerical Example: Free Vibration Analysis of A Composite Laminated Conical Shell 222

8 Miscellaneous Applications of Differential Quadrature Method 224
8.1 Introduction 224
8.2 Application to Heat Transfer 224
8.3 Application to Chemical Reactor 228
8.4 Application to Lubrication Problems 232
8.5 Application to Waveguide Analysis 235
8.6 Solution of the Helmholtz Equation 239
8.7 Effect of Mesh Point Distribution on Accuracy of DQ Results 242

9 Application of Differential Quadrature to Complex Problems 245
9.1 Introduction 245
9.2 Multi-domain DQ Method 245
9.2.1 Topology of Interface 246
9.2.1.1 Patched Interface 246
9.2.1.2 Overlapped Interface 248
9.2.2 Multi-domain DQ Application in Fluid Mechanics 249
9.2.3 Multi-domain DQ Application in Solid Mechanics 251
9.2.4 Multi-domain DQ Application in Waveguide Analysis 252
9.3 DQ Application in Curvilinear Coordinate System 254
9.3.1 Coordinate Transformation 254
9.3.2 DQ Simulation of Incompressible Flows in Irregular Domains 256
9.3.3 DQ Vibration Analysis of Irregular Plates 260
9.3.3.1 Partial Transformation 260

9.3.3.2 Complete Transformation ... 261
9.3.3.3 Implementation of Boundary Conditions ... 262
9.3.3.4 Sample Application ... 264
9.4 Differential Cubature Method for Complex Problems ... 266

10 Generalized Integral Quadrature (GIQ) And Its Application to Solve Boundary Layer Equations ... 267
10.1 Introduction ... 267
10.2 Generalized Integral Quadrature (GIQ) ... 268
10.2.1 One-dimensional Generalized Integral Quadrature ... 268
10.2.2 Error Analysis ... 271
10.2.3 Extension to Multi-dimensional Cases ... 272
10.2.4 Sample Application of GIQ Method ... 273
10.3 DQ-GIQ Algorithm to Solve Boundary Layer Equations ... 275
10.3.1 Stream Function as Dependent Variable ... 275
10.3.2 Primitive Variables as Dependent Variables ... 277
10.4 Steady Boundary Layer Solutions ... 279
10.4.1 One-dimensional Case ... 279
10.4.2 Two-dimensional Case ... 280
10.4.3 Three-dimensional Case ... 281
10.5 Unsteady Boundary Layer Solutions ... 285

Appendices
A. A Fortran Program for Simulation of Natural Convection in A Square Cavity ... 288
B. A Fortran Program for Vibration Analysis of Rectangular Plates ... 305
C. A Fortran Program for *L*-Shaped Waveguide Analysis by Multi-domain DQ Method ... 315

References ... 324

Index ... 336

Chapter 1

Mathematical Fundamentals of Differential Quadrature Method: Linear Vector Space Analysis and Function Approximation

1.1 Introduction

Most engineering problems are governed by a set of partial differential equations (PDEs) with proper boundary conditions. For example, Newtonian fluid flows are modeled by the Navier-Stokes equations; the vibration of thin plates is governed by a fourth order partial differential equation; whereas acoustic waves and microwaves can be simulated by the Helmholtz equation. In general, it is very difficult for us to obtain the closed-form solution of these equations. On the other hand, the solution of these PDEs is always demanded due to practical interests. For example, when we design an aircraft, we need to know the curve of c_l (lift coefficient) versus c_d (drag coefficient) for a given airfoil shape. The values of c_l and c_d can be obtained from the solution of Navier-Stokes equations. Therefore, it is important for us to develop some approximate solutions to the given PDEs.

In most cases, the approximate solution is represented by functional values at certain discrete points (grid points or mesh points). At this stage, one may question the relationship between the derivatives in the partial differential equation and the functional values at the grid points. It seems that there exists a bridge which links them. The numerical discretization technique is such a bridge, and the corresponding approximate solution is termed the numerical solution.

Currently, there are many available numerical discretization techniques. Among them, the finite difference (FD), finite element (FE), and finite volume (FV) methods fall under the category of low order methods, whereas spectral and pseudospectral methods are considered global methods. The FD method is based on the Taylor series expansion or the polynomial approximation while the FE method is based on the variational principle or the principle of weighted residuals. The FV method applies the physical conservation law directly to a finite cell. The spectral method may be viewed as an extreme development of the class of discretization schemes known as the methods of weighted residuals. The key elements of spectral methods are the base functions and the weighting functions. There is a close relationship between FE and spectral methods in the sense that both methods use a set of base functions to approximate the solution. The choice of the base functions is one of the features that distinguishes the spectral method from the FE method. The base functions for spectral methods are infinitely differential functions with global properties. In the case of FE methods, the domain is subdivided into small elements, and a base function is specified in each element. The base functions are thus local in character, and well suited for handling complex geometries. Spectral methods may be considered an extension of FE methods, and can be viewed as a whole space approximation technique.

Most numerical simulations of engineering problems can be carried out by the low order FD, FE, and FV methods using a large number of grid points. In some practical applications, however, the numerical solutions of PDEs are required at only a few specified points in the physical domain. To achieve an acceptable degree of accuracy, low order methods still require the use of a large number of grid points to obtain accurate solutions at these specified points. An example can be found in vibration analysis. Upon numerical discretization of the governing PDEs, the eigenvalues of the resultant algebraic equation system provide the vibrational frequencies of the problem. Usually, the number of interior grid points is equal to the dimension of the resultant algebraic equation system, thus giving the same number of eigenfrequencies. Among all the computed eigenfrequencies, only the low frequencies are of practical interest. However, since the computed eigenfrequencies have the same order of accuracy, a large number of grid points are still necessary in order to obtain such low frequencies accurately. As a result, a lot of virtual storage and computational effort are required. It seems that the above drawbacks of low order methods can be improved by using high order and global methods. In general, high order methods have a high order of truncation error. Thus, to achieve the same order of accuracy, the mesh spacing used by high order methods can be much larger than that used by low order methods. As a consequence, high order methods are capable of yielding accurate numerical solutions using very few grid points. The spectral method is the natural choice for this purpose. At present, the spectral method is highly successful in several areas: turbulence modeling, weather prediction, nonlinear waves, seismic modeling, etc. Application of the spectral method demands substantial mathematical knowledge of the theory.

On the other hand, in seeking an efficient discretization technique to obtain accurate numerical solutions using a considerably small number of grid points, Bellman et al. (1971, 1972) introduced the method of differential quadrature (DQ), where a partial derivative of a function with respect to a coordinate direction is expressed as a linear weighted sum of all the functional values at all mesh points along that direction. The DQ method was initiated from the idea of integral quadrature. The key to DQ is to determine the weighting coefficients for the discretization of a derivative of any order. Bellman et al. (1972) suggested two methods to determine the weighting coefficients of the first order derivative. The first method solves an algebraic equation system. The second uses a simple algebraic formulation, but with the coordinates of grid points chosen as the roots of the shifted Legendre polynomial. Most early applications of DQ in engineering (Bellman et al. 1971, 1972, 1974, 1975*a*, *b*, 1986, Kashef and Bellman 1974, Hu and Hu 1974, Mingle 1977, Wang 1982, Civan and Sliepcevich 1983*a, b*, 1984*a, b*, Naadimuthu et al. 1984, Bert et al. 1988, 1989, Jang et al. 1989) used Bellman's approaches to compute the weighting coefficients. Among Bellman's two methods, the first method is usually applied because it allows the coordinates of the grid points to be chosen arbitrarily. Unfortunately, when the order of the algebraic equation system is large, its matrix is ill-conditioned. Thus, it is very difficult to obtain the weighting coefficients for a large number of grid points using this method. This is probably the reason why early applications of this scheme only used the number of grid points less than or equal to 13. To overcome the drawbacks of Bellman's approaches to compute the weighting coefficients, some effort has been made. Civan (1989) realized that the difficulty in computing the weighting coefficients for a large number of grid points by Bellman's first approach is actually due to the ill-conditioned character of the Vandermonde matrix. The Vandermonde matrix commonly appears in many engineering problems. There are some special algorithms to solve the Vandermonde system of equations. One such efficient algorithm was presented by Björck and Pereyra (1970). Using the BP algorithm, he successfully computed the weighting coefficients accurately by using up to 31 grid points. To further improve the computation of weighting coefficients, Quan and Chang (1989*a, b*) applied Lagrange interpolated polynomials as test functions, and obtained explicit formulations to calculate the weighting coefficients for the discretization of the first and second order derivatives. A major breakthrough in computing the weighting coefficients was made by Shu and Richards (1990), and Shu (1991), in which all the current methods for determination of the weighting coefficients are generalized under the analysis of a high order polynomial approximation and the analysis of a linear vector space. In Shu's approach, the weighting coefficients of the first order derivative are determined by a simple algebraic formulation without any restriction on the choice of grid points, whereas the weighting coefficients of the second and higher order derivatives are determined by a recurrence relationship. Clearly, all the above work is based on the polynomial approximation, and accordingly, the related DQ method can be considered as the polynomial-based differential quadrature (PDQ) method. Recently, Shu and Chew (1997), and Shu and Xue (1997) have further developed some simple algebraic formulations to compute the weighting coefficients of the first and second order derivatives in the DQ approach when the function or the solution of a PDE is approximated by a Fourier series expansion. These formulations are different from those in PDQ, and the approach can be termed the Fourier expansion-based differential quadrature (FDQ) method. In the development of FDQ, a linear vector space analysis is also employed. From these processes, it is evident that the mathematical fundamentals of PDQ and FDQ lie in the analysis of a linear vector space and the analysis of a function approximation. These fundamentals will be discussed in the following sections.

1.2 Derivative Approximation by Differential Quadrature (DQ) Method

The differential quadrature (DQ) method was presented by R. E. Bellman and his associates in the early 1970's. The DQ method is a numerical discretization technique for the approximation of derivatives. The DQ method was initiated from the idea of conventional integral quadrature.

1.2.1 Integral Quadrature

One problem which frequently arises in science and engineering is the evaluation of $\int_a^b f(x)dx$ over a finite interval $[a, b]$. If a function F exists such that $dF/dx = f$, then the value of this integral is $F(b) - F(a)$. Unfortunately, in practical problems, it is extremely difficult, if not impossible, to obtain an explicit expression for F. Indeed, the values of f may only be known at a discrete set of points and in this situation, a numerical approach is essential.

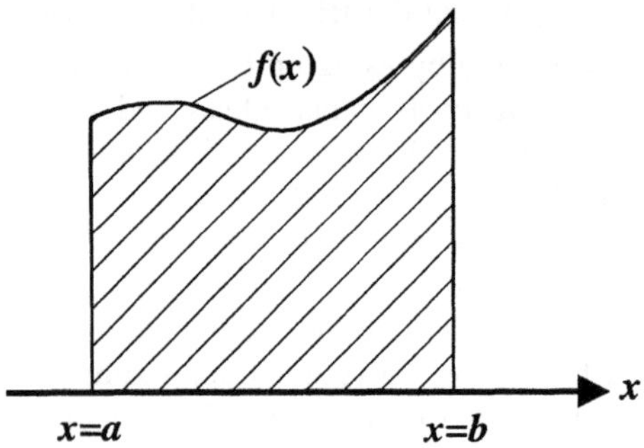

Figure 1.1 Integral of $f(x)$ over an interval

On the other hand, it was found that the integral $\int_a^b f(x)dx$ represents the area under the curve $f(x)$ as shown in Figure1.1. Thus, evaluating the integral is equivalent to the approximation of the area. Using this principle, many numerical techniques were developed. In general, the integral $\int_a^b f(x)dx$ can be approximated by

$$\int_a^b f(x)dx = w_1 f_1 + w_2 f_2 + \ldots + w_n f_n = \sum_{k=1}^{n} w_k f_k \tag{1.1}$$

where $w_1, w_2, \ldots, w_n$ are the weighting coefficients, $f_1, f_2, \ldots, f_n$ are the functional values at the discrete points $a = x_1, x_2, \ldots, x_n = b$. Equation 1.1 is called the integral quadrature, which uses all the functional values in the whole integral domain to approximate an integral over a finite interval.

In general, the discrete points are selected so as to give a uniform distribution, i.e., $x_i = x_{i-1} + h$, i = 2, 3, ..., n, where h is called the step size. All the conventional quadrature rules can be written in the form of Equation 1.1. The following lists two special cases of Equation 1.1:

(1) Trapezoidal rule: for this case, $w_1, w_2, ..., w_n$ are taken as

$$w_1 = \frac{h}{2}, w_2 = w_3 = ... = w_{n-1} = h, w_n = \frac{h}{2}$$

(2) Simpson's one-third rule: for this case, $w_1, w_2, ..., w_n$ are taken as

$$w_k = \begin{cases} \frac{h}{3}, & \text{when } k = 1, n \\ \frac{4h}{3}, & \text{when } k = 2, 4, 6, ..., n-2 \\ \frac{2h}{3}, & \text{when } k = 3, 5, 7, ..., n-1 \end{cases}$$

1.2.2 Differential Quadrature

As shown in Figure 1.2, we consider a one-dimensional problem. It is assumed that a function $f(x)$ is sufficiently smooth over the whole domain.

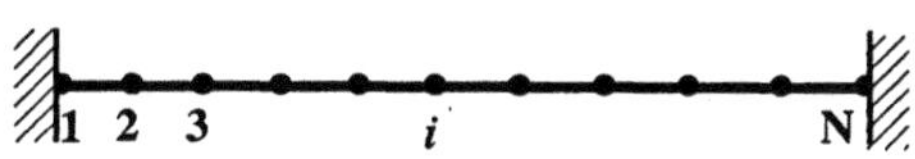

Figure 1.2 A one-dimensional problem

Following the idea of integral quadrature (Equation 1.1), Bellman et al. (1972) suggested that the first order derivative of the function $f(x)$ with respect to x at a grid point x_i, is approximated by a linear sum of all the functional values in the whole domain, that is,

$$f_x(x_i) = \left. \frac{df}{dx} \right|_{x_i} = \sum_{j=1}^{N} a_{ij} \cdot f(x_j), \quad \text{for } i = 1, 2, \ldots, N \tag{1.2}$$

where a_{ij} represent the weighting coefficients, and N is the number of grid points in the whole domain. Equation 1.2 is called differential quadrature (DQ). It should be noted that the weighting coefficients a_{ij} are different at different locations of x_i. The key procedure in DQ approximation is to determine the weighting coefficients a_{ij}. In Chapters 2 and 3, we will show the determination of the weighting coefficients in detail from the analysis of a linear vector space and the analysis of function approximation.

1.3 Analysis of A Linear Vector Space

In this section, we will review a number of concepts and properties in a linear vector space, which will be used in the following chapters to compute the weighting coefficients in the DQ approximation.

1.3.1 Definition of A Linear Vector Space

A linear vector space is defined over a field. The formal definition of a field is given as follows.

Definition of a field: A field, denoted by F, consists of a set of elements called scalars and two operations called addition "+" and multiplication "×" (or "·"). The two operations are defined to satisfy the following conditions:

(1) To every pair of elements a and b in F, there is a corresponding element $a + b$ called the sum of a and b, and an element $a{\cdot}b$ or ab in F, called the product of a and b.

(2) Addition and multiplication are respectively commutative: for any a and b in F,

$$a+b=b+a\ ,\ a\cdot b=b\cdot a$$

(3) Addition and multiplication are respectively associative: for any a, b and c in F,

$$(a+b)+c=a+(b+c)\ ,\ (a\cdot b)\cdot c=a\cdot(b\cdot c)$$

(4) Multiplication is distributive with respect to addition: for any a, b and c in F,

$$a\cdot(b+c)=(a\cdot b)+(a\cdot c)$$

(5) F contains two elements, denoted by 0 and 1 respectively, such that $a + 0 = a$, $1{\cdot}a = a$ for every a in F.

(6) To every a in F, there is an element b in F such that $a+b=0$.

(7) To every a in F which is not the element 0, there is an element c in F such that $a\cdot c=1$.

From the above features, it can be seen that the set of objects that form a field could be anything as long as the two operations can be defined for these objects. The fields we will encounter in this book are the most familiar ones: the field of real numbers and the field of rational functions with real coefficients. The addition and multiplication of these fields are defined in the usual ways.

Before we introduce the concept of vector spaces, let us consider a two-dimensional geometric plane. When the origin is chosen, every point in the plane can be represented by a vector. The vector has direction as well as magnitude, and a vector can be extended. Any two vectors can be added, but the product of two vectors is not defined. Such a plane, using mathematical terminology, is called a linear space, or a vector space, or a linear vector space. The formal definition of a linear vector space is given as follows.

Definition of a linear vector space: A linear vector space, denoted by V, consists of a set of elements called vectors, a field F, and two operations called vector addition and scalar multiplication. The two operations are defined such that the following conditions are satisfied:

(1) To every pair of vectors α and β in V, there is a corresponding vector $\alpha + \beta$ called the sum of α and β.

(2) Addition is commutative: For any α and β in V, $\alpha + \beta = \beta + \alpha$.

(3) Addition is associative: For any α, β and γ in V, $(\alpha + \beta) + \gamma = \alpha + (\beta + \gamma)$.

(4) V contains a vector, denoted by 0, such that $0 + \alpha = \alpha$, for every α in V. The vector 0 is called the zero vector or the origin.

(5) To every α in V, there is a vector β in V such that $\alpha + \beta = 0$.

(6) To every c in F, and every α in V, there is a corresponding vector $c\alpha$ in V called the scalar product of c and α.

(7) Scalar multiplication is associative: For any a, b in F and any γ in V, $a(b\gamma) = (ab)\gamma$.

(8) Scalar multiplication is distributive with respect to vector addition: For any a in F, and any β, γ in V, $a(\beta + \gamma) = a\beta + a\gamma$.

(9) Scalar multiplication is distributive with respect to scalar addition: For any a, b in F, and any γ in V, $(a + b)\gamma = a\gamma + b\gamma$.

(10) For any α in V, $1\alpha = \alpha$ where 1 is the element 1 in F.

From the above features, it is known that a field forms a linear vector space over itself with vector addition and scalar multiplication.

Now, let us consider the set $P_n(x)$ of all polynomials of degree less than n with real coefficients c_i

$$P_n(x) = \sum_{i=0}^{n-1} c_i x^i \tag{1.3}$$

Let the vector addition and scalar multiplication be defined as

$$\sum_{i=0}^{n-1} c_i x^i + \sum_{i=0}^{n-1} d_i x^i = \sum_{i=0}^{n-1} (c_i + d_i) x^i \tag{1.4a}$$

$$a\left(\sum_{i=0}^{n-1} c_i x^i\right) = \sum_{i=1}^{n-1} (ac_i) x^i \tag{1.4b}$$

It is easy to verify that $P_n(x)$ constitutes a linear vector space.

1.3.2 Properties of A Linear Vector Space

Some properties of a linear vector space will be shown in this section. These properties will be used in the following chapters to develop PDQ and FDQ formulations.

Linear independence: A set of vectors, $\alpha_1, \alpha_2, ..., \alpha_n$, in a linear vector space V over a field F, is said to be linearly independent if and only if the equation

$$c_1\alpha_1 + c_2\alpha_2 + ... + c_n\alpha_n = 0 \tag{1.5}$$

implies $c_1 = c_2 = ... = c_n = 0$, where $c_1, c_2, ..., c_n$ are the elements in a field F.

From the above definition, it can be seen that the linear independence depends not only on the set of vectors but also on the field. It is also clear from the definition of linear independence that if the vectors $\alpha_1, \alpha_2, ..., \alpha_n$ are linearly dependent, then at least one of them can be written as a linear combination of the others.

Dimension of a linear vector space: The maximum number of linearly independent vectors in a linear vector space V is called the dimension of the linear vector space.

Basis or base vectors: A set of linearly independent vectors in a linear vector space V is said to be a basis of V if every vector in V can be expressed as a unique linear combination of these vectors.

When the dimension of the linear vector space is n, the above property can be stated as: in an n-dimensional linear vector space V_n, any set of linearly independent vectors qualifies as a basis.

It is easy to show this property. Let $\alpha_1, \alpha_2, ..., \alpha_n$ be any n linearly independent vectors in V_n, and let α be an arbitrary vector in V_n. Then the set of $(n+1)$ vectors $\alpha, \alpha_1, \alpha_2, ..., \alpha_n$ is linearly dependent since, from the definition of dimension for a vector space, n is the maximum number of linearly independent vectors we can have in the space. Consequently, there exist $c_0, c_1, c_2, ..., c_n$ in F, not all zero, such that

$$c_0\alpha + c_1\alpha_1 + c_2\alpha_2 + ... + c_n\alpha_n = 0 \tag{1.6}$$

Note that $c_0 \neq 0$. If $c_0 = 0$, Equation 1.6 is reduced to Equation 1.5, which, together with the linear independence assumption of $\alpha_1, \alpha_2, ..., \alpha_n$, implies that $c_1 = c_2 = ... = c_n = 0$. This contradicts the assumption that not all $c_0, c_1, c_2, ..., c_n$ are zero. Therefore, $c_0 \neq 0$. By defining $d_i = -c_i / c_0$, for i = 1, 2, ..., n, then Equation 1.6 becomes

$$\alpha = d_1\alpha_1 + d_2\alpha_2 + ... + d_n\alpha_n \tag{1.7}$$

This shows that every vector α in V_n can be expressed as a linear combination of $\alpha_1, \alpha_2, ..., \alpha_n$. Now we show that this combination is unique. Suppose that there exists another linear combination, say

$$\alpha = \bar{d}_1\alpha_1 + \bar{d}_2\alpha_2 + ... + \bar{d}_n\alpha_n \tag{1.8}$$

Then by subtracting Equation 1.8 from Equation 1.7, we obtain

$$0 = (d_1 - \bar{d}_1)\alpha_1 + (d_2 - \bar{d}_2)\alpha_2 + ... + (d_n - \bar{d}_n)\alpha_n \tag{1.9}$$

Which, together with the linear independence of $\alpha_1, \alpha_2, ..., \alpha_n$, implies that

$$d_i = \bar{d}_i, \quad i = 1, 2, ..., n$$

This property has a very important implication. In an n-dimensional linear vector space V_n, if a basis is chosen, then every vector in V_n can be uniquely represented by a set of scalars $d_1, d_2, ..., d_n$ in F.

Change of basis: In an n-dimensional linear vector space V_n, there exist many sets of basis, and each set of basis (base vectors) can be uniquely expressed by another set of basis (base vectors). This property is obvious, and can be easily derived from the property of basis.

Example: Consider two sets of base vectors $\alpha_1, \alpha_2, ..., \alpha_n$ and $\beta_1, \beta_2, ..., \beta_n$ in V_n. This property indicates that

$$\alpha_i = \sum_{j=1}^{n} c_{i,j} \cdot \beta_j, \quad i = 1, 2, ..., n \tag{1.10}$$

$$\beta_i = \sum_{j=1}^{n} d_{i,j} \cdot \alpha_j, \quad i = 1, 2, ..., n \tag{1.11}$$

where $c_{i,j}$ and $d_{i,j}$ are the elements in a field F.

Linear operator: A function is said to be a linear operator if and only if

$$L(c_1\alpha_1 + c_2\alpha_2) = c_1 L(\alpha_1) + c_2 L(\alpha_2) \tag{1.12}$$

for any vectors α_1, α_2 in V_n and any scalars c_1, c_2 in F.

It is easy to show that Equation 1.2 is a linear operator since

$$\left.\frac{d(c_1 f_1 + c_2 f_2)}{dx}\right|_{x=x_i} = c_1 \sum_{j=1}^{N} a_{ij} \cdot f_1(x_j) + c_2 \sum_{j=1}^{N} a_{ij} \cdot f_2(x_j) \tag{1.13}$$

where $f_1(x), f_2(x)$ are two functions.

Linear operator and a set of base vectors in a linear vector space: In an n-dimensional linear vector space V_n, $\alpha_1, \alpha_2, ..., \alpha_n$ is a set of base vectors. If all the base vectors $\alpha_1, \alpha_2, ..., \alpha_n$ satisfy a linear operator, so does any vector in the space V_n. Let $\alpha_1, \alpha_2, ..., \alpha_n$ be a set of base vectors in V_n, which satisfies a linear equation

$$L(\alpha_i) = 0, \quad i = 1, 2, ..., n \tag{1.14}$$

and α be an arbitrary vector in V_n. Using Equation 1.7 and Equation 1.12, we have

$$L(\alpha) = L\left(\sum_{i=1}^{n} d_i \alpha_i\right) = \sum_{i=1}^{n} L(d_i \alpha_i) = \sum_{i=1}^{n} d_i L(\alpha_i) = 0 \tag{1.15}$$

Linear operator and different sets of base vectors in a linear vector space: In an n-dimensional linear vector space, if one set of base vectors satisfies a linear operator, so does another set of base vectors.

Let $\alpha_1, \alpha_2, ..., \alpha_n$ and $\beta_1, \beta_2, ..., \beta_n$ be two sets of base vectors in V_n. Suppose $\alpha_1, \alpha_2, ..., \alpha_n$ satisfy the linear Equation 1.14. Then, using Equation 1.11 and Equation 1.12, we have

$$L(\beta_i) = L\left(\sum_{j=1}^{n} d_{i,j} \alpha_j\right) = \sum_{j=1}^{n} L(d_{i,j} \alpha_j) = \sum_{j=1}^{n} d_{i,j} L(\alpha_j) = 0 \tag{1.16}$$

Equation 1.16 indicates that the set of base vectors $\beta_1, \beta_2, ..., \beta_n$ also satisfies the linear equation. The above properties will be used to determine the weighting coefficients in the DQ approximation.

1.4 Solution of Partial Differential Equations (PDEs) and Function Approximation

The DQ method is a numerical discretization technique, which is used to solve well-posed partial differential equations and to obtain the numerical solution. The solution or approximate solution of a partial differential equation has a close relationship with the function approximation.

1.4.1 Two Basic Types of Solution for PDEs

First, let us consider the following two examples.

Example 1: Consider a boundary layer flow over a flat plate as shown in Figure 1.3.

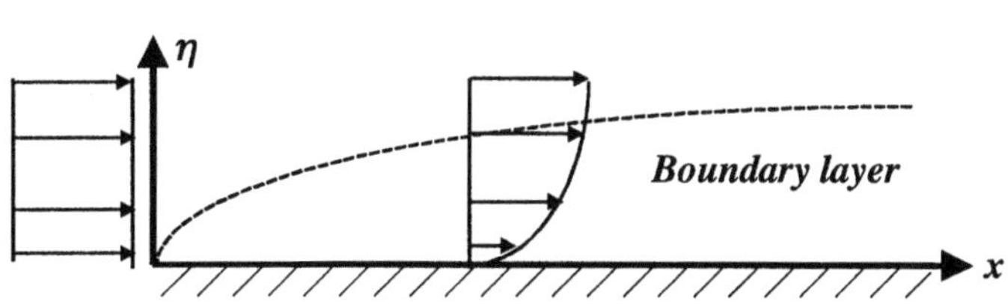

Figure 1.3 Boundary layer flow over a flat plate

The governing equation of this problem is

$$f''' + ff'' = 0 \tag{1.17}$$

and the two boundary conditions are

$$\begin{aligned} \eta = 0: &\quad f(0) = f'(0) = 0 \\ \eta \to \infty: &\quad f'(\eta) = 2.0 \end{aligned} \tag{1.18}$$

where f is the non-dimensional stream function, f', f'' and f''' represent the first, second, and third order derivatives of $f(\eta)$ with respect to η.

Equation 1.17 is a nonlinear equation. There is no simple way to solve it. Blasius first solved Equation 1.17 using a polynomial series expansion. In other words, the solution of Equation 1.17 is assumed to be approximated by a polynomial series expansion

$$f(\eta) = A_0 + A_1\eta + \frac{1}{2!}A_2\eta^2 + \frac{1}{3!}A_3\eta^3 + \ldots \tag{1.19}$$

The coefficients $A_0, A_1, A_2, \ldots$ can be determined by substituting Equation 1.19 into Equation 1.18. Blasius finally obtained the solution as

$$\begin{aligned} f(\eta) = A_2^{1/3}\Bigg[&\frac{1}{2!}(A_2^{1/3}\eta)^2 - \frac{1}{5!}(A_2^{1/3}\eta)^5 + \\ &\frac{11}{8!}(A_2^{1/3}\eta)^8 - \frac{375}{11!}(A_2^{1/3}\eta)^{11} + \ldots \Bigg] \end{aligned} \tag{1.20}$$

where $A_2 = 1.32824$. From this example, it can be seen that the solution of a PDE (Equation 1.17) can be approximated by a high order polynomial.

Example 2: Consider the solution of the following two-dimensional Helmholtz equation

$$\frac{\partial^2\phi}{\partial x^2} + \frac{\partial^2\phi}{\partial y^2} + K\phi = f(x, y) \tag{1.21}$$

over a rectangular domain as shown in Figure 1.4. The Helmholtz equation is the governing equation for many engineering problems such as waveguides in electromagnetic fields, vibration of membranes, and water wave diffraction in offshore structure engineering. In this example, L_x and L_y are taken as 1, $K = 0.5$, and

$f(x, y) = (-2\pi^2 + 0.5)\sin \pi x \cdot \sin \pi y$. The boundary condition is $\phi = 0$ on all boundaries.

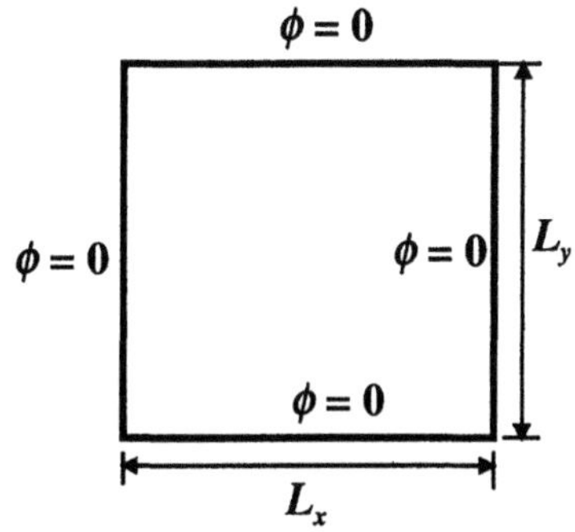

Figure 1.4 A two-dimensional Helmholtz problem

The exact solution of this problem can be given by

$$\phi = \sin \pi x \cdot \sin \pi y \tag{1.22}$$

The above two examples show that the solution of PDEs can be expressed or approximated by a polynomial, or by a harmonic function. Although the solution of PDEs may be expressed in other forms, the polynomial and harmonic function are two basic types of the solution.

For a general case, it may not be possible to express the solution of a PDE in a closed-form. However, from mathematical theories, the solution function of a well-posed PDE exists. This solution function can be approximated by the following two types of function approximation: high order polynomial approximation and Fourier series expansion (harmonic functions).

1.4.2 High Order Polynomial Approximation

Weierstrass' first theorem: Let $f(x)$ be a real valued continuous function defined in the closed interval $[a, b]$. Then there exists a sequence of polynomials $P_n(x)$ which converges to $f(x)$ uniformly as n goes to infinity.

The Weierstrass' first theorem can also be stated as: if $f(x)$ is a real valued function which is continuous in the closed interval $[a, b]$, then for every $\varepsilon > 0$, there exists a polynomial $P_n(x)$ of degree $n = n(\varepsilon)$ such that the inequality

$$|f(x) - P_n(x)| \le \varepsilon \tag{1.23}$$

holds through the interval $[a, b]$.

The proof of Weierstrass' first theorem can be found in many textbooks (Bellman and Roth 1986, Hyman 1956). In practical applications, if $f(x)$ represents the solution of a partial differential equation, then it can be approximated by a polynomial of degree less than N. The conventional form of this approximation is

$$f(x) \approx P_N(x) = \sum_{k=0}^{N-1} c_k x^k \tag{1.24}$$

where c_k is a constant. It has been shown in the above section that $P_N(x)$ constitutes an N-dimensional linear vector space V_N with respect to the operation of vector addition and scalar multiplication. Obviously, in the linear vector space V_N, a set of vectors (polynomials), $1, x, x^2, \ldots, x^{N-1}$, is linearly independent. Thus,

$$s_k(x) = x^{k-1}, \quad k = 1, 2, \ldots, N \tag{1.25}$$

is a basis of V_N.

For the numerical solution of a PDE, we need to find out the functional values at certain discrete points. Now, it is supposed that in a closed interval $[a, b]$, there are N mesh points with coordinates $a = x_1, x_2, \ldots, x_N = b$, and the functional value at a mesh point x_i is $f(x_i)$. Then the constants in Equation 1.24 can be determined from the following equation system

$$\begin{cases} c_0 + c_1 x_1 + c_2 x_1^2 + \ldots + c_{N-1} x_1^{N-1} = f(x_1) \\ c_0 + c_1 x_2 + c_2 x_2^2 + \ldots + c_{N-1} x_2^{N-1} = f(x_2) \\ \cdots\cdots\cdots\cdots\cdots\cdots\cdots\cdots\cdots\cdots \\ c_0 + c_1 x_N + c_2 x_N^2 + \ldots + c_{N-1} x_N^{N-1} = f(x_N) \end{cases} \tag{1.26}$$

The matrix of Equation 1.26 is of Vandermonde form, which is not singular. Thus, Equation 1.26 can give unique solutions for constants $c_0, c_1, c_2, \ldots, c_{N-1}$. Once the constants are determined, the approximated polynomial is obtained. On the other hand, when N is large, the matrix is highly ill-conditioned and its inversion is very difficult. Then for this case, it is hard to determine the constants $c_0, c_1, c_2, \ldots, c_N$ from Equation 1.26.

The difficulty of determining the approximated polynomial in Equation 1.24 can be removed by the Lagrange interpolated polynomial

$$P_N(x) = \sum_{i=1}^{N} f(x_i) \cdot l_i(x) \tag{1.27}$$

where

$$l_i(x) = \frac{M(x)}{M^{(1)}(x_i) \cdot (x - x_i)}$$

$$M(x) = (x - x_1)(x - x_2) \cdots (x - x_N)$$

$$M^{(1)}(x_i) = (x_i - x_1) \cdots (x_i - x_{i-1})(x_i - x_{i+1}) \cdots (x_i - x_N) = \prod_{k=1,k \neq i}^{N} (x_i - x_k)$$

It is obvious from Equation 1.27 that, once the functional values at the mesh points are given, the approximated polynomial can be obtained. The degree of the polynomial decided by Equation 1.27 is less than N, and the operation of vector addition and scalar multiplication are also satisfied by Equation 1.27. Thus, we can say that the polynomial given by Equation 1.27 is in the polynomial vector space V_N. We can show that $l_k(x), k = 1, 2, ..., N$ is another basis of V_N. Actually, $l_k(x)$ possesses the following property

$$l_k(x_i) = \begin{cases} 1, & \text{when } k = i \\ 0, & \text{others} \end{cases} \tag{1.28}$$

Let $d_1, d_2, ..., d_n$ be the elements in a field F, and

$$d_1 l_1(x) + d_2 l_2(x) + \cdots + d_N l_N(x) = \sum_{k=1}^{N} d_k l_k(x) = 0 \tag{1.29}$$

be exact in the closed interval [a, b]. Using Equation 1.28, it is observed that application of Equation 1.29 at a point x_i gives $d_i = 0$, i = 1, 2, ..., N. Thus, we can say that $l_1(x), l_2(x), \cdots, l_N(x)$ are linearly independent, which can be taken as a basis in V_N. The approximated polynomial $P_N(x)$ can also be determined by the Newton interpolation polynomial

$$P_N(x) = a_1 + \sum_{k=2}^{N} a_k (x - x_1)(x - x_2) \cdots (x - x_{k-1}) \tag{1.30}$$

Clearly, Equation 1.30 satisfies the operation of vector addition and scalar multiplication, and the degree of polynomial given by Equation 1.30 is less than N. So, the Newton's interpolation polynomial is also in the polynomial vector space V_N. Obviously,

$$n_1(x) = 1,\ n_k(x) = n_{k-1}(x) \cdot (x - x_{k-1}),\ k = 2, 3, ..., N \tag{1.31}$$

is a set of vectors (polynomials) in V_N. We can easily show that this set of vectors is linearly independent. Let $d_1, d_2, ..., d_n$ be the elements in a field *F*, and

$$d_1 n_1(x) + d_2 n_2(x) + \cdots + d_N n_N(x) = \sum_{k=1}^{N} d_k n_k(x) = 0 \tag{1.32}$$

Equation 1.32 should be satisfied at any point in the closed interval [*a*, *b*]. If we let Equation 1.32 be satisfied respectively at mesh points, $x_1, x_2, ..., x_N$, we then obtain $d_i = 0$, *i* = 1, 2, ..., *N*. So, $n_k(x)$, *k* = 1, 2, ..., *N*, are linearly independent, and can be considered as a basis in V_N.

The constants in Equation 1.30 can be expressed in terms of the functional values at the mesh points. Note that a_N in Equation 1.30 is identical to the coefficient of the highest power of the polynomial $P_N(x)$ (Equation 1.27) and that

$$a_1 + \sum_{k=2}^{m} a_k (x - x_1)(x - x_2) \cdots (x - x_{k-1})$$

is the polynomial of degree (*m*-1) which satisfies the functional values at the points $x_1, x_2, ..., x_m$. We find directly from Equation 1.27 that

$$a_m = \sum_{k=1}^{m} \frac{f(x_k)}{M_m^{(1)}(x_k)}, \qquad m = 1, 2, ..., N \tag{1.33}$$

where

$$M_m^{(1)}(x_k) = (x_k - x_1) \cdots (x_k - x_{k-1})(x_k - x_{k+1}) \cdots (x_k - x_m) = \prod_{j=1, j \neq k}^{m} (x_k - x_j)$$

From the above discussion, it can be seen that in a linear polynomial vector space V_N, there are many sets of base vectors. The three typical sets of base vectors are

$$r_k(x) = x^{k-1}, \quad k = 1, 2, ..., N \tag{1.34a}$$

$$r_k(x) = \frac{M(x)}{M^{(1)}(x_k) \cdot (x - x_k)}, \quad k = 1, 2, ..., N \tag{1.34b}$$

$$r_1(x) = 1,\ r_k(x) = r_{k-1}(x) \cdot (x - x_{k-1}),\ k = 2, 3, ..., N \tag{1.34c}$$

When the approximated polynomial is expressed using other forms, we may have different sets of base vectors. For example, when the approximated polynomial is expressed using the Legendre expansion

$$P_N(x) = \sum_{k=0}^{N-1} c_k \cdot L_k(x) \tag{1.35}$$

where $L_k(x)$ is the Legendre polynomial of degree k, and c_k is a constant, then $L_k(x)$, k = 0, 1, ..., N-1, is a set of base vectors. $L_k(x)$, k = 0, 1, ..., N-1, are the eigenfunctions of the singular Sturm-Liouville problem

$$(1-x^2)L_k^{(2)}(x) - 2xL_k^{(1)}(x) + k(k+1)L_k(x) = 0 \tag{1.36}$$

where $L_k^{(1)}(x)$ and $L_k^{(2)}(x)$ represent the first and second order derivatives of $L_k(x)$ respectively. The Legendre polynomials satisfy the recurrence relation

$$L_{k+1}(x) = \frac{2k+1}{k+1} x L_k(x) - \frac{k}{k+1} L_{k-1}(x) \tag{1.37}$$

where $L_0(x) = 1$ and $L_1(x) = x$.

Similarly, when the approximated polynomial is expressed using the Chebyshev expansion

$$P_N(x) = \sum_{k=0}^{N-1} c_k \cdot T_k(x) \tag{1.38}$$

where $T_k(x)$ is the Chebyshev polynomial of degree k, and c_k is a constant, then $T_k(x)$, k = 0, 1, ..., N-1, is a set of base vectors. $T_k(x)$, k = 0, 1, ..., N-1, are the eigenfunctions of the following singular Sturm-Liouville problem

$$\frac{d\left(\sqrt{1-x^2}\,T_k^{(1)}(x)\right)}{dx} + \frac{k^2}{\sqrt{1-x^2}} T_k(x) = 0 \tag{1.39}$$

where $T_k^{(1)}(x)$ represents the first order derivative of $T_k(x)$. The Chebyshev polynomials satisfy the recurrence relation

$$T_{k+1}(x) = 2xT_k(x) - T_{k-1}(x) \tag{1.40}$$

where $T_0(x) = 1$ and $T_1(x) = x$. $T_k(x)$ can also be expressed in the following form

$$T_k(x) = \cos k\theta, \quad \theta = \arccos x \tag{1.41}$$

On the other hand, it can be observed from Equation 1.34*b* that the set of base vectors also depends on the distribution of the mesh points (coordinates). When the coordinates of the mesh points are taken as the roots of the Legendre polynomial of degree *N* in a closed interval [-1, 1], Equation 1.34*b* can be reduced to

$$r_k(x) = \frac{L_N(x)}{L_N^{(1)}(x_k)\cdot(x - x_k)}, \quad k = 0,1,\ldots,N \tag{1.42}$$

1.4.3 Fourier Series Expansion

When the solution of a partial differential equation is periodic over an interval, it can be approximated by a Fourier series expansion.

1.4.3.1 General Function

For a complex-valued function $f(x)$ defined on [0, 2π], the Fourier series expansion can be written as

$$f(x) = \sum_{k=-\infty}^{\infty} c_k e^{ikx} \tag{1.43}$$

where

$$c_k = \frac{1}{2\pi}\int_0^{2\pi} f(x)e^{-ikx}dx$$

For a real-valued function $f(x)$ defined on [0, 2π], the Fourier series expansion is usually written as

$$f(x) = a_0 + \sum_{k=1}^{\infty} a_k \cos kx + \sum_{k=1}^{\infty} b_k \sin kx \tag{1.44}$$

where

$$a_0 = \frac{1}{2\pi}\int_0^{2\pi} f(x)dx$$

$$a_k = \frac{1}{\pi}\int_0^{2\pi} f(x)\cos kx dx$$

$$b_k = \frac{1}{\pi}\int_0^{2\pi} f(x)\sin kx \cdot dx$$

In most engineering problems, the solution of PDEs involves real functions. So, only Equation 1.44 is considered. For practical applications, the truncated Fourier series expansion is usually used. Thus,

$$f(x) \approx F_{N+1}(x) = a_0 + \sum_{k=1}^{N/2} a_k \cos kx + \sum_{k=1}^{N/2} b_k \sin kx \tag{1.45}$$

The convergence of the truncated Fourier series expansion to $f(x)$ as N tends to ∞ is guaranteed by Weierstrass' second theorem.

Weierstrass' second theorem: If $f(x)$ is a continuous function of period 2π, then, corresponding to every positive number ε, there exists a trigonomeric sum

$$S_n(x) = a_0 + \sum_{k=1}^{n} a_k \cos kx + \sum_{k=1}^{n} b_k \sin kx \tag{1.46}$$

($n = n(\varepsilon)$) such that the inequality

$$|f(x) - S_n(x)| \leq \varepsilon \tag{1.47}$$

is satisfied for all values of x.

The proof of Weierstrass' second theorem can be found in the book of Hyman (1956). It is easy to show that $F_{N+1}(x)$ given by Equation 1.45 satisfies the operation of vector addition and scalar multiplication. So, $F_{N+1}(x)$ constitutes a linear vector space V_{N+1}. In the linear vector space V_{N+1}, the following set of vectors

$$1, \cos x, \sin x, ..., \cos(Nx/2), \sin(Nx/2) \tag{1.48}$$

is linearly independent. Thus, it constitutes a set of base vectors in V_{N+1}. In the following, we will show another set of base vectors in V_{N+1}. Let $C_m(\cos x)$ represent a set of cosine functions of the form

$$C_m(\cos x) = a_0 + \sum_{k=1}^{m} a_k \cos^k x \tag{1.49}$$

and $S_m(\sin x)$ be a set of sine functions of the form

$$S_m(\sin x) = \sin x \sum_{k=1}^{m} b_k \cos^{k-1} x \tag{1.50}$$

Then, we can see that $\cos x$ is in the set of $C_1(\cos x)$, $\sin x$ is in the set of $S_1(\sin x)$, and $\cos 2x = 2\cos^2 x - 1$ is in the set of $C_2(\cos x)$, $\sin 2x = 2\sin x \cos x$ is in the set of $S_2(\sin x)$. Now, we assume that, for $k \le n$ (n is an arbitrary integer only if $n \ge 2$), $\cos kx$ is in the set of $C_k(\cos x)$ and $\sin kx$ is in the set of $S_k(\sin x)$, and we want to prove that $\cos[(n+1)x]$ is in the set of $C_{n+1}(\cos x)$ and $\sin[(n+1)x]$ is in the set of $S_{n+1}(\sin x)$.

For $\cos[(n+1)x]$, we have

$$\begin{aligned}
\cos[(n+1)x] &= \cos x \cos(nx) - \sin x \sin(nx) \\
&= \cos x C_n(\cos x) - \sin x S_n(\sin x) \\
&= \cos x \left[a_0 + \sum_{k=1}^{n} a_k \cos^k x \right] - \sin x \sin x \sum_{k=1}^{n} b_k \cos^{k-1} x \\
&= a_0 \cos x + \sum_{k=1}^{n} a_k \cos^{k+1} x + (\cos^2 x - 1) \sum_{k=1}^{n} b_k \cos^{k-1} x \\
&= \bar{a}_0 + \sum_{k=1}^{n+1} \bar{a}_k \cos^k x
\end{aligned} \tag{1.51}$$

So, $\cos[(n+1)x]$ is in the set of $C_{n+1}(\cos x)$. Similarly, for $\sin[(n+1)x]$, we have

$$\begin{aligned}
\sin[(n+1)x] &= \sin x \cos(nx) + \cos x \sin(nx) \\
&= \sin x C_n(\cos x) + \cos x S_n(\sin x) \\
&= \sin x \left[a_0 + \sum_{k=1}^{n} a_k \cos^k x \right] + \cos x \sin x \sum_{k=1}^{n} b_k \cos^{k-1} x \\
&= \sin x \sum_{k=1}^{n+1} \bar{b}_k \cos^{k-1} x
\end{aligned} \tag{1.52}$$

Thus, sin[$(n+1)x$] is in the set of $S_{n+1}(\sin x)$. Therefore, $F_{N+1}(x)$ can be expressed as

$$F_{N+1}(x) = \bar{a}_0 + \sum_{k=1}^{N/2} \bar{a}_k \cos^k x + \sin x \sum_{k=1}^{N/2} \bar{b}_k \cos^{k-1} x \tag{1.53}$$

and

$$1, \cos x, \cos^2 x, \ldots, \cos^{(N/2)} x, \sin x, \sin x \cos x, \ldots, \sin x \cos^{(N/2-1)} x$$

is a set of linearly independent vectors, which can be considered as a basis of V_{N+1}.

For the numerical solution of a PDE, we need to determine the functional values at discrete mesh points. For this case, the Fourier series expansion should be used in the discrete form. Let $x_0, x_1, \ldots, x_N$ be the coordinates of (N+1) mesh points in the interval [0,2π], and the functional value at a mesh point x_i be $f(x_i)$. Then $F_{N+1}(x)$ can be expressed using the following Lagrange interpolation formula for trigonometric polynomials

$$F_{N+1}(x) = \sum_{i=0}^{N} f(x_i) \cdot g_i(x) \tag{1.54}$$

where

$$g_i(x) = \frac{G(x)}{G^{(1)}(x_i) \cdot \sin[(x - x_i)/2]}$$

$$G(x) = \sin\frac{x - x_0}{2} \sin\frac{x - x_1}{2} \cdots \sin\frac{x - x_N}{2}$$

$$G^{(1)}(x_i) = \prod_{k=0,k\neq i}^{N} \sin\frac{x_i - x_k}{2}$$

$g_k(x_i)$ has the same property as $l_k(x_i)$ shown in Equation 1.28. Similar to $l_i(x)$, it is easy to show that $g_i(x), i = 0, 1, \ldots, N$, is a set of linearly independent vectors in V_{N+1}. Thus

$$g_i(x) = \frac{G(x)}{G^{(1)}(x_i) \cdot \sin[(x - x_i)/2]}, \quad i = 0, 1, \ldots, N \tag{1.55}$$

is a basis of V_{N+1}.

1.4.3.2 Even Function

For the even function $f(x)$ defined on [0, π], the Fourier series expansion can be written as

$$f(x) = a_0 + \sum_{k=1}^{\infty} a_k \cos kx \tag{1.56}$$

where

$$a_0 = \frac{1}{\pi}\int_0^{\pi} f(x)dx$$

$$a_k = \frac{2}{\pi}\int_0^{\pi} f(x)\cos kx \cdot dx\ , \ k > 0$$

And accordingly, the truncated Fourier series expansion of Equation 1.56 is

$$f(x) \approx F_{N+1}(x) = a_0 + \sum_{k=1}^{N} a_k \cos kx \tag{1.57}$$

From the definition of a linear vector space, $F_{N+1}(x)$ given by Equation 1.57 constitutes a linear vector space V_{N+1}. And one obvious set of base vectors in V_{N+1} is

$$1, \cos x, ..., \cos(Nx) \tag{1.58}$$

On the other hand, using Equation 1.51, $F_{N+1}(x)$ can also be expressed as

$$F_{N+1}(x) = \overline{a}_0 + \sum_{k=1}^{N} \overline{a}_k \cos^k x \tag{1.59}$$

If we set $\xi = \cos x$, then Equation 1.59 can be reduced to

$$F_{N+1}(x) = \overline{a}_0 + \sum_{k=1}^{N} \overline{a}_k \xi^k = \sum_{k=0}^{N} \overline{a}_k \xi^k, \quad \xi = \cos x \tag{1.60}$$

The form of Equation 1.60 is exactly the same as Equation 1.24. Thus,

$$\xi^i = \cos^i x, \quad i = 0, 1, ..., N \tag{1.61}$$

is a set of base vectors in V_{N+1}. The Lagrange interpolation formula for $F_{N+1}(x)$ can be written as

$$F_{N+1}(x) = \sum_{i=0}^{N} f(x_i) \cdot c_i(x) \tag{1.62}$$

where

$$c_i(x) = \frac{C(x)}{C^{(1)}(x_i) \cdot (\cos x - \cos x_i)}$$

$$C(x) = (\cos x - \cos x_0)(\cos x - \cos x_1) \cdots (\cos x - \cos x_N)$$

$$C^{(1)}(x_i) = \prod_{k=0, k \neq i}^{N} (\cos x_i - \cos x_k)$$

$c_k(x_i)$ has the same property as $l_k(x_i)$ shown in Equation 1.28. Similar to $l_i(x)$, it is easy to show that $c_i(x)$, $i = 0, 1, ..., N$, is a set of linearly independent vectors in V_{N+1}. Thus

$$c_i(x) = \frac{C(x)}{C^{(1)}(x_i) \cdot (\cos x - \cos x_i)}, \quad i = 0, 1, ..., N \tag{1.63}$$

is a basis of V_{N+1}.

1.4.3.3 Odd Function

For the odd function $f(x)$ defined on $(0, \pi)$, the Fourier series expansion can be written as

$$f(x) = \sum_{k=1}^{\infty} b_k \sin kx \tag{1.64}$$

where

$$b_k = \frac{2}{\pi} \int_0^{\pi} f(x) \sin kx \cdot dx$$

The truncated Fourier series expansion of Equation 1.64 is

$$f(x) \approx F_N(x) = \sum_{k=1}^{N} b_k \sin kx \tag{1.65}$$

$F_N(x)$ given by Equation 1.65 satisfies the operation of vector addition and scalar multiplication. Thus it constitutes a linear vector space V_N. A set of base vectors in V_N is

$$\sin x, \sin 2x, ..., \sin(Nx) \tag{1.66}$$

On the other hand, using Equation 1.52, $F_N(x)$ can be expressed as

$$F_N(x) = \sin x \sum_{k=1}^{N} \overline{b}_k \cos^{k-1} x \tag{1.67}$$

Thus,

$$\sin x \cos^{i-1} x, \quad i = 1, 2, ..., N \tag{1.68}$$

is another set of base vectors in V_N. Furthermore, the Lagrange interpolation formula of $F_N(x)$ can be written from Equation 1.67 as

$$F_N(x) = \sum_{i=1}^{N} f(x_i) \cdot s_i(x) \tag{1.69}$$

where

$$s_i(x) = \frac{\sin x \cdot S(x)}{S^{(1)}(x_i) \cdot \sin x_i (\cos x - \cos x_i)}$$

$$S(x) = (\cos x - \cos x_1)(\cos x - \cos x_2) \cdots (\cos x - \cos x_N)$$

$$S^{(1)}(x_i) = \prod_{k=1, k \neq i}^{N} (\cos x_i - \cos x_k)$$

$s_k(x_i)$ has the same property as $l_k(x_i)$ shown in Equation 1.28. Similar to $l_i(x)$, it can be seen that $s_i(x),\ i = 1, 2, ..., N,$ is a set of linearly independent vectors in V_N. Thus

$$s_i(x) = \frac{\sin x \cdot S(x)}{S^{(1)}(x_i) \cdot \sin x_i (\cos x - \cos x_i)}, \quad i = 1, 2, ..., N \tag{1.70}$$

is another basis of V_N. In the application of Equation 1.70, $\sin x_i$ must not be zero. This is guaranteed by the application in the open interval (0, π). From Equation 1.65, it is observed that when x = 0 and π, the functional values are zero. Thus, Equation 1.69 does not involve the points x = 0 and π. In other words, the coordinates of $x_1, x_2, ..., x_N$ should be taken from the open interval (0, π).

Chapter 2

Polynomial-based Differential Quadrature (PDQ)

2.1 Introduction

The fundamentals of the differential quadrature (DQ) method, that is, linear vector space analysis and function approximation, have been described in the last chapter. In this chapter, we will show the details of determining the weighting coefficients in the DQ approximation when the solution of a partial differential equation (PDE) is approximated by a polynomial of high degree. Since for this case, the DQ approximation is related to the polynomial approximation to the solution of a PDE, for simplicity, it is termed the polynomial-based differential quadrature (PDQ) method.

The computation of the weighting coefficients in the PDQ method was initially made by Bellman and his associates (Bellman et al. 1971, 1972). They suggested two ways to determine the weighting coefficients for the first order derivative. The first way involves the solution of an algebraic equation system. The second way uses a simple algebraic formulation, but with the coordinates of the grid points chosen as the roots of the shifted Legendre polynomial. Most early applications of PDQ in engineering (Bellman et al. 1972, Mingle 1977, Civan and Sliepcevich 1983*a*, *b*, 1984*a*, *b*) used Bellman's first approach to obtain the weighting coefficients because it allows the coordinates of the grid points to be chosen arbitrarily. Unfortunately, when the order of the algebraic equation system, which is equal to the number of grid points, is large, the resulting matrix is ill-conditioned. Thus it is very difficult to obtain the weighting coefficients for a large number of grid points. To overcome the difficulties of Bellman's approaches in computing the weighting coefficients, many attempts have been made. Using the Lagrange interpolation polynomials as the test functions, Quan and Chang (1989*a*, *b*) obtained explicit formulations to compute the weighting coefficients for the first and second order derivatives. At this stage, two questions may arise. First, why should there be several ways to compute the weighting coefficients? Next, how can we be sure that the weighting coefficients computed using different approaches are the same? These two questions can be easily answered by employing linear vector space analysis. Using the properties of a linear vector space, Shu and Richards (1990), and Shu (1991) found that all the methods of computing the weighting coefficients in the PDQ can be generalized by proper choice of base vectors in the linear vector space. As a general case, Shu (1991) presented a simple algebraic formulation to compute the weighting coefficients of the first order derivative without any restriction on the choice of the grid points, and a recurrence relationship to compute the weighting coefficients of the second and higher order derivatives. In this chapter, the details of this process will be shown. Some specific results for typical grid point distributions, the error estimates of function and derivative approximations, the relationship between PDQ and Chebyshev collocation method as well as the highest order finite difference scheme, are also provided in this chapter. For multi-dimensional problems, two approaches will be introduced. One is based on the structured grid, and the other is based on the unstructured grid. In addition, this chapter will show how explicit PDQ formulations can be applied to generate the low order finite difference schemes.

2.2 Computation of Weighting Coefficients for the First Order Derivative

As indicated in the last chapter, the DQ approximation of derivatives was initiated from the idea of integral quadrature. Consider a one-dimensional problem over a closed interval $[a, b]$. It is supposed that there are N grid points with coordinates as $a = x_1, x_2, \ldots, x_N = b$. Bellman et al. (1972) assumed that a function $f(x)$ is sufficiently smooth over the interval $[a, b]$ so that its first order derivative $f^{(1)}(x)$ at any grid point can be approximated by the following formulation

$$f_x^{(1)}(x_i) = \sum_{j=1}^{N} a_{ij} \cdot f(x_j) \quad , \quad \text{for } i = 1, 2, \cdots, N \tag{2.1}$$

where $f(x_j)$ represents the functional value at a grid point x_j, $f_x^{(1)}(x_i)$ indicates the first order derivative of $f(x)$ at x_i, and a_{ij} is the weighting coefficient of the first order derivative. It was shown in the last chapter that $f^{(1)}(x)$ is a linear operator. The determination of weighting coefficients a_{ij} in Equation 2.1 is a key procedure in the DQ approximation. Once the weighting coefficients are determined, the bridge to link the derivatives in the governing differential equation and the functional values at the mesh points is established. In other words, with the weighting coefficients, one can easily use the functional values to compute the derivatives. In the following, we shall show that the weighting coefficients can be efficiently computed by employing some explicit formulations.

2.2.1 Bellman's Approaches

Bellman et al. (1972) proposed two approaches to compute the weighting coefficients a_{ij} in Equation 2.1. The two approaches are based on the use of two different test functions.

Bellman's first approach: In this approach, the test functions are chosen as

$$g_k(x) = x^k, k = 0, 1, \cdots, N-1 \tag{2.2}$$

Obviously, Equation 2.2 gives N test functions. For the weighting coefficients a_{ij} in Equation 2.1, i and j are taken from 1 to N. Thus, the total number of weighting coefficients is $N \times N$. To obtain these weighting coefficients, the N test functions should be applied at N grid points $x_1, x_2, ..., x_N$. As a consequence, the following $N \times N$ algebraic equations for a_{ij} are obtained

$$\begin{cases} \sum_{j=1}^{N} a_{ij} = 0 \\ \sum_{j=1}^{N} a_{ij} \cdot x_j = 1 \\ \sum_{j=1}^{N} a_{ij} \cdot x_j^k = k \cdot x_i^{k-1}, k = 2, 3, ..., N-1 \end{cases} \tag{2.3}$$

for $i = 1, 2, \cdots, N$

Equation System 2.3 has a unique solution because its matrix is of Vandermonde form. Unfortunately, when N is large, the matrix is ill-conditioned and its inversion is difficult. In the practical application of this approach, N is usually chosen to be less than 13.

Bellman's second approach: This approach is similar to the first approach, but uses different test functions

$$g_k(x) = \frac{L_N(x)}{(x - x_k) \cdot L_N^{(1)}(x_k)}, k = 1, 2, \cdots, N \tag{2.4}$$

where $L_N(x)$ is the Legendre polynomial of degree N and $L_N^{(1)}(x)$ is the first order derivative of $L_N(x)$. By choosing x_k to be the roots of the shifted Legendre polynomial and applying Equation 2.4 at N grid points $x_1, x_2, ..., x_N$, Bellman et al. (1972) obtained a simple algebraic formulation to compute a_{ij}

$$a_{ij} = \frac{L_N^{(1)}(x_i)}{(x_i - x_j)L_N^{(1)}(x_j)}, \quad \text{for } j \neq i \tag{2.5a}$$

$$a_{ii} = \frac{1-2x_i}{2x_i(x_i - 1)} \tag{2.5b}$$

Using Equation 2.5, the computation of the weighting coefficients is a simple task. However, this approach is not as flexible as the first approach because the coordinates of the grid points in this approach cannot be chosen arbitrarily. Instead, they should be chosen as the roots of the Legendre polynomial of degree N. So, Equation 2.5 only reflects a special case. Due to the inflexibility associated with the second approach in selecting the grid points, the first approach is usually adopted in practical applications.

2.2.2 Quan and Chang's Approach

To improve Bellman's approaches in computing the weighting coefficients, many attempts have been made by researchers. One of the most useful approaches is the one introduced by Quan and Chang (1989*a, b*). Quan and Chang used the following Lagrange interpolation polynomials as the test functions,

$$g_k(x) = \frac{M(x)}{(x - x_k)\cdot M^{(1)}(x_k)}, \quad k = 1, 2, \cdots, N \tag{2.6}$$

where

$$M(x) = (x - x_1)(x - x_2)\cdots(x - x_N) \tag{2.7}$$

$$M^{(1)}(x_i) = \prod_{k=1,k\neq i}^{N}(x_i - x_k) \tag{2.8}$$

Subsequently, by applying Equation 2.6 at N grid points, they obtained the following algebraic formulations to compute the weighting coefficients a_{ij},

$$a_{ij} = \frac{1}{x_j - x_i}\prod_{k=1,k\neq i,j}^{N}\frac{x_i - x_k}{x_j - x_k}, \text{ for } j \neq i \tag{2.9a}$$

$$a_{ii} = \sum_{k=1,k\neq i}^{N}\frac{1}{x_i - x_k} \tag{2.9b}$$

When Equation 2.9 is used, there is no restriction on the choice of the grid points.

2.2.3 Shu's General Approach

Shu's general approach was inspired from Bellman's approaches. It covers all the approaches, including Quan and Chang's approach.

Starting from Bellman's two approaches, Shu raised two questions. The first one is why we can use two approaches to compute the weighting coefficients. The second is whether these two approaches give the same weighting coefficients. If so, there may exist some other approaches to compute the weighting coefficients. It was found that these questions could be answered by polynomial approximation and linear vector space analysis. Both polynomial approximation and linear vector space analyses have been described in the last chapter. Here, we shall only quote the relevant results.

It was shown in Chapter 1 that the solution of a PDE can be accurately approximated by a polynomial of high degree. Now, we suppose that the degree of the approximated polynomial is N-1. Chapter 1 shows that this approximated polynomial constitutes an N-dimensional linear vector space V_N with the operation of vector addition and scalar multiplication, and can be expressed in different forms. One popular form is

$$f(x) = \sum_{k=1}^{N} c_k \cdot x^{k-1} \tag{2.10}$$

where c_k is a constant. As shown in Chapter 1, there exist many sets of base vectors in the linear vector space V_N. In this case, the vector is actually the polynomial. So, the base vectors are also called the base polynomials. Four typical sets of the base polynomials are listed as follows,

$$r_k(x) = x^{k-1}, \quad k = 1, 2, ..., N \tag{2.11a}$$

$$r_k(x) = \frac{L_N(x)}{(x - x_k) \cdot L_N^{(1)}(x_k)}, \quad k = 1, 2, \cdots, N \tag{2.11b}$$

$$r_k(x) = \frac{M(x)}{(x - x_k) \cdot M^{(1)}(x_k)}, \quad k = 1, 2, \cdots, N \tag{2.11c}$$

$$r_1(x) = 1, \ r_k(x) = (x - x_{k-1}) \cdot r_{k-1}(x), \ k = 2, 3, ..., N \tag{2.11d}$$

where $L_N(x)$ is the Legendre polynomial of degree N and $M(x)$ is defined in Equation 2.7. Among the four sets of base polynomials, Equations 2.11b and 2.11c are from the Lagrange interpolation polynomials while Equation 2.11d is from the Newton interpolation polynomials. The difference between Equation 2.11b and 2.11c lies in the distribution of the grid points. Equation 2.11b is a special case of Equation 2.11c since it is only valid at the Legendre collocation points.

It can be seen that Equation 2.11*a* is the same as the test functions of Bellman's first approach while Equation 2.11*b* is the same as the test functions of Bellman's second approach. In other words, the test functions of Bellman's two approaches are actually two sets of base polynomials in V_N. Note that Equation 2.1 is a linear operator. Then from the properties of a linear vector space, we know that if one set of base polynomials satisfies a linear operator such as Equation 2.1, so do other sets of base polynomials. This means that every set of base polynomials would give the same weighting coefficients. Hence, the weighting coefficients do not depend on the choice of the test functions. On the other hand, we notice that the difference in Bellman's two approaches only lies in the use of different test functions. The test functions are equivalent to the base polynomials. Thus, the use of different sets of base polynomials will result in different approaches to compute the weighting coefficients. Since there are many sets of base polynomials in the linear vector space V_N, we have many approaches to compute the weighting coefficients.

The property of a linear vector space also gives us the ability to apply the weighting coefficients for the discretization of a PDE. Note that the solution of the PDE is approximated by a polynomial of degree (N-1) which constitutes the N-dimensional vector space V_N. In general, the actual expression of the polynomial is unknown. For example, the constant c_k in Equation 2.10 is to be determined. On the other hand, we find that, in the linear vector space V_N, the set of base polynomials can be chosen to be independent of the solution. This was revealed by Equation 2.11. From the property of a linear vector space, if a set of base polynomials satisfies a linear operator such as Equation 2.1, so does any polynomial in the space. This indicates that the solution of the PDE (Equation 2.10) also satisfies Equation 2.1. In other words, the weighting coefficients obtained from the base polynomials can be used to discretize the derivatives in the PDE.

When the set of base polynomials is given by Equation 2.11*a*, we can obtain the same Equation System 2.3 as given by Bellman's first approach, and when the set of base polynomials is obtained from Equation 2.11*b*, we can get the same algebraic formulation 2.5 as given by Bellman's second approach. Here, for generality, we use two sets of base polynomials. The Lagrange interpolation polynomials (Equation 2.11*c*) are taken as the first set of base polynomials. For simplicity, we set

$$M(x) = N(x, x_k) \cdot (x - x_k), \quad k = 1, 2, ..., N \tag{2.12}$$

with $N(x_i, x_j) = M^{(1)}(x_i) \cdot \delta_{ij}$, where δ_{ij} is the Kronecker operator. Using Equation 2.12, Equation 2.11*c* can be simplified to

$$r_k(x) = \frac{N(x, x_k)}{M^{(1)}(x_k)}, \quad k = 1, 2, ..., N \tag{2.13}$$

Substituting Equation 2.13 into Equation 2.1 gives

$$a_{ij} = \frac{N^{(1)}(x_i, x_j)}{M^{(1)}(x_j)} \tag{2.14}$$

In Equation 2.14, $M^{(1)}(x_j)$ can be easily computed from Equation 2.8. To evaluate $N^{(1)}(x_i, x_j)$, we successively differentiate Equation 2.12 with respect to x and obtain the following recurrence formulation

$$M^{(m)}(x) = N^{(m)}(x, x_k) \cdot (x - x_k) + m \cdot N^{(m-1)}(x, x_k) \tag{2.15}$$
$$\text{for } m = 1, 2, \ldots, N\text{-}1;\ k = 1, 2, \ldots, N$$

where $M^{(m)}(x)$ and $N^{(m)}(x, x_k)$ indicate the mth order derivative of $M(x)$ and $N(x, x_k)$. From Equation 2.15, we can obtain the expression for $N^{(1)}(x_i, x_j)$ as

$$N^{(1)}(x_i, x_j) = \frac{M^{(1)}(x_i)}{x_i - x_j}, \quad \text{for } i \neq j \tag{2.16a}$$

$$N^{(1)}(x_i, x_i) = \frac{M^{(2)}(x_i)}{2} \tag{2.16b}$$

Substituting Equation 2.16 into Equation 2.14, we obtain

$$a_{ij} = \frac{M^{(1)}(x_i)}{(x_i - x_j) \cdot M^{(1)}(x_j)}, \quad \text{for } i \neq j \tag{2.17a}$$

$$a_{ii} = \frac{M^{(2)}(x_i)}{2M^{(1)}(x_i)} \tag{2.17b}$$

It is observed from Equation 2.17 that, if x_i is given, it is easy to compute $M^{(1)}(x_i)$ from Equation 2.8, and hence a_{ij} for $i \neq j$. However, the calculation of a_{ii} is based on the computation of the second order derivative $M^{(2)}(x_i)$ which is not an easy task. This difficulty can be eliminated by using the second set of base polynomials. According to the property of a linear vector space, if one set of base polynomials satisfies a linear operator, say Equation 2.1, so does another set of base polynomials. As a consequence, the equation system for determination of a_{ij} derived from the Lagrange interpolation polynomials (Equation 2.11*c*) should be equivalent to that derived from another set of base polynomials $x^{k-1}, k = 1, 2, ..., N$ (Equation 2.11*a*). Thus a_{ij} satisfies the following equation which is obtained by the base polynomial x^{k-1} when $k = 1$

$$\sum_{j=1}^{N} a_{ij} = 0 \text{ or } a_{ii} = -\sum_{j=1, j\neq i}^{N} a_{ij} \tag{2.18}$$

Equations 2.17*a* and 2.18 are two formulations to compute the weighting coefficients a_{ij}. It is noted that in the development of these two formulations, two sets of base polynomials were used in the linear polynomial vector space V_N.

2.3 Computation of Weighting Coefficients for the Second and Higher Order Derivatives

In this section, we show that the weighting coefficients of the second and higher order derivatives can be computed from a recurrence relationship.

2.3.1 Weighting Coefficients of the Second Order Derivative

For the discretization of the second order derivative, we introduce a similar approximation form given by

$$f_x^{(2)}(x_i) = \sum_{j=1}^{N} b_{ij} \cdot f(x_j), \quad \text{for } i = 1, 2, \cdots, N \tag{2.19}$$

where $f_x^{(2)}(x_i)$ is the second order derivative of $f(x)$ at x_i, b_{ij} is the weighting coefficient of the second order derivative. Obviously, Equation 2.19 is a linear operator. Note that Equation 2.19 has the same form as Equation 2.1. The only difference between Equation 2.19 and Equation 2.1 is the use of different weighting coefficients. The following two approaches can be used to determine b_{ij}.

Quan and Chang's approach: In this approach, Quan and Chang used the Lagrange interpolation polynomials as the test functions and then derived

$$b_{ij} = \frac{2}{x_j - x_i}\left(\prod_{k=1,k\neq i,j}^{N} \frac{x_i - x_k}{x_j - x_k}\right)\left(\sum_{l=1,l\neq i,j}^{N} \frac{1}{x_i - x_l}\right), \text{ for } i \neq j \tag{2.20a}$$

$$b_{ii} = 2\sum_{k=1,k\neq i}^{N-1}\left[\frac{1}{x_i - x_k}\left(\sum_{l=k+1,l\neq i}^{N} \frac{1}{x_i - x_l}\right)\right] \tag{2.20b}$$

Shu's general approach: Similar to the first order derivative, Shu's general approach is also based on polynomial approximation and linear vector space analyses. Two sets of base polynomials (Equations 2.11*c* and 2.11*a*) are used. Substituting Equation 2.13 into Equation 2.19 gives

$$b_{ij} = \frac{N^{(2)}(x_i, x_j)}{M^{(1)}(x_j)} \tag{2.21}$$

On the other hand, from Equation 2.15, we obtain

$$N^{(2)}(x_i, x_j) = \frac{M^{(2)}(x_i) - 2N^{(1)}(x_i, x_j)}{x_i - x_j}, \quad \text{for } i \neq j \tag{2.22a}$$

$$N^{(2)}(x_i, x_i) = \frac{M^{(3)}(x_i)}{3} \tag{2.22b}$$

Substituting Equation 2.22 into Equation 2.21 yields

$$b_{ij} = \frac{M^{(2)}(x_i) - 2N^{(1)}(x_i, x_j)}{(x_i - x_j)\cdot M^{(1)}(x_j)}, \quad \text{for } i \neq j \tag{2.23a}$$

$$b_{ii} = \frac{M^{(3)}(x_i)}{3M^{(1)}(x_i)} \tag{2.23b}$$

Finally, by substituting Equation 2.17 into Equation 2.23*a*, we obtain

$$b_{ij} = 2a_{ij}\left(a_{ii} - \frac{1}{x_i - x_j}\right), \quad \text{for } i \neq j \tag{2.24}$$

When $i \neq j$, b_{ij} can be easily computed from Equation 2.24. However, from Equation 2.23*b*, it can be seen that the computation of b_{ii} involves the third order derivative $M^{(3)}(x_i)$ which cannot be computed easily. This difficulty can be eliminated by employing the properties of a linear vector space. Similar to the analysis for the case of the first order derivative, the equation system for b_{ij} derived from the Lagrange interpolation polynomials (a set of base polynomials) is equivalent to that derived from another set of base polynomials $x^{k-1}, k = 1, 2, ..., N$. Thus b_{ij} should also satisfy the following formulation derived from the base polynomial x^{k-1} when $k = 1$

$$\sum_{j=1}^{N} b_{ij} = 0 \quad \text{or} \quad b_{ii} = -\sum_{j=1, j\neq i}^{N} b_{ij} \tag{2.25}$$

For the application of Shu's general approach, b_{ij} is firstly computed from Equation 2.24 when $i \neq j$. Subsequently, b_{ii} is computed from Equation 2.25.

2.3.2 Shu's Recurrence Formulation for Higher Order Derivatives

For the discretization of higher order derivatives, the following two linear operators are applied

$$f_x^{(m-1)}(x_i) = \sum_{j=1}^{N} w_{ij}^{(m-1)} \cdot f(x_j) \tag{2.26}$$

$$f_x^{(m)}(x_i) = \sum_{j=1}^{N} w_{ij}^{(m)} \cdot f(x_j) \tag{2.27}$$

for $i = 1, 2, \ldots, N;\ m = 2, 3, \ldots, N\text{-}1$

where $f_x^{(m-1)}(x_i)$, $f_x^{(m)}(x_i)$ indicate the $(m\text{-}1)$th and mth order derivatives of $f(x)$ with respect to x at x_i. $w_{ij}^{(m-1)}, w_{ij}^{(m)}$ are the weighting coefficients related to $f_x^{(m-1)}(x_i)$ and $f_x^{(m)}(x_i)$. Two sets of base polynomials will also be used to derive explicit formulations for $w_{ij}^{(m)}$. The first set of base polynomials is given by Equation 2.13. Substituting Equation 2.13 into Equations 2.26 and 2.27 gives

$$w_{ij}^{(m-1)} = \frac{N^{(m-1)}(x_i, x_j)}{M^{(1)}(x_j)} \tag{2.28}$$

$$w_{ij}^{(m)} = \frac{N^{(m)}(x_i, x_j)}{M^{(1)}(x_j)} \tag{2.29}$$

By rewriting Equation 2.28, we obtain

$$N^{(m-1)}(x_i, x_j) = w_{ij}^{(m-1)} M^{(1)}(x_j) \tag{2.30}$$

Equation 2.30 is valid for any i and j. On the other hand, from the recurrence formulation 2.15, we have

$$N^{(m-1)}(x_i, x_i) = \frac{M^{(m)}(x_i)}{m} \tag{2.31}$$

$$N^{(m)}(x_i, x_j) = \frac{M^{(m)}(x_i) - mN^{(m-1)}(x_i, x_j)}{x_i - x_j}, \quad \text{for } i \neq j \tag{2.32}$$

$$N^{(m)}(x_i, x_i) = \frac{M^{(m+1)}(x_i)}{m+1} \tag{2.33}$$

Substituting Equation 2.31 into Equation 2.32 leads to

$$N^{(m)}(x_i, x_j) = \frac{m[N^{(m-1)}(x_i, x_i) - N^{(m-1)}(x_i, x_j)]}{x_i - x_j}, \quad \text{for } i \neq j \tag{2.34}$$

Equation 2.34 can be further simplified using Equation 2.30,

$$N^{(m)}(x_i, x_j) = \frac{m[w_{ii}^{(m-1)} M^{(1)}(x_i) - w_{ij}^{(m-1)} M^{(1)}(x_j)]}{x_i - x_j}, \quad \text{for } i \neq j \tag{2.35}$$

Substituting Equation 2.35 into Equation 2.29, and using Equation 2.17a, a recurrence formulation is obtained as follows

$$w_{ij}^{(m)} = m\left(a_{ij} w_{ii}^{(m-1)} - \frac{w_{ij}^{(m-1)}}{x_i - x_j} \right) \tag{2.36}$$

for $i, j = 1, 2, \ldots, N;\ m = 2, 3, \ldots, N\text{-}1$

where a_{ij} is the weighting coefficient of the first order derivative described above. The formulation for $w_{ii}^{(m)}$ can be obtained by substituting Equation 2.33 into Equation 2.29, which gives

$$w_{ii}^{(m)} = \frac{M^{(m+1)}(x_i)}{(m+1) \cdot M^{(1)}(x_i)} \tag{2.37}$$

for $i, j = 1, 2, \ldots, N;\ m = 2, 3, \ldots, N\text{-}1$

Obviously, Equation 2.36 offers an easy way to compute the weighting coefficients $w_{ij}^{(m)}$ for $j \neq i$. However, it is very difficult to apply Equation 2.37 to compute the weighting coefficients $w_{ii}^{(m)}$. Again, this difficulty can be overcome by the properties of a linear vector space. In terms of the analysis of the N-dimensional linear vector space, the equation system for $w_{ij}^{(m)}$ derived from the Lagrange interpolation polynomials should be equivalent to that derived from the base polynomials $x^{k-1}, k = 1, 2, \ldots, N$. Thus $w_{ij}^{(m)}$ should satisfy the following equation obtained from the base polynomial x^{k-1} when $k = 1$

$$\sum_{j=1}^{N} w_{ij}^{(m)} = 0 \quad \text{or} \quad w_{ii}^{(m)} = -\sum_{j=1, j\neq i}^{N} w_{ij}^{(m)} \tag{2.38}$$

From this formulation, $w_{ii}^{(m)}$ can be determined from $w_{ij}^{(m)}$ $(i \neq j)$.

2.3.3 Matrix Multiplication Approach

From the definition of the differential operator, we have

$$\frac{\partial^2 f}{\partial x^2} = \frac{\partial}{\partial x}\left(\frac{\partial f}{\partial x}\right) \tag{2.39}$$

For simplicity, $\partial^2 f / \partial x^2$ is represented by $f_x^{(2)}$ while $\partial f / \partial x$ is denoted by $f_x^{(1)}$. When the DQ approximation is applied to the left side of Equation 2.39, we obtain

$$f_x^{(2)}(x_i) = \sum_{j=1}^{N} b_{ij} f(x_j), \; i = 1, 2, ..., N \tag{2.40}$$

and when the DQ method is applied to the right side of Equation 2.39 twice, we get

$$\begin{aligned} f_x^{(2)}(x_i) &= \sum_{k=1}^{N} a_{ik} f_x^{(1)}(x_k) = \sum_{k=1}^{N} a_{ik} \sum_{j=1}^{N} a_{kj} f(x_j) \\ &= \sum_{j=1}^{N} \left[\sum_{k=1}^{N} a_{ik} \cdot a_{kj} \right] \cdot f(x_j) \end{aligned} \quad , \; i = 1, 2, ..., N \tag{2.41}$$

By comparing Equations 2.40 and 2.41, we obtain

$$b_{ij} = \sum_{k=1}^{N} a_{ik} \cdot a_{kj} \tag{2.42}$$

Now, if we define two matrices by

$$[A^{(1)}] = \begin{bmatrix} a_{11} & a_{12} & \cdot & \cdot & \cdot & a_{1N} \\ a_{21} & a_{22} & \cdot & \cdot & \cdot & a_{2N} \\ \cdot & \cdot & \cdot & \cdot & \cdot & \cdot \\ \cdot & \cdot & \cdot & \cdot & \cdot & \cdot \\ \cdot & \cdot & \cdot & \cdot & \cdot & \cdot \\ a_{N1} & a_{N2} & \cdot & \cdot & \cdot & a_{NN} \end{bmatrix}, \quad [A^{(2)}] = \begin{bmatrix} b_{11} & b_{12} & \cdot & \cdot & \cdot & b_{1N} \\ b_{21} & b_{22} & \cdot & \cdot & \cdot & b_{2N} \\ \cdot & \cdot & \cdot & \cdot & \cdot & \cdot \\ \cdot & \cdot & \cdot & \cdot & \cdot & \cdot \\ \cdot & \cdot & \cdot & \cdot & \cdot & \cdot \\ b_{N1} & b_{N2} & \cdot & \cdot & \cdot & b_{NN} \end{bmatrix}$$

then Equation 2.42 gives

$$[A^{(2)}] = [A^{(1)}][A^{(1)}] \tag{2.43}$$

Equation 2.43 indicates that the weighting coefficients for the second order derivative can be computed by the matrix multiplication of the weighting coefficient matrix for the first order derivative. Similarly, the mth order derivative of $f(x)$ can be written as

$$\frac{\partial^{(m)} f}{\partial x^{(m)}} = \frac{\partial}{\partial x}\left(\frac{\partial^{(m-1)} f}{\partial x^{(m-1)}} \right) = \frac{\partial^{(m-1)}}{\partial x^{(m-1)}}\left(\frac{\partial f}{\partial x} \right) \tag{2.44}$$

Let $[A^{(m-1)}]$, $[A^{(m)}]$ be the weighting coefficient matrices of the (m-1)th and mth order derivatives respectively. Then the application of DQ approximation to Equation 2.44 results in the following recurrence relationship

$$[A^{(m)}]=[A^{(1)}][A^{(m-1)}]=[A^{(m-1)}][A^{(1)}]\,,\quad m=2,3,...,N-1 \tag{2.45}$$

Although Equation 2.45 looks simple, it involves more arithmetic operations as compared to Equations 2.36 and 2.38. It is noted that the calculation of each weighting coefficient by Equation 2.43 or 2.45 involves N multiplications and (N-1) additions, i.e., a total of (2N-1) arithmetic operations. On the other hand, Equation 2.24 or the recurrence relationship 2.36 only involves two multiplications, one division, and one subtraction, i.e., a total of four arithmetic operations for the calculation of each off-diagonal weighting coefficient, which is independent of the number of grid points N. The calculation of each diagonal weighting coefficient from Equation 2.25 or 2.38 involves (N-2) subtractions. Thus, the number of arithmetic operations for Equations 2.36 and 2.38 is substantially smaller than that for Equation 2.45. Consequently, with increasing number of grid points, the calculation of the weighting coefficients from Equations 2.36 and 2.38 ought to be relatively inexpensive as well as more accurate due to smaller rounding-off errors, than that using Equation 2.45. In addition, the conditions on Equation 2.25 and 2.38 are not guaranteed by Equation 2.45. Practical applications showed that this may affect the accuracy of the numerical solution.

2.4 Error Analysis

It is interesting to analyze the errors resulting from the approximation of a function and its derivatives. Shu (1991) has given a thorough error analysis in his PhD thesis. In this section, we will start with the error estimate for the function approximation, followed by the error estimate for the derivative approximation. The error distributions for some derivative approximations given by Chen (1996) are also shown in this section.

2.4.1 The Function Approximation

Firstly, we discuss the approximation error when $f(x)$ is approximated by a polynomial of degree (N-1), particularly by the Lagrange interpolation polynomial,

$$P_N f=\sum_{i=1}^{N} f(x_i)\cdot r_i(x) \tag{2.46}$$

where $r_i(x)$ is the Lagrange interpolation polynomial given by Equation 2.11*c*. The approximation error of $f(x)$ is defined as

$$E(f) = f(x) - P_N f \tag{2.47}$$

If the *N*th order derivative of the function $f(x)$ is assumed to be a constant, say *K*, then $f(x)$ can be expressed as

$$f(x) = m_0 + m_1 x + m_2 x^2 + \cdots + m_{N-1} x^{N-1} + \frac{Kx^N}{N!} \tag{2.48}$$

Since Equation 2.46 is exactly satisfied for a polynomial of degree less than or equal to (*N*-1), we have

$$E(x^k) = 0, \ k = 0, 1, ..., N-1 \tag{2.49}$$

Substituting Equation 2.48 into Equation 2.47 and using Equation 2.49, we obtain

$$E(f) = K \cdot E(x^N) / N! \tag{2.50}$$

where

$$E(x^N) = x^N - \sum_{i=1}^{N} x_i^N \cdot r_i(x)$$

On the other hand, substituting the polynomial of degree (*N*-1), $g(x) = x^N - M(x)$, into Equation 2.47, where *M*(*x*) is defined by Equation 2.7, we obtain

$$E(g) = x^N - M(x) - \sum_{i=1}^{N} \left[x_i^N - M(x_i) \right] \cdot r_i(x) = 0 \tag{2.51}$$

Since $M(x_i) = 0$, Equation 2.51 can be further reduced to

$$x^N - \sum_{i=1}^{N} x_i^N \cdot r_i(x) = M(x) \tag{2.52}$$

Finally, substituting Equation 2.52 into Equation 2.50, we get

$$E(f) = K \cdot M(x) / N! \tag{2.53}$$

In most cases, the *N*th order derivative of $f(x)$ is not a constant, but it may be bounded. In this case, we can adopt another method to analyze *E(f)*. For simplicity, we set $\phi(x) = P_N f$ and define a function *F*(*z*) as

$$F(z) = f(z) - \phi(z) - c \cdot M(z) \tag{2.54}$$

Clearly, when $z = x_1, x_2, ..., x_N$, $F(z) = 0$. If we set $F(x) = 0$, we can obtain

$$E(f) = f(x) - P_N f = f(x) - \phi(x) = c \cdot M(x) \tag{2.55}$$

Since $F(z)$ has (*N*+1) roots, $x, x_1, x_2, ..., x_N$, in the domain, by repeated applications of Rolle's theorem, the *N*th order derivative of $F(z)$, $F^{(N)}(z)$, is found to have at least one root lying between x_1 and x_N. Denoting this root by ξ, we have

$$F^{(N)}(\xi) = 0 \tag{2.56}$$

Note that ϕ (*z*) is a polynomial of degree (*N*-1). So from Equation 2.54, we obtain

$$c = f^{(N)}(\xi) / N! \tag{2.57}$$

Hence,

$$E(f) = f^{(N)}(\xi) \cdot M(x) / N! \tag{2.58}$$

In general, ξ is a function of *x*.

2.4.2 The Derivative Approximation

The error for the *m*th order derivative approximation can be defined as

$$E_D^{(m)}(f) = \frac{\partial^m f}{\partial x^m} - \frac{\partial^m (P_N f)}{\partial x^m} = \frac{\partial^m f}{\partial x^m} - \frac{\partial^m \phi}{\partial x^m} \tag{2.59}$$

where $m = 1, 2, ..., N-1$. Using Equation 2.58, Equation 2.59 can be written as

$$E_D^{(m)}(f) = \frac{1}{N!} \frac{\partial^m \left[f^{(N)}(\xi) \cdot M(x) \right]}{\partial x^m} \tag{2.60}$$

Since ξ is an unknown function of x, it is difficult to estimate $E_D^{(m)}(f)$ using Equation 2.60. As a special case, if we assume that the Nth order derivative of $f(x)$ is a constant, say K, Equation 2.60 can be simplified to

$$E_D^{(m)}(f) = \frac{K \cdot M^{(m)}(x)}{N!} \tag{2.61}$$

Then for this case, the error distribution of derivative approximation can be easily obtained from Equation 2.61 as

$$E_D^{(m)}[f(x_i)] = \frac{K \cdot M^{(m)}(x_i)}{N!} \tag{2.62}$$

Using Equation 2.37, we have

$$M^{(m)}(x_i) = m w_{ii}^{(m-1)} M^{(1)}(x_i) \tag{2.63}$$

where $w_{ii}^{(m-1)}$ is the diagonal weighting coefficient of the (m-1)th order derivative, and $M^{(1)}(x_i)$ is given by Equation 2.8. So, substituting Equation 2.63 into Equation 2.62, we obtain

$$E_D^{(m)}[f(x_i)] = \frac{m K w_{ii}^{(m-1)} M^{(1)}(x_i)}{N!} \tag{2.64}$$

$$\text{for } m = 1, 2, \ldots, N-1$$

For the general case where $f^{(N)}(\xi)$ is not a constant, we can use a similar method as in the analysis of the function approximation to conduct error analysis of the derivative approximation. Since $g(z) = f(z) - \phi(z)$ has N roots in the interval, according to Rolle's theorem, its mth order derivative $g^{(m)}(z)$ has at least (N-m) roots in the interval, namely, $\bar{x}_1, \bar{x}_2, \ldots, \bar{x}_{N-m}$. Thus, the function

$$F^{(m)}(z) = g^{(m)}(z) - \bar{c} \cdot \overline{M}(z) = f^{(m)}(z) - \phi^{(m)}(z) - \bar{c} \cdot \overline{M}(z) \tag{2.65}$$

where $\overline{M}(z) = (z - \overline{x}_1)(z - \overline{x}_2)\cdots(z - \overline{x}_{N-m})$, would vanish at $\overline{x}_1, \overline{x}_2, \ldots, \overline{x}_{N-m}$. Now, if we set $F^{(m)}(\overline{x}) = 0$, where $\overline{x}$ is different from $\overline{x}_1, \overline{x}_2, \ldots, \overline{x}_{N-m}$, then $F^{(m)}(z)$ has $(N - m + 1)$ roots, and

$$E_D^{(m)}[f(\overline{x})] = f^{(m)}(\overline{x}) - \phi^{(m)}(\overline{x}) = \overline{c} \cdot \overline{M}(\overline{x}) \tag{2.66}$$

Using Rolle's theorem repeatedly ((*N*-*m*) times), the (*N*-*m*)th order derivative of $F^{(m)}(z)$ is found to have at least one root $\overline{\xi}$, i.e.,

$$f^{(N)}(\overline{\xi}) - \overline{c} \cdot (N - m)! = 0$$

so

$$\overline{c} = f^{(N)}(\overline{\xi}) / (N - m)! \tag{2.67}$$

Using Equation 2.67, Equation 2.66 can be reduced to

$$E_D^{(m)}[f(\overline{x})] = \frac{f^{(N)}(\overline{\xi}) \cdot \overline{M}(\overline{x})}{(N - m)!} \tag{2.68}$$

Equation 2.68 can be used to estimate the error of the derivative approximation. If it is assumed that all the coordinates are in the interval *h*, and the *N*th order derivative of the function $f(x)$ is bounded, then

$$\left| f^{(N)}(\overline{\xi}) \right| \le C \text{, where } C \text{ is a positive constant}$$

$$\left| M^{(m)}(x) \right| \le N(N-1)\cdots(N-m+1)h^{N-m}$$

$$\left| \overline{M}(\overline{x}) \right| \le h^{N-m}$$

So, Equations 2.62 and 2.68 can be simplified to

$$\left| E_D^{(m)}(f) \right| \le \frac{C \cdot h^{N-m}}{(N-m)!}, \quad \text{for } m = 1, 2, \ldots, N-1 \tag{2.69}$$

For the general case, the error distribution of the derivative approximation can also be studied using Equation 2.60. The error distributions of the first, second, third, and fourth order derivative approximation have been studied by Chen (1996) in his PhD work. For the first order derivative, Equation 2.60 gives

$$E_D^{(1)}(f) = \frac{f^{(N+1)}(\xi)\cdot\xi_x\cdot M(x) + f^{(N)}(\xi)\cdot M^{(1)}(x)}{N!} \tag{2.70}$$

Let $K_1 = \max\left\{\left|f^{(N)}(\xi)\right|\right\}$ and note that $M(x_i)=0$. Applying Equation 2.70 at the grid point x_i gives

$$E_D^{(1)}[f(x_i)] = \frac{f^{(N)}(\xi)\cdot M^{(1)}(x_i)}{N!} \le \frac{K_1\cdot M^{(1)}(x_i)}{N!} = K_1 e^{(1)}(x_i) \tag{2.71}$$

$$\text{for } i = 1, 2, \ldots, N$$

where

$$e^{(1)}(x_i) = \frac{M^{(1)}(x_i)}{N!} \tag{2.72}$$

is the error distribution of the first order derivative approximation. For the second order derivative, we obtain from Equation 2.60

$$E_D^{(2)}[f(x_i)] = \frac{2\xi_x f^{(N+1)}(\xi)M^{(1)}(x_i)}{N!} + \frac{f^{(N)}(\xi)M^{(2)}(x_i)}{N!} \tag{2.73}$$

$$\text{for } i = 1, 2, \ldots, N$$

On the other hand, using Equation 2.63, we have

$$M^{(2)}(x_i) = 2\cdot w_{ii}^{(1)} M^{(1)}(x_i)$$

Let $K_2 = \max\{\left|f^{(N)}(\xi)\right|, \left|\xi_x f^{(N+1)}(\xi)\right|\}$, then Equation 2.73 can be reduced to

$$E_D^{(2)}[f(x_i)] \le 2K_2\left(1+\left|w_{ii}^{(1)}\right|\right)\frac{M^{(1)}(x_i)}{N!} = K_2 e^{(2)}(x_i) \tag{2.74}$$

where

$$e^{(2)}(x_i) = 2\left(1+\left|w_{ii}^{(1)}\right|\right)\frac{M^{(1)}(x_i)}{N!} \tag{2.75}$$

is the error distribution of the second order derivative approximation. Similarly, we can obtain

$$E_D^{(3)}[f(x_i)] \le 3K_3\left(2+2\left|w_{ii}^{(1)}\right|+\left|w_{ii}^{(2)}\right|\right)\frac{M^{(1)}(x_i)}{N!} = K_3 e^{(3)}(x_i) \tag{2.76}$$

$$E_D^{(4)}[f(x_i)] \le 4K_4\left(5+6\left|w_{ii}^{(1)}\right|+3\left|w_{ii}^{(2)}\right|+\left|w_{ii}^{(3)}\right|\right)\frac{M^{(1)}(x_i)}{N!} = K_4 e^{(4)}(x_i) \tag{2.77}$$

for the third and fourth order derivatives, respectively, where K_3 and K_4 are the maximum values of the composite derivatives of ξ and $f(x)$ up to order (N+3), and

$$e^{(3)}(x_i) = 3\left(2+2\left|w_{ii}^{(1)}\right|+\left|w_{ii}^{(2)}\right|\right)\frac{M^{(1)}(x_i)}{N!} \tag{2.78}$$

$$e^{(4)}(x_i) = 4\left(5+6\left|w_{ii}^{(1)}\right|+3\left|w_{ii}^{(2)}\right|+\left|w_{ii}^{(3)}\right|\right)\frac{M^{(1)}(x_i)}{N!} \tag{2.79}$$

are the error distributions of the third and fourth order derivatives.

2.5 Relationship Between PDQ and Other Approaches

As shown above, PDQ is based on the polynomial approximation. Some other methods such as the finite difference scheme and the spectral collocation method are also based on the polynomial approximation. The relationship between PDQ and these methods will be discussed in this section.

2.5.1 Relationship Between PDQ and Finite Difference Scheme

2.5.1.1 Generation of Finite Difference Scheme

Finite difference scheme is a most popular approach used in engineering. Basically, the finite difference scheme can be generated from two ways. One way is to use Taylor series expansion. The other way is based on the polynomial approximation. Like the PDQ approach, the finite difference scheme is a numerical discretization technique which uses the functional values at local and neighbouring points to approximate the derivatives in a PDE. Thus, an important step for the finite difference scheme is to determine how many neighbouring points are involved in the scheme before one starts to generate it.

Consider a function $f(x)$. Using Taylor series expansion, the functional value at a point $x_i + s$ can be approximated by the function and its derivatives at the point x_i, that is,

$$f(x_i + s) = f(x_i) + f^{(1)}(x_i) \cdot s + \frac{1}{2!} f^{(2)}(x_i) \cdot s^2 + \cdots + \frac{1}{k!} f^{(k)}(x_i) \cdot s^k + \cdots \quad (2.80)$$

where $f^{(k)}(x_i)$ represents the kth order derivative at the point x_i. It is noted that in the finite difference scheme, the functional values at the grid points are considered as known values, which are used to approximate the unknown derivatives. So, in Taylor series expansion, the derivatives $f^{(1)}(x_i), f^{(2)}(x_i), \ldots f^{(k)}(x_i), \ldots$ are taken as unknowns. Then, the critical step is to determine which term the Taylor series expansion should be truncated to. Once the truncated term is decided, the number of derivatives (unknowns) is determined. For example, if Taylor series expansion is truncated after the term $f^{(K)}(x_i) \cdot s^K / K!$, the number of unknowns is K (unknowns are $f^{(1)}(x_i)$, $f^{(2)}(x_i)$,..., $f^{(K)}(x_i)$). To solve for these unknowns, we need the same number of equations. On the other hand, we note that the application of Taylor series expansion at a grid point provides one equation. Since the number of unknowns is K, Taylor series expansion should be applied at K neighbouring points, giving K equations. The above is a basic principle for the generation of a difference scheme. The value of K depends on the order of the given derivative and the required order of accuracy.

Consider an nth order derivative of a function $f(x)$. We want to generate a finite difference scheme with pth order of accuracy at a grid point x_i. In other words, the required difference scheme is

$$f^{(n)}(x_i) = \text{difference expression} + \mathrm{O}(s^p) \quad (2.81)$$

For the general case, we assume that the mesh is non-uniform. To solve this problem, we notice that the term involving $f^{(n)}(x_i)$ in Taylor series expansion 2.80 is $f^{(n)}(x_i) \cdot s^n / n!$. Thus, to get the expression for $f^{(n)}(x_i)$, each term in Taylor series expansion should be divided by s^n. To meet the requirement of Equation 2.81, the truncation error term in the Taylor series expansion should be the order s^{n+p}. Clearly, the term of order s^{n+p} divided by s^n is in the order of s^p, which is the required order of accuracy. From this analysis, we know that the Taylor series expansion should include terms up to and including $f^{(n+p-1)}(x_i) \cdot s^{n+p-1} /(n+p-1)!$. As a result, the number of unknowns (derivatives) in the Taylor series expansion is $(n + p - 1)$. Therefore, Taylor series expansion has to be applied at $(n + p - 1)$ neighbouring points, providing $(n + p - 1)$ equations. In other words, $(n + p - 1)$ neighbouring points will be involved in the difference expression in Equation 2.81. Next, the difference expression can be generated by either the Taylor series expansion approach or the polynomial approximation technique, which will be demonstrated in the following examples.

Example 1 (Taylor series expansion approach): For a one-dimensional problem, discretize the first order derivative $\partial f / \partial x$ at the left boundary with second order of accuracy. The mesh is shown in the following figure.

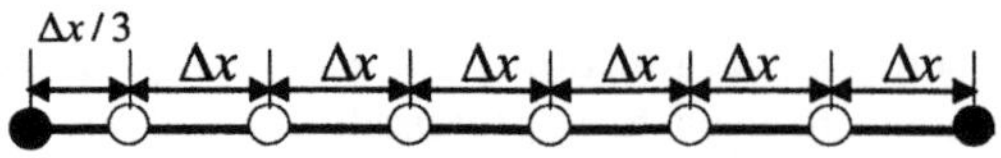

Figure 2.1 A non-uniform mesh

It is supposed that the number of grid points is N. In other words, x_1 and x_N are the left and right boundary points respectively. Following the above-mentioned principle, the number of neighbouring points involved in the difference scheme is $n + p - 1 = 1 + 2 - 1 = 2$. Thus, applying Taylor series expansion at two neighbouring points x_2 and x_3 gives

$$f_2 = f_1 + \left(\frac{\partial f}{\partial x}\right)_1 \cdot \frac{\Delta x}{3} + \frac{1}{2}\left(\frac{\partial^2 f}{\partial x^2}\right)_1 \cdot \left(\frac{\Delta x}{3}\right)^2 + O(\Delta x^3),\ s = \frac{\Delta x}{3} \tag{2.82}$$

$$f_3 = f_1 + \left(\frac{\partial f}{\partial x}\right)_1 \cdot \frac{4\Delta x}{3} + \frac{1}{2}\left(\frac{\partial^2 f}{\partial x^2}\right)_1 \cdot \left(\frac{4\Delta x}{3}\right)^2 + O(\Delta x^3),\ s = \frac{4\Delta x}{3} \tag{2.83}$$

$16 \times$ Equation 2.82 - Equation 2.83 gives

$$16 f_2 - f_3 = 15 f_1 + 4\left(\frac{\partial f}{\partial x}\right)_1 \cdot \Delta x + O(\Delta x^3)$$

So,

$$\left(\frac{\partial f}{\partial x}\right)_1 = \frac{16 f_2 - f_3 - 15 f_1}{4\Delta x} + O(\Delta x^2)$$

Example 2 (Polynomial approximation technique): As shown in the following figure, it is supposed that a function $f(x)$ is approximated by a second degree polynomial between the points W and E. Derive the expressions for the first order derivative approximation at the point P using the functional values at the three points.

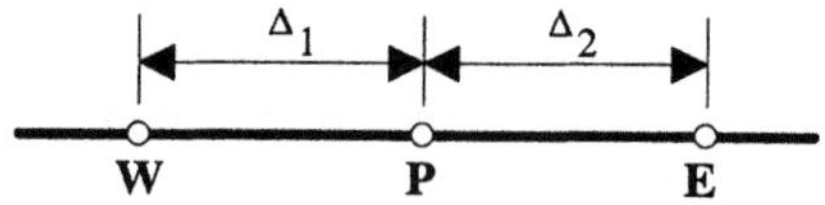

Figure 2.2 Illustration of three mesh points

Let the origin be the point P and the coordinates of the three points be $x_W = -\Delta_1$, $x_P = 0$, and $x_E = \Delta_2$. According to the problem, $f(x)$ is approximated by

$$f(x) = a + bx + cx^2 \tag{2.84}$$

Let Equation 2.84 be satisfied at the three points *W*, *P*, and *E*. We have

$$\begin{cases} f_W = a - b\Delta_1 + c\Delta_1^2 \\ f_P = a \\ f_E = a + b\Delta_2 + c\Delta_2^2 \end{cases}$$

Solving the above equation system, we obtain

$$a = f_P$$

$$b = \frac{(f_E - f_P)\Delta_1^2 - (f_W - f_P)\Delta_2^2}{\Delta_1\Delta_2(\Delta_1 + \Delta_2)}$$

$$c = \frac{(f_E - f_P)\Delta_1 + (f_W - f_P)\Delta_2}{\Delta_1\Delta_2(\Delta_1 + \Delta_2)}$$

Thus, at the point *P*,

$$\left(\frac{df}{dx}\right)_P = b = \frac{(f_E - f_P)\Delta_1^2 - (f_W - f_P)\Delta_2^2}{\Delta_1\Delta_2(\Delta_1 + \Delta_2)} \tag{2.85}$$

Obviously, when $\Delta_1 = \Delta_2$, Equation 2.85 yields a central difference scheme.

Example 3: It is known that the following finite difference scheme is of second order accuracy,

$$\left(\frac{d^n u}{dx^n}\right)_i = a_1 u_{i-3} + a_2 u_{i-2} + a_3 u_{i-1} + a_4 u_i + a_5 u_{i+1} + a_6 u_{i+2} \tag{2.86}$$

where $a_1, a_2, a_3, a_4, a_5, a_6$ are constants. Determine the order of the approximated derivative, n. The mesh is assumed to be non-uniform.

To solve this problem, we notice that, according to the principle described previously, the number of neighbouring points in the difference scheme is (n+p-1). For this problem, the number of neighbouring points is 5 and $p = 2$. So,

$$n + p - 1 = n + 2 - 1 = 5 \quad \Rightarrow \quad n = 4$$

2.5.1.2 Relationship Between PDQ and Highest Order Finite Difference Scheme

This relationship has been shown in the works of Shu (1991), Shu and Chew (1998). For the general one-dimensional case, we suppose that the mesh is non-uniform. It is known from the above analysis that n grid points (local point plus the neighbouring points) can generate an (n-m)th order finite difference scheme for the mth order derivative where m = 1, 2, ..., n-1. Therefore, if there are only N grid points in the whole domain, the (N-m)th order finite difference scheme for the mth order derivative should be the highest order finite difference scheme. For the first order derivative, the difference scheme of order (N-1) is the highest order finite difference scheme which can be written as a linear sum of all the functional values at N grid points, taking the same form as Equation 2.1, i.e.,

$$f_x^{(1)}(x_i) = \sum_{j=1}^{N} a_{ij} \cdot f(x_j) \quad , \quad \text{for } i = 1, 2, \cdots, N \tag{2.87}$$

where $f_x^{(1)}(x_i)$ indicates the first order derivative of $f(x)$ at x_i, and a_{ij} are the weighting coefficients. Obviously, the highest order finite difference scheme is a global approximation approach since it uses all the functional values in the whole computational domain. As we know, the underlying mathematical tool for finite difference schemes is the Taylor series expansion. Therefore, the weighting coefficients in Equation 2.87 could be determined from Taylor series expansion. Using Taylor series expansion, $f(x_j)$ can be expressed as

$$\begin{aligned} f(x_j) = f(x_i) + f^{(1)}(x_i)(x_j - x_i) + \cdots + f^{(k)}(x_i)(x_j - x_i)^k / k! \\ + \cdots + f^{(N-1)}(x_i)(x_j - x_i)^{N-1} / (N-1)! + R_N \end{aligned} \tag{2.88}$$

where $f^k(x_i)$ is the kth order derivative of $f(x)$ at x_i. R_N is the truncation error, which can be written as

$$R_N = f^{(N)}(\xi) \cdot (x_j - x_i)^N / N!, \quad \xi \in [x_i, x_j] \tag{2.89}$$

Substituting Equation 2.88 into Equation 2.87 and retaining the (N-1)th order accuracy lead to the following algebraic equations

$$\begin{cases} \sum_{j=1}^{N} a_{ij} = 0 \\ \sum_{j=1}^{N} a_{ij} \cdot (x_j - x_i) = 1 \\ \sum_{j=1}^{N} a_{ij} \cdot (x_j - x_i)^k = 0, \quad k = 2, 3, ..., N-1 \end{cases} \tag{2.90}$$

for $i = 1, 2, \ldots, N$

Equation 2.90 is an equation system for the determination of the weighting coefficients of the highest order finite difference scheme for the first order derivative. Similarly, the highest order finite difference scheme for the *m*th order derivative of the function $f(x)$ at x_i, $f_x^{(m)}(x_i)$, can be written as

$$f_x^{(m)}(x_i) = \sum_{j=1}^{N} w_{ij}^{(m)} \cdot f(x_j) \tag{2.91}$$

for $i = 1, 2, \ldots, N;\ m = 2, 3, \ldots, N\text{-}1$

where $w_{ij}^{(m)}$ are the weighting coefficients. Substituting Equation 2.88 into Equation 2.91, and retaining (*N*-*m*)th order accuracy, we obtain

$$\begin{cases} \sum_{j=1}^{N} w_{ij}^{(m)} = 0 \\ \sum_{j=1}^{N} w_{ij}^{(m)} \cdot (x_j - x_i)^m = m! \\ \sum_{j=1}^{N} w_{ij}^{(m)} \cdot (x_j - x_i)^k = 0, \quad k = 1, 2, ..., N-1, \text{but } k \neq m \end{cases} \tag{2.92}$$

Equation set 2.92 is an algebraic equation system for determining the weighting coefficients of the second and higher order derivatives in the highest order finite difference scheme. In the following, we will show that all these equation systems are equivalent to the PDQ equation systems. In other words, the weighting coefficients in the highest order finite difference schemes can be computed by the PDQ formulations, and the PDQ method can be considered as the highest order finite difference scheme.

As stated above, from the analysis of high order polynomial approximation in a linear polynomial vector space, the equation system for the determination of the weighting coefficients derived from one set of base vectors is equivalent to that derived from another set of base vectors. Thus, without loss of generality, we will choose only one Equation System 2.3 obtained from the base vectors $x^k, k = 0, 1, \ldots, N-1$ and prove that this PDQ equation system is equivalent to the Equation System 2.90 given by the highest order finite difference scheme.

It is obvious that the first equation of Equation sets 2.3 and 2.90 are the same, i.e.

$$\sum_{j=1}^{N} a_{ij} = 0 \tag{2.93}$$

Furthermore, it can be shown that the second equation of the two systems is the same, i.e.,

$$\sum_{j=1}^{N} a_{ij} \cdot (x_j - x_i) - 1 = \sum_{j=1}^{N} a_{ij} \cdot x_j - 1 = 0 \tag{2.94}$$

Now, assuming that the first $p+1$ equations of the two systems are the same, that is

$$\sum_{j=1}^{N} a_{ij} \cdot (x_j - x_i)^k = \sum_{j=1}^{N} a_{ij} \cdot x_j^k - k \cdot x_i^{k-1} = 0 \tag{2.95}$$
$$\text{for } k = 0, 1, \ldots, p; \; i = 1, 2, \ldots, N$$

and using the binomial expression

$$(a-b)^p = a^p - c_p^1 \cdot a^{p-1} b + \cdots + (-1)^k c_p^k a^{p-k} b^k + \cdots + (-1)^p b^p \tag{2.96}$$

where c_p^k is the combination of p terms taken k at a time, and setting $a = b = 1$, the following expression can be obtained

$$c_p^1 - c_p^2 + \cdots + (-1)^{k+1} c_p^k + \cdots + (-1)^{p+1} = 1 \tag{2.97}$$

Using Equation 2.96, the $(p+2)$th equation of Equation set 2.90 can be written as

$$\begin{aligned} \sum_{j=1}^{N} a_{ij} \cdot (x_j - x_i)^{p+1} = \sum_{j=1}^{N} a_{ij} \cdot x_j^{p+1} - c_{p+1}^1 \cdot x_i \Bigg[\sum_{j=1}^{N} a_{ij} (x_j^p \\ - \frac{1}{2} c_p^1 \cdot x_j^{p-1} \cdot x_i + \cdots + \frac{(-1)^p x_i^p}{p+1}) \Bigg] \end{aligned} \tag{2.98}$$

Substituting Equations 2.95, 2.97 into Equation 2.98 leads to

$$\sum_{j=1}^{N} a_{ij} \cdot (x_j - x_i)^{p+1} = \sum_{j=1}^{N} a_{ij} \cdot x_j^{p+1} - (p+1) \cdot x_i^p \qquad (2.99)$$

Equation 2.99 demonstrates the equivalence of the (p+2)th equation of the two systems. Since p is an arbitrary integer only if $p \le N$-2, it has thus been proven that the two Equation Systems 2.3 and 2.90 are the same.

Next, we intend to show that Equations 2.36 and 2.38 given by the PDQ method satisfy Equation set 2.92. Obviously, the first equation of Equation 2.92 is exactly the same as Equation 2.38. To prove that $w_{ij}^{(m)}$, for $2 \le m \le N$-1, satisfies the other equations of Equation 2.92, we suppose that $w_{ij}^{(m-1)}$ satisfies those equations first, that is,

$$\sum_{j=1}^{N} w_{ij}^{(m-1)} \cdot (x_j - x_i)^k = \begin{bmatrix} (m-1)! & \text{when } k = m-1 \\ 0 & \text{others} \end{bmatrix} \qquad (2.100)$$

Using Equation 2.36, we have, for $1 \le k \le N$-1

$$\sum_{j=1}^{N} w_{ij}^{(m)} (x_j - x_i)^k = m w_{ii}^{(m-1)} \sum_{j=1}^{N} a_{ij} (x_j - x_i)^k + m \sum_{j=1}^{N} w_{ij}^{(m-1)} (x_j - x_i)^{k-1} \qquad (2.101)$$

Substituting Equations 2.90 and 2.100 into Equation 2.101 leads to

$$\sum_{j=1}^{N} w_{ij}^{(m)} \cdot (x_j - x_i)^k = \begin{bmatrix} m! & \text{when } k = m \\ 0 & \text{others} \end{bmatrix} \qquad (2.102)$$

Since m is an arbitrary integer for $2 \le m \le N$-1, we have proven that $w_{ij}^{(m)}$ given from the PDQ method satisfies exactly Equation set 2.92 derived from the highest order finite difference scheme. So, it can be concluded that the PDQ method is an extension of finite difference methods, and is actually the highest order finite difference scheme. It should be noted that Equations 2.87 and 2.91 bear the same form, which can be applied to both interior points and boundary points, and can also be applied to a uniform mesh or a non-uniform mesh. Actually, the implementation of these equations is much easier than that for low order finite difference schemes.

2.5.2 Relationship Between PDQ and Chebyshev Collocation Method

The spectral method is a global approach, which uses a set of base functions to approximate the solution of a PDE. The approximate error in the spectral method is controlled by a weighted integral. Thus, the key elements of the spectral method are the base functions and the weighting functions. According to the choice of the weighting functions, there are three most commonly used spectral schemes, namely, the Galerkin scheme where the weighting functions are taken to be the same as the base functions; the collocation scheme where the weighting functions are the delta functions centered at the collocation points; and the Tau scheme which is similar to the Galerkin scheme in the way that the differential equation is enforced, but the weighting functions do not need to satisfy the boundary conditions. The spectral method has been extensively applied in many engineering problems. For details of the spectral method, the reader is advised to consult the books of Canuto et al. (1988), and Gottlieb and Orszag (1977). Among spectral methods, the collocation method, especially the Chebyshev collocation method, is the most popular scheme. The relationship between PDQ and the collocation method has been discussed by Shu (1991) and Bellomo (1997). In the following, we will show that when the grid points are chosen to be the Chebyshev collocation points, the PDQ method is identical to the Chebyshev collocation method. Some comments on the two methods are also given.

First, we consider the Chebyshev collocation method. Following the work of Canuto et al. (1988), a smooth function $f(x)$ in the interval $-1 \le x \le 1$ can be expanded by the Chebyshev polynomial as

$$f(x) = \sum_{k=0}^{N} c_k \cdot T_k(x) \tag{2.103}$$

where c_k is a constant, $T_k(x)$ is the Chebyshev polynomial of degree k defined by

$$T_k(x) = \cos k\theta, \qquad \theta = \arccos x \tag{2.104}$$

The Chebyshev collocation points are chosen as the roots of $|T_N(x)| = 1$, which are given by

$$x_k = \cos(k\pi / N), \qquad 0 \le k \le N \tag{2.105}$$

Using Equation 2.103, the function value of $f(x)$ at a collocation point x_i can be expressed by

$$f(x_i) = \sum_{k=0}^{N} c_k \cdot T_k(x_i) \tag{2.106}$$

and accordingly, the first order derivative of $f(x)$ can be approximated by

$$f_x^{(1)}(x_i) = \sum_{k=0}^{N} d_{ik} \cdot f(x_k) \tag{2.107}$$

where

$$d_{ik} = \frac{(-1)^{i+k}\bar{c}_i}{(x_i - x_k)\bar{c}_k}, \qquad \text{for } 0 \le i,k \le N,\ i \ne k \tag{2.108a}$$

$$d_{ii} = -\frac{x_i}{2(1-x_i^2)}, \qquad \text{for } 1 \le i \le N-1 \tag{2.108b}$$

$$d_{00} = -d_{NN} = \frac{2N^2+1}{6} \tag{2.108c}$$

$$\bar{c}_0 = \bar{c}_N = 2$$
$$\bar{c}_k = 1, 1 \le k \le N-1$$

Now, we consider the PDQ method. When the grid points are chosen as the Chebyshev collocation points, that is

$$x_i = \cos(\theta_i), \quad \theta_i = i\pi / N, \quad i = 0, 1, \ldots, N \tag{2.109}$$

then the function $M(x)$ in Equation 2.7 takes the following form

$$M(x) = (1-x^2)T_N^{(1)}(x) \tag{2.110}$$

where $T_N^{(1)}(x)$ is the first order derivative of $T_N(x)$. Using the following formulations

$$x = \cos(\theta), \qquad T_N(x) = \cos(N\theta)$$

Equation 2.110 can be reduced to

$$M(x) = M(\theta) = N\sin(\theta)\sin(N\theta) \tag{2.111}$$

Differentiating Equation 2.111 successively, we obtain

$$M^{(1)}(x) = -\frac{N\cos(\theta)\sin(N\theta) + N^2\sin(\theta)\cos(N\theta)}{\sin(\theta)} \tag{2.112}$$

$$M^{(2)}(x) = \frac{N\cos(\theta)[N\sin(\theta)\cos(N\theta) - \cos(\theta)\sin(N\theta)]}{\sin^3(\theta)} - \frac{N\sin(\theta)\sin(N\theta)(1+N^2)}{\sin^2(\theta)} \tag{2.113}$$

Since $N\theta_i = i\pi$, when $\sin(\theta_i) \neq 0$, i.e., $i \neq 0, N$, Equations 2.112 and 2.113 reduce to

$$M^{(1)}(x_i) = (-1)^{i+1} N^2 \tag{2.114a}$$

$$M^{(2)}(x_i) = (-1)^i N^2 \cdot \frac{x_i}{1 - x_i^2} \tag{2.114b}$$

When $\sin(\theta_i) = 0$, application of L'Hospital's rule gives

$$M^{(1)}(x_0) = -2N^2 \tag{2.115a}$$

$$M^{(1)}(x_N) = (-1)^{N+1} \cdot 2N^2 \tag{2.115b}$$

$$M^{(2)}(x_0) = -\frac{2}{3} N^2 (1 + 2N^2) \tag{2.115c}$$

$$M^{(2)}(x_N) = (-1)^N \frac{2}{3} N^2 (1 + 2N^2) \tag{2.115d}$$

Substituting Equations 2.114 and 2.115 into Equation 2.17, we obtain the following simplified forms

$$a_{ij} = \frac{(-1)^{i+j}\bar{c}_i}{(x_i - x_j)\bar{c}_j}, \qquad \text{for } 0 \le i, j \le N,\ i \neq j \tag{2.116a}$$

$$a_{ii} = -\frac{x_i}{2(1 - x_i^2)}, \qquad \text{for } 1 \le i \le N-1 \tag{2.116b}$$

$$a_{00} = -a_{NN} = \frac{2N^2 + 1}{6} \tag{2.116c}$$

where

$$\bar{c}_k = \begin{cases} 2 & \text{when k = 0, N} \\ 1 & \text{otherwise} \end{cases}$$

Equation 2.116 derived from the PDQ method is exactly the same as Equation 2.108 given by the Chebyshev collocation method.

The equivalence between the PDQ and Chebyshev collocation method is natural. As we know, the spectral method uses a set of orthogonal base functions to approximate the solution of a PDE, and the coordinates of the grid points are chosen as the roots of the base functions. Obviously, in this procedure, one has to choose the base functions first, and then determine the grid points from the base functions. In contrast, the PDQ adopts another procedure. In the PDQ, we firstly specify a distribution of the grid points, and then construct the base polynomials from the coordinates of the specified grid points. Definitely, this process is more flexible. In general, the base polynomials in the PDQ are not orthogonal. From the above processes, it can be easily deduced that when the grid points are taken to be the same for both methods, their polynomial approximation form should be identical, leading to identical derivative approximations. For practical applications, we recommend the use of the PDQ method because it allows the use of an arbitrary grid distribution.

2.6 Extension to the Multi-dimensional Case

In practice, most problems are two-dimensional or three-dimensional. Thus, it is necessary to extend the DQ approximation from the one-dimensional case to the multi-dimensional cases. In this section, we will only demonstrate the extension of the one-dimensional case to the two-dimensional case. Extension to the three-dimensional case can be carried out in the same manner as the extension to the two-dimensional case.

As discussed in the previous sections, the DQ approximation is actually based on polynomial approximations. Since polynomial approximations for regular and irregular domains are quite different, the extension of the DQ approximation to regular and irregular domains would obviously be different.

2.6.1 Direct Extension for Regular Domain

As shown by Shu (1991), the one-dimensional PDQ formulations can be directly extended to the multi-dimensional case if the discretization domain is regular. The regular domain could be a rectangle or other regular shapes such as a circle. Here, for simplicity, we only consider a rectangular domain for demonstration.

Consider a two-dimensional function $f(x, y)$ defined on a rectangular domain, as shown in Figure 2.3. It is clearly seen from the figure that along each horizontal line, the x interval is the same, and along each vertical line, the y interval is the same. Thus, we can use the same x coordinate distribution for each horizontal line, and the same y coordinate distribution for each vertical line.

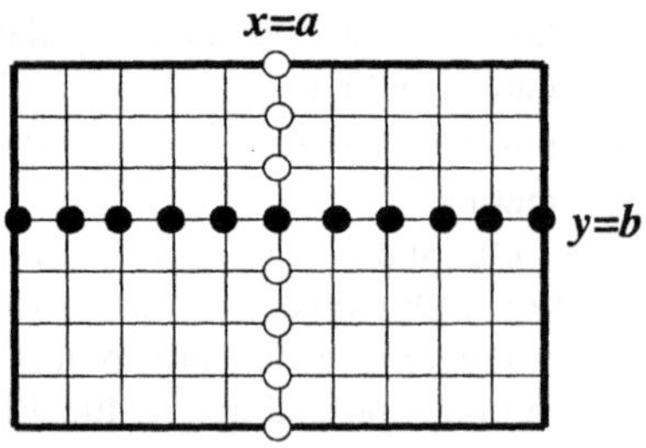

Figure 2.3 Mesh distribution on a regular domain

It is supposed that along a horizontal line $y = b$, where b is a constant, the values of $f(x,b)$, denoted by the filled circles, can be approximated by a polynomial of degree (N-1), $P_N(x)$, which constitutes an N-dimensional linear vector space V_N with N base vectors (polynomials). Along a vertical line $x = a$, where a is a constant, the values of $f(a,y)$, denoted by the open circles, can be approximated by a polynomial of degree (M-1), $P_M(y)$, which constitutes a M-dimensional linear vector space V_M with M base vectors (polynomials) $s_j(y), j = 1, 2, ..., M$. The value of the function $f(x,y)$ at any location in the domain can then be approximated by a polynomial $P_{N\times M}(x,y)$ of the form

$$f(x,y) \approx P_{N\times M}(x,y) = \sum_{i=1}^{N}\sum_{j=1}^{M} \bar{c}_{ij} \cdot x^{i-1} \cdot y^{j-1} \tag{2.117}$$

where $\bar{c}_{ij}$ is a coefficient. It is clear that $P_{N\times M}(x,y)$ constitutes an $N \times M$ dimensional linear polynomial vector space $V_{N\times M}$ with respect to the operation of vector addition and scalar multiplication. We will show that $\Phi_{ij}(x,y) = r_i(x)s_j(y)$ constitutes the base vectors (polynomials) in the linear vector space $V_{N\times M}$. Since $r_i(x), s_j(y)$ are the base vectors in V_N and V_M, they must be linearly independent, that is

$$\sum_{i=1}^{N} c_i \cdot r_i(x) = 0 \quad \text{only if} \quad c_i = 0\ , \ i = 1, 2, \cdots, N \tag{2.118a}$$

$$\sum_{j=1}^{M} d_j \cdot s_j(y) = 0 \quad \text{only if} \quad d_j = 0\ , \ j = 1, 2, \cdots, M \tag{2.118b}$$

Now we see that if

$$\sum_{i=1}^{N}\sum_{j=1}^{M} e_{ij} \cdot \Phi_{ij}(x,y) = 0 \quad , \text{ i.e., } \sum_{i=1}^{N}\left[\sum_{j=1}^{M} e_{ij} \cdot s_j(y)\right] \cdot r_i(x) = 0 \qquad (2.119)$$

then from Equation 2.118*a*, the following equation can be obtained

$$\sum_{j=1}^{M} e_{ij} \cdot s_j(y) = 0 \qquad (2.120)$$

Finally using Equation 2.118*b*, Equation 2.120 leads to $e_{ij} = 0$. This indicates that $\Phi_{ij}(x,y)$ is a set of base vectors in $V_{N\times M}$.

Now it is assumed that the following linear formulations are satisfied for the function $f(x,y)$ and its first order derivatives

$$f_x^{(1)}(x_i, y_j) = \sum_{k=1}^{N} a_{ik}^{x} \cdot f(x_k, y_j) \qquad (2.121a)$$

$$f_y^{(1)}(x_i, y_j) = \sum_{k=1}^{M} a_{jk}^{y} \cdot f(x_i, y_k) \qquad (2.121b)$$

$$\text{for } i = 1, 2, \cdots, N; \;\; j = 1, 2, \cdots, M$$

where $f_x^{(1)}, f_y^{(1)}$ are the first order derivatives of the function $f(x,y)$ with respect to x and y respectively, and a_{ik}^{x}, a_{jk}^{y} are the corresponding weighting coefficients. From the properties of a linear vector space, we know that if all the base polynomials $\Phi_{ij}(x,y)$ satisfy the linear Equation 2.121, so does any polynomial in $V_{N\times M}$. Substituting $\Phi_{ij}(x,y)$ into Equation. 2.121 leads to

$$\sum_{k=1}^{N} a_{ik}^{x} \cdot r_j(x_k) = r_j^{(1)}(x_i), \; i, j = 1,2,\ldots,N \qquad (2.122a)$$

$$\sum_{k=1}^{M} a_{ik}^{y} \cdot s_j(y_k) = s_j^{(1)}(y_i), \; i, j = 1,2,\ldots,M \qquad (2.122b)$$

where $r_j^{(1)}(x_i)$ represents the first order derivative of $r_j(x)$ at x_i and $s_j^{(1)}(y_i)$ represents the first order derivative of $s_j(y)$ at y_i. From Equation 2.122, it is obvious that a_{ik}^{x} or a_{jk}^{y} is only related to $r_i(x)$ or $s_j(y)$. Hence the formulation of the one dimensional case can be directly extended to the two dimensional case, that is

$$a_{ij}^x = \frac{M^{(1)}(x_i)}{(x_i - x_j)M^{(1)}(x_j)}, \quad \text{for } j \neq i \tag{2.123a}$$

$$a_{ii}^x = -\sum_{j=1, j\neq i}^{N} a_{ij}^x \tag{2.123b}$$

for $i, j = 1, 2, ..., N$

$$a_{ij}^y = \frac{P^{(1)}(y_i)}{(y_i - y_j)P^{(1)}(y_j)}, \quad \text{for } j \neq i \tag{2.124a}$$

$$a_{ii}^y = -\sum_{j=1, j\neq i}^{M} a_{ij}^y \tag{2.124b}$$

for $i, j = 1, 2, ..., M$

where

$$M^{(1)}(x_i) = \prod_{j=1, j\neq i}^{N} (x_i - x_j), \quad P^{(1)}(y_i) = \prod_{j=1, j\neq i}^{M} (y_i - y_j)$$

Similarly, for the second and higher order derivatives, the recurrence relationships of the weighting coefficients are obtained as follows

$$w_{ij}^{(n)} = n\left(a_{ij}^x \cdot w_{ii}^{(n-1)} - \frac{w_{ij}^{(n-1)}}{x_i - x_j}\right), \quad \text{for } j \neq i \tag{2.125a}$$

$$w_{ii}^{(n)} = -\sum_{j=1, j\neq i}^{N} w_{ij}^{(n)} \tag{2.125b}$$

for $n = 2, 3, ..., N$-1; $i, j = 1, 2, ..., N$

$$\overline{w}_{ij}^{(m)} = m\left(a_{ij}^y \cdot \overline{w}_{ii}^{(m-1)} - \frac{\overline{w}_{ij}^{(m-1)}}{y_i - y_j}\right), \quad \text{for } j \neq i \tag{2.126a}$$

$$\overline{w}_{ii}^{(m)} = -\sum_{j=1, j\neq i}^{M} \overline{w}_{ij}^{(m)} \tag{2.126b}$$

for $m = 2, 3, ..., M$-1; $i, j = 1, 2, ..., M$

where $w_{ij}^{(n)}$ are the weighting coefficients of the nth order derivative of $f(x, y)$ with respect to x at (x_i, y_j), namely $f_x^{(n)}(x_i, y_j)$ and $\overline{w}_{ij}^{(m)}$ are the weighting coefficients of the mth order derivative of $f(x, y)$ with respect to y at (x_i, y_j), namely $f_y^{(m)}(x_i, y_j)$. They satisfy

$$f_x^{(n)}(x_i,y_j) = \sum_{k=1}^{N} w_{ik}^{(n)} \cdot f(x_k,y_j) \tag{2.127a}$$

$$f_y^{(m)}(x_i,y_j) = \sum_{k=1}^{M} \overline{w}_{jk}^{(m)} \cdot f(x_i,y_k) \tag{2.127b}$$

for $i = 1, 2, ..., N;\ j = 1, 2, ..., M;\ n = 1, 2, ..., N\text{-}1;\ m = 1, 2, ..., M\text{-}1$

If the functional values at all grid points are obtained, it is easy to determine the functional value at any location in the domain using the Lagrange interpolation polynomials. Along a horizontal grid line $y = y_j$, the functional value can be computed from

$$f(x, y_j) = \sum_{i=1}^{N} f(x_i, y_j) r_i(x), \quad j = 1, 2, ..., M \tag{2.128}$$

and along a vertical grid line $x = x_i$, the functional value can be calculated from

$$f(x_i, y) = \sum_{j=1}^{M} f(x_i, y_j) s_j(y), \quad i = 1, 2, ..., N \tag{2.129}$$

At any location in the domain, $f(x, y)$ can be computed from

$$f(x, y) = \sum_{i=1}^{N} \sum_{j=1}^{M} f(x_i, y_j) \cdot r_i(x) \cdot s_j(y) \tag{2.130}$$

$r_i(x)$ and $s_j(y)$ in Equations 2.128-2.130 are the Lagrange interpolation polynomials given by

$$r_i(x) = \prod_{k=1, k\neq i}^{N} \frac{x - x_k}{x_i - x_k} \tag{2.131a}$$

$$s_j(y) = \prod_{k=1, k\neq j}^{M} \frac{y - y_k}{y_j - y_k} \tag{2.131b}$$

It should be indicated that the use of Equations 2.128-2.130 will not result in the loss of accuracy of the numerical solution since the PDQ discretization is also based on these polynomial approximations. Equations 2.128-2.130 are very useful for post-processing. In general, accurate DQ results can be obtained by using a coarse mesh. After we have obtained the converged solution on the coarse mesh, we then interpolate the results on a fine mesh. The results on the fine mesh can subsequently be used to give a smooth contour plot or curve, and can also be used to compute certain interesting statistical parameters such as the maximum or minimum value attained by a variable.

2.6.2 Differential Cubature Method

For the regular domain, the formulations shown in Section 2.6.1 are very useful because each direction can be treated in the same way as in the one-dimensional case. The weighting coefficients in each direction can be computed easily by using the explicit formulations. However, for an irregular domain, these formulations cannot be applied directly. One has to perform a coordinate transformation to map the irregular physical domain onto a regular computational domain. Then in the computational domain, the formulations in Section 2.6.1 can be applied. In Chapter 9, we will show the application of this approach. In this section, we will introduce the differential cubature (DC) method proposed by Civan (1989, 1991), and Malik and Civan (1995), which has potential in treatment of irregular geometries.

We will consider a two-dimensional case to demonstrate the DC method. In the DC method, the partial derivative at any grid point in a two-dimensional domain is approximated by a linear weighted sum of all the functional values in the whole two-dimensional domain. Let $\partial^n f/\partial x^n$ and $\partial^m f/\partial y^m$ be the nth and mth order derivatives of a function $f(x, y)$ with respect to x and y respectively. At any grid point, the DC approximation gives

$$\left.\frac{\partial^n f}{\partial x^n}\right|_i = \sum_{j=1}^{NM} c_{ij}^{x,n} \cdot f_j, \quad \text{for } i = 1,2,\ldots, NM \tag{2.132a}$$

$$\left.\frac{\partial^m f}{\partial y^m}\right|_i = \sum_{j=1}^{NM} c_{ij}^{y,m} \cdot f_j, \quad \text{for } i = 1,2,\ldots, NM \tag{2.132b}$$

where $c_{ij}^{x,n}, c_{ij}^{y,m}$ are the weighting coefficients related to $\partial^n f/\partial x^n$ and $\partial^m f/\partial y^m$, and NM is the total number of the grid points in the whole two-dimensional domain. It should be noted that the index i in Equation 2.132 denotes the one-dimensional indexing of the two-dimensional grid points. So, before Equation 2.132 is applied, one has to covert the two-dimensional grid point distribution into a one-dimensional array. Like the DQ method, the key procedure in the DC method is to determine the weighting coefficients. It is noted that the number of the weighting coefficients involved in Equation 2.132 is $NM \times NM$. Thus, to determine the weighting coefficients, we need $NM \times NM$ algebraic equations. This can be done by selecting NM test functions and NM grid points. Applying each test function at each grid point provides one algebraic equation. So, applying NM test functions at NM grid points would give $NM \times NM$ equations. In the DC method, the function $f(x, y)$ is approximated by

$$f(x,y) \approx \sum_{i=0}^{N-1} \sum_{j=0}^{i} d_{ij} \cdot x^{i-j} \cdot y^{j} \tag{2.133}$$

This approximation form also constitutes a linear vector space. The test functions (base vectors in the linear vector space) can be chosen as

$$r_{ij}(x,y) = x^{i-j} y^{j}, \quad i = 0,1,\ldots,N-1;\ j = 0,1,\ldots,i \tag{2.134}$$

The number of test functions given by Equation 2.134 is $N(N+1)/2$. On the other hand, we find that the form of the approximation given by Equation 2.134 is different from Equation 2.117. When M is taken to be the same as N, Equation 2.117 provides N^2 test functions, which not only covers the test functions given by Equation 2.134, but also includes some higher order test functions. When the test functions given by Equation 2.134 are substituted into Equation 2.132, we can get a set of algebraic equations to compute the weighting coefficients. Clearly, the procedure for determining the weighting coefficients in the DC method is exactly the same as Bellman's first approach in the DQ method. So, the DC method suffers from the same difficulties as Bellman's first approach in computing the weighting coefficients. That is, when N is relatively large, the obtained algebraic equation system is highly ill-conditioned. Like the DQ method, the ideal way in the DC method is to develop some explicit formulations to compute the weighting coefficients. Unfortunately, this problem has not yet been solved. There is another difficulty in applying the DC method. As discussed previously in the determination of the weighting coefficients in Equation 2.132, the number of test functions should be equal to the number of grid points. However, for some practical applications, the number of test functions given by Equation 2.134, i.e. $N(N+1)/2$, may not be equal to the number of grid points NM. To ensure that the number of test functions is exactly NM, one has to add some test functions selected from the higher degree polynomials. However, this selection is not arbitrary. A good example was given by Malik and Civan (1995). Consider a rectangular domain as shown in Figure 2.4. There are 61 grid points. On the other hand, the number of test functions provided by Equation 2.134 is $N(N+1)/2 = 55$ when N is taken as 10, and $N(N+1)/2 = 66$ when N is taken as 11. Clearly, these two values do not match 61 (number of grid points). To get 61 test functions, we first take N as 10, and obtain 55 test functions from Equation 2.134. Another 6 test functions should be obtained from a set of basic polynomials of degree 10 as shown below

$$x^{10}, x^9 y, x^8 y^2, x^7 y^3, x^6 y^4, x^5 y^5, x^4 y^6, x^3 y^7, x^2 y^8, xy^9, y^{10} \tag{2.135}$$

It was found by Malik and Civan (1995) that, to successfully obtain a solution, the 6 additional test functions should be chosen as

$$x^5y^5, x^4y^6, x^3y^7, x^2y^8, xy^9, y^{10} \tag{2.136}$$

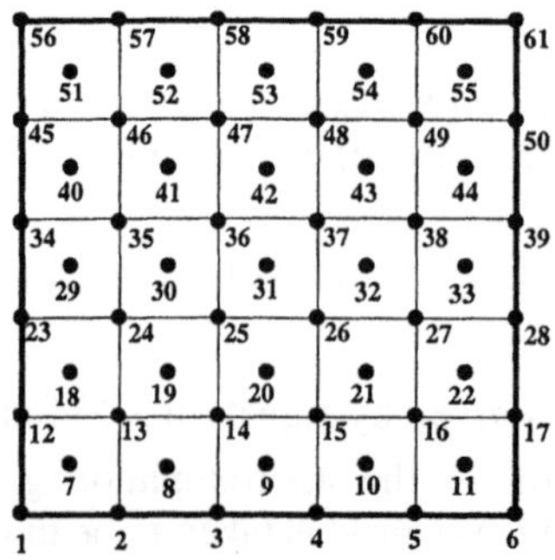

Figure 2.4 An example of cubature grid

The idea of the DC method is very good. It can be applied to Equation 2.132 for the discretization of partial derivatives. By nature, the DC method has the potential of treating problems with irregular geometries. However, due to the above-mentioned difficulties, the application of DC method has so far been limited to simple geometries using a small number of grid points in the whole two-dimensional domain.

2.7 Specific Results for Typical Grid Point Distributions

In this section, three specific formulations for the weighting coefficients will be given for three typical grid point distributions: uniform grid; Chebyshev-Gauss-Lobatto grid; and a grid with coordinates chosen as the roots of Chebyshev polynomial. Since the complete weighting coefficients of the second and higher order derivatives can be calculated from those of the first order derivative, and for the multi-dimensional cases, each direction can be treated using the same way as in the one-dimensional case, only the weighting coefficients of the first order derivative in the one-dimensional case are considered. Since the diagonal weighting coefficients can be easily computed from Equation 2.18, we will only show the formulation for the non-diagonal weighting coefficients in the following.

2.7.1 Uniform Grid

By a uniform grid, we mean that the grid has the same step sizes. Thus setting

$$\Delta x = x_2 - x_1 = x_i - x_{i-1} = x_N - x_{N-1}\text{, etc.}$$

one can obtain

$$x_j - x_i = (j-i)\cdot \Delta x$$
$$M^{(1)}(x_i) = (-1)^{N-1}\cdot(\Delta x)^{N-1}\cdot(i-1)!\cdot(N-i)!, \text{ for } i = 1, 2, \ldots, N$$

So, for this case, Equation 2.17*a* reduces to

$$a_{ij} = (-1)^{i+j}\cdot\frac{(i-1)!\cdot(N-i)!}{\Delta x\cdot(i-j)\cdot(j-1)!\cdot(N-j)!}, \quad \text{for } j \neq i \tag{2.137}$$

2.7.2 Chebyshev-Gauss-Lobatto Grid

For this case, the coordinates of the grid points are chosen as $x_i = \cos(i\pi/N)$, $i = 0, 1, \ldots, N$, and the simplified formulation for the weighting coefficients has been shown in Section 2.5.2, which is written as

$$a_{ij} = \frac{(-1)^{i+j}\bar{c}_i}{\bar{c}_j(x_i - x_j)}, \qquad \text{for } j \neq i \tag{2.138}$$

where $\bar{c}_i$ has been defined in Section 2.5.2. Equation 2.138 is valid for the interval [1, -1]. If the physical domain is not [1, -1], but [*a*, *b*], then we need to use the following coordinate transformation

$$x = \frac{b-a}{2}(1-\xi) + a \tag{2.139}$$

to map the interval [*a*, *b*] in the *x* domain onto the interval [1, -1] in the ξ domain. In the ξ domain, we can use Equation 2.138 to compute the weighting coefficients. Using the chain rule for differentiation, we have,

$$\frac{\partial f}{\partial x} = \frac{\partial f}{\partial \xi}\frac{\partial \xi}{\partial x} = \frac{2}{a-b}\frac{\partial f}{\partial \xi} \tag{2.140}$$

With Equation 2.140, Equation 2.138 can be modified to compute the weighting coefficients in the interval [*a*, *b*] as

$$a_{ij} = \frac{(-1)^{i+j}2\bar{c}_i}{(a-b)\bar{c}_j(\xi_i - \xi_j)}, \qquad \text{for } j \neq i \tag{2.141}$$

2.7.3 Coordinates of Grid Points Chosen as the Roots of Chebyshev Polynomial

For this case, the coordinates of grid points are chosen as

$$x_i = \cos[(2i-1)\pi/2N] \quad , \quad i = 1, 2, \cdots, N \tag{2.142}$$

It should be indicated that x_i in Equation 2.142 is in the domain $[x_1, x_N]$, where $x_1 = \cos(\pi/2N)$, $x_N = -x_1$. In this case, the function $M(x)$ in Equation 2.7 takes the following form

$$M(x) = T_N(x) \tag{2.143}$$

Since $x = \cos(\theta), T_N(x) = \cos(N\theta)$, Equation 2.143 can be reduced to

$$M(x) = M(\theta) = \cos(N\theta) \tag{2.144}$$

Differentiating Equation 2.144 with respect to x, one obtains

$$M^{(1)}(x) = N\sin(N\theta)/\sin(\theta) \tag{2.145}$$

Since $N\theta_i = \pi(2i-1)/2$, Equation 2.145 can be further reduced to

$$M^{(1)}(x_i) = (-1)^{i+1} N/\sqrt{1-x_i^2} \tag{2.146}$$

Substituting Equation 2.146 into Equation 2.17a, we obtain the following simplified form

$$a_{ij} = \frac{(-1)^{i+j}\sqrt{1-x_j^2}}{(x_i - x_j)\sqrt{1-x_i^2}}, \quad \text{for } j \neq i \tag{2.147}$$

Similarly, if the physical domain is $[a, b]$, using the following transformation

$$\xi = \frac{d_2}{b-a}(x-a) + d_1 \tag{2.148}$$

where $d_1 = \cos(\pi/2N)$, $d_2 = -2d_1$, we can modify Equation 2.147 to compute the weighting coefficients in the interval $[a, b]$ as

$$a_{ij} = \frac{(-1)^{i+j} d_2 \sqrt{1-\xi_j^2}}{(b-a)(\xi_i - \xi_j)\sqrt{1-\xi_i^2}}, \quad \text{for } j \neq i \tag{2.149}$$

2.8 Generation of Low Order Finite Difference Schemes by PDQ

As an example, we will use the PDQ formulations 2.17*a* and 2.18 to generate a second order finite difference scheme for the first order derivative on a non-uniform mesh. Following the discussion in Section 2.5.1, the second order finite difference scheme for the first order derivative involves the functional values at two neighbouring points. Supposing at a grid point x_i, the two neighbouring points are x_{i-1} and x_{i+1}. Then in the interval $[x_{i-1}, x_{i+1}]$, we can apply the PDQ formulations. For this case, $N = 3$. Using Equation 2.8, we have

$$M^{(1)}(x_{i-1}) = (x_{i-1} - x_i)(x_{i-1} - x_{i+1}) = \Delta_1(\Delta_1 + \Delta_2) \tag{2.150a}$$

$$M^{(1)}(x_i) = (x_i - x_{i-1})(x_i - x_{i+1}) = -\Delta_1\Delta_2 \tag{2.150b}$$

$$M^{(1)}(x_{i+1}) = (x_{i+1} - x_i)(x_{i+1} - x_{i-1}) = \Delta_2(\Delta_1 + \Delta_2) \tag{2.150c}$$

where $\Delta_1 = x_i - x_{i-1}$, $\Delta_2 = x_{i+1} - x_i$. Substituting Equation 2.150 into Equation 2.17*a*, we obtain

$$a_{i,i-1} = -\Delta_2 / [\Delta_1(\Delta_1 + \Delta_2)] \tag{2.151a}$$

$$a_{i,i+1} = \Delta_1 / [\Delta_2(\Delta_1 + \Delta_2)] \tag{2.151b}$$

Using Equation 2.18, $a_{i,i}$ can be computed by

$$a_{i,i} = -(a_{i,i-1} + a_{i,i+1}) = (\Delta_2 - \Delta_1)/(\Delta_1\Delta_2) \tag{2.152}$$

Finally, using Equation 2.1, the first order derivative of a function $f(x)$ at a grid point x_i is approximated by

$$f_x^{(1)}(x_i) = a_{i,i-1}f_{i-1} + a_{i,i}f_i + a_{i,i+1}f_{i+1}$$
$$= -\frac{\Delta_2}{\Delta_1(\Delta_1+\Delta_2)}f_{i-1} + \frac{\Delta_2-\Delta_1}{\Delta_1\Delta_2}f_i + \frac{\Delta_1}{\Delta_2(\Delta_1+\Delta_2)}f_{i+1} \tag{2.153}$$

For the uniform mesh, $\Delta_1 = \Delta_2 = \Delta$, Equation 2.153 can be simplified to

$$f_x^{(1)}(x_i) = (f_{i+1} - f_{i-1})/(2\Delta) \tag{2.154}$$

which is a central difference scheme on the uniform mesh. In a similar manner, the discretization of the first order derivative at x_{i-1} and x_{i+1} can be written as

$$f_x^{(1)}(x_{i-1}) = -\frac{2\Delta_1+\Delta_2}{\Delta_1(\Delta_1+\Delta_2)}f_{i-1} + \frac{\Delta_2+\Delta_1}{\Delta_1\Delta_2}f_i - \frac{\Delta_1}{\Delta_2(\Delta_1+\Delta_2)}f_{i+1} \tag{2.155}$$

$$f_x^{(1)}(x_{i+1}) = \frac{\Delta_2}{\Delta_1(\Delta_1+\Delta_2)}f_{i-1} - \frac{\Delta_2+\Delta_1}{\Delta_1\Delta_2}f_i + \frac{2\Delta_2+\Delta_1}{\Delta_2(\Delta_1+\Delta_2)}f_{i+1} \tag{2.156}$$

For a one-dimension problem, we can divide the whole domain into (*N*-1) elements with grid points as $x_1, x_2, ..., x_N$. At location x_i, $i = 2, 3, ..., N-1$, the first order derivative of a function can be approximated by Equation 2.153 in the element $[x_{i-1}, x_{i+1}]$. It is noted that for this case, the two neighbouring elements $[x_{i-1}, x_{i+1}]$ and $[x_i, x_{i+2}]$, used for the discretization of the first order derivative at the grid points x_i and x_{i+1}, are overlapped with the region of $[x_i, x_{i+1}]$. This behavior is different from the standard finite element approach where the neighbouring elements are patched. Similarly, at x_1 and x_N, the discretization of the first order derivative of the function can be made by Equation 2.155 in the element $[x_1, x_3]$ and by Equation 2.156 in the element $[x_{N-2}, x_N]$.

The feature of overlapped elements for the finite difference scheme may give us some hint regarding the stability analysis. In the overlapped region, the solution can be given from one element, and can also be determined from the neighbouring elements. During the computation, the solutions given from the neighbouring elements in the overlapped region may be different. The inconsistency of the solutions in the overlapped region may be the cause of instability for a difference scheme.

For the general case, the PDQ method can be used to generate a *p*th order finite difference scheme for the *n*th order derivative of a function $f(x)$. Using the principle shown in Section 2.5.1, we know that the number of grid points involved in the difference scheme is *n* + *p* - 1. For simplicity, we set $K = n + p - 1$. Consider a location x_i, the total number of grid points involved in the difference scheme should be $K+1$ (the *K* neighbouring points plus the local point x_i). Among the *K* neighbouring points, it is supposed that K_1 points are on the left side of x_i, and K_2 points are on the right side of x_i, where $K = K_1 + K_2$. For the left boundary point x_1, $K_1 = 0, K_2 = K$, while for the right boundary point x_N, $K_1 = K, K_2 = 0$. When the PDQ method is used to generate a difference scheme, we need to start from the first order derivative, and then use the recurrence relationship 2.36 to generate the required difference scheme. Using the K+1 grid points, $x_{i-K_1}, x_{i-K_1+1}, \ldots, x_{i-1}, x_i, x_{i+1}, \ldots, x_{i+K_2}$, the first order derivative of the function $f(x)$ at x_i is approximated by

$$f_x^{(1)}(x_i) = \sum_{j=-K_1}^{K_2} a_{i,i+j} f(x_{i+j}) \tag{2.157}$$

The weighting coefficients $a_{i,i+j}$ in Equation 2.157 are given by Equation 2.17*a* as

$$a_{i,i+j} = \frac{M^{(1)}(x_i)}{(x_i - x_{i+j}) \cdot M^{(1)}(x_{i+j})}, \quad \text{for } j \neq 0 \tag{2.158}$$

where

$$M^{(1)}(x_i) = \prod_{k=-K_1, k\neq 0}^{K_2} (x_i - x_{i+k})$$

$$M^{(1)}(x_{i+j}) = \prod_{k=-K_1, k\neq j}^{K_2} (x_{i+j} - x_{i+k})$$

The diagonal weighting coefficients $a_{i,i}$ are given by Equation 2.18 as

$$a_{i,i} = -\sum_{k=-K_1, k\neq 0}^{K_2} a_{i,i+k} \tag{2.159}$$

It is noted that the order of accuracy for the difference scheme 2.157 is $K = n + p - 1$. Similar to Equation 2.157, the *p*th order difference scheme of the *n*th order derivative can be written as

$$f_x^{(n)}(x_i) = \sum_{j=-K_1}^{K_2} w_{i,i+j}^{(n)} f(x_{i+j}) \tag{2.160}$$

The weighting coefficients $w_{i,i+j}^{(n)}$ can be computed from Equation 2.36 as

$$w_{i,i+j}^{(n)} = n\left(a_{i,i+j} \cdot w_{i,i}^{(n-1)} - \frac{w_{i,i+j}^{(n-1)}}{x_i - x_{i+j}} \right), \text{ for } j = -K_1, ..., K_2, j \neq 0 \quad (2.161)$$

and the diagonal weighting coefficients $w_{i,i}^{(n)}$ are given by

$$w_{i,i}^{(n)} = -\sum_{k=-K_1, k\neq 0}^{K_2} w_{i,i+k}^{(n)} \quad (2.162)$$

It is noted that in Equations 2.161 and 2.162, $w_{i,i+j}^{(1)} = a_{i,i+j}$. Equations 2.161, 2.162 can be repeatedly applied from the second order derivative to the nth order derivative.

In particular, when the required difference scheme is on a uniform mesh, the formulations for the weighting coefficients given in Equations 2.158 and 2.161 can be simplified to

$$a_{i,i+j} = \frac{(-1)^{j+1} K_1! \cdot K_2!}{j\Delta x \cdot (K_1 + j)! \cdot (K_2 - j)!}, \quad \text{for } j \neq 0 \quad (2.163)$$

$$w_{i,i+j}^{(n)} = n\left(a_{i,i+j} \cdot w_{i,i}^{(n-1)} + \frac{w_{i,i+j}^{(n-1)}}{j\Delta x} \right), \text{ for } j = -K_1, ..., K_2, j \neq 0 \quad (2.164)$$

where Δx is the mesh spacing between two neighbouring grid points, that is,

$$\Delta x = x_i - x_{i-1} = x_{i+1} - x_i$$

Chapter 3

Fourier Expansion-based Differential Quadrature (FDQ)

3.1 Introduction

In Chapter 1, we have shown that the solution of a PDE can be approximated by a polynomial of high degree or by the Fourier series expansion, depending on the feature of the problem. It is noted that the expressions for polynomial approximation and Fourier series expansion are quite different. Thus, when the DQ approximation is applied to these two cases, the formulations to compute the weighting coefficients in the DQ approximation may be different. In Chapter 2, the computation of the weighting coefficients for polynomial-based differential quadrature (PDQ) was described in detail. In PDQ, the solution of a PDE is approximated by a high degree polynomial. The polynomial approximation is suitable for most engineering problems. However, for some problems, especially for those with periodic behaviors such as the Helmholtz problems, polynomial approximation may not be the best choice for the true solution. In contrast, Fourier series expansion could be the best approximation. In this chapter, we will demonstrate that using Fourier series expansion and linear vector space analysis, the weighting coefficients in the Fourier expansion-based differential quadrature (FDQ) can also be calculated by explicit formulations.

For simplicity, the one-dimensional problem is chosen to demonstrate the FDQ method. As shown in Chapter 2, when a structured grid is used, the one-dimensional results can be directly extended to the multi-dimensional cases. The FDQ approximation for the discretization of derivatives is exactly the same as the PDQ approximation. The only difference between FDQ and PDQ is in the way the weighting coefficients are computed. Consider a one-dimensional function $f(x)$ over an interval $[a, b]$ and suppose that there are N grid points in the whole domain with coordinates $x_1, x_2, ..., x_N$. From Chapter 2, it is known that the DQ method approximates the derivative at any grid point by a linear weighted sum of all the functional values in the whole domain. For example, the first and second order derivatives of $f(x)$ at a point x_i are approximated by

$$f_x^{(1)}(x_i) = \sum_{j=1}^{N} a_{ij} \cdot f(x_j), \qquad \text{for } i = 1, 2, \cdots, N \tag{3.1}$$

$$f_x^{(2)}(x_i) = \sum_{j=1}^{N} b_{ij} \cdot f(x_j), \qquad \text{for } i = 1, 2, \cdots, N \tag{3.2}$$

where $f_x^{(1)}(x_i), f_x^{(2)}(x_i)$ represent the first and second order derivatives respectively at x_i, and a_{ij}, b_{ij} are the corresponding weighting coefficients. Generally, for the mth order derivative $f_x^{(m)}(x_i)$, we have a similar approximation form

$$f_x^{(m)}(x_i) = \sum_{j=1}^{N} w_{ij}^{(m)} \cdot f(x_j), \quad \text{for } i = 1, 2, ..., N; m = 1,2,..., N-1 \tag{3.3}$$

where $w_{ij}^{(m)}$ is the weighting coefficient with $w_{ij}^{(1)} = a_{ij}$ and $w_{ij}^{(2)} = b_{ij}$. It is shown in Chapter 2 that when the function $f(x)$ is approximated by a polynomial of degree (N-1), the weighting coefficients can be computed respectively by Equations 2.17a, 2.18, 2.24, 2.25, 2.36, 2.38. In the following, we will show the relevant formulations for computing the weighting coefficients when the function $f(x)$ is approximated by Fourier series expansion.

3.2 Cosine Expansion-based Differential Quadrature (CDQ) for Even Functions

As shown in Chapter 1, for an even function $f(x)$ defined on the interval [0, π], it can be approximated by a Fourier series expansion given by

$$f(x) = d_0 + \sum_{k=1}^{N} d_k \cos kx \tag{3.4}$$

It is demonstrated in Chapter 1 that $f(x)$ in Equation 3.4 constitutes a (N+1) dimensional linear vector space V_{N+1}. In the linear vector space V_{N+1}, the two typical sets of base vectors are

$$c_k(x) = \cos(kx), \; k = 0, 1, ..., N \tag{3.5}$$

$$c_k(x) = \frac{C(x)}{p(x_k) \cdot (\cos x - \cos x_k)}, \quad k = 0, 1, ..., N \tag{3.6}$$

where

$$C(x) = \prod_{k=0}^{N} (\cos x - \cos x_k) \tag{3.7}$$

$$p(x_i) = \prod_{k=0, k \neq i}^{N} (\cos x_i - \cos x_k) \tag{3.8}$$

Note that Equations 3.1, 3.2 and 3.3 are linear operators. It should be noted that for this problem, there are (N+1) grid points with coordinates $0 = x_0, x_1, ..., x_N = \pi$. From the properties of a linear vector space, we know that if a set of base vectors given by Equation 3.5 satisfies the linear Equations 3.1, 3.2 and 3.3, so does another set of base vectors given by Equation 3.6. And if all the base vectors satisfy the linear Equations 3.1, 3.2 and 3.3, so does any vector in the space. This indicates that when the weighting coefficients are determined from the base vectors, they can be used to discretize the derivatives in a PDE when the solution is approximated by Equation 3.4.

Similar to PDQ, two sets of base vectors given by Equations 3.5 and 3.6 will be used in deriving the explicit formulations for the weighting coefficients in FDQ. The first set of base vectors is given by Equation 3.6. For simplicity, we set

$$C(x) = \prod_{k=0}^{N} (\cos x - \cos x_k) = N(x, x_k) \cdot (\cos x - \cos x_k) \tag{3.9}$$

where

$$N(x_i, x_i) = \prod_{k=0, k \neq i}^{N} (\cos x_i - \cos x_k) = p(x_i)$$

$N(x_i, x_j) = N(x_i, x_i) \cdot \delta_{ij} = p(x_i)\delta_{ij}$, δ_{ij} is the Kronecker operator.

Equation 3.6 can then be reduced to

$$c_k(x) = \frac{N(x, x_k)}{p(x_k)}, \quad k = 0, 1, ..., N \tag{3.10}$$

Using the same methodology as in PDQ, we let all the base vectors given by Equation 3.10 satisfy linear Equations 3.1, 3.2 and 3.3, and then obtain

$$a_{ij} = \frac{N^{(1)}(x_i, x_j)}{p(x_j)} \tag{3.11}$$

$$b_{ij} = \frac{N^{(2)}(x_i, x_j)}{p(x_j)} \tag{3.12}$$

$$w_{ij}^{(m)} = \frac{N^{(m)}(x_i, x_j)}{p(x_j)}, \quad m = 1, 2, ..., N \tag{3.13}$$

where $N^{(m)}(x, x_k)$ is the mth order derivative of $N(x, x_k)$. It is observed from Equations 3.11-3.13 that the computation of a_{ij}, b_{ij} and $w_{ij}^{(m)}$ is equivalent to the evaluation of $N^{(1)}(x, x_k)$, $N^{(2)}(x, x_k)$ and $N^{(m)}(x, x_k)$ since $p(x_j)$ can be easily calculated from Equation 3.8. To evaluate $N^{(1)}(x, x_k)$, $N^{(2)}(x, x_k)$ and $N^{(m)}(x, x_k)$, we successively differentiate Equation 3.9 and obtain

$$C^{(1)}(x) = N^{(1)}(x, x_k) \cdot (\cos x - \cos x_k) - N(x, x_k) \cdot \sin x \tag{3.14}$$

$$\begin{aligned} C^{(2)}(x) = {} & N^{(2)}(x, x_k) \cdot (\cos x - \cos x_k) - \\ & 2N^{(1)}(x, x_k) \cdot \sin x - N(x, x_k) \cdot \cos x \end{aligned} \tag{3.15}$$

$$\begin{aligned} C^{(3)}(x) = {} & N^{(3)}(x, x_k) \cdot (\cos x - \cos x_k) - 3N^{(2)}(x, x_k) \cdot \sin x - \\ & 3N^{(1)}(x, x_k) \cdot \cos x + N(x, x_k) \cdot \sin x \end{aligned} \tag{3.16}$$

$$\begin{aligned} C^{(4)}(x) = {} & N^{(4)}(x, x_k) \cdot (\cos x - \cos x_k) - 4N^{(3)}(x, x_k) \cdot \sin x - \\ & 6N^{(2)}(x, x_k) \cdot \cos x + 4N^{(1)}(x, x_k) \cdot \sin x + N(x, x_k) \cdot \cos x \end{aligned} \tag{3.17}$$

$$C^{(m)}(x) = N^{(m)}(x, x_k) \cdot (\cos x - \cos x_k) + \sum_{k=1}^{m} A_{m,k}, \quad m = 1, 2, ..., N \tag{3.18}$$

where

$$A_{m,k} = \begin{cases} c_m^k \cdot (-1)^{\frac{k}{2}} \cdot N^{(m-k)}(x, x_k) \cdot \cos x, & \text{when } k = 2, 4, 6, ... \\ c_m^k \cdot (-1)^{\frac{k+1}{2}} \cdot N^{(m-k)}(x, x_k) \cdot \sin x, & \text{when } k = 1, 3, 5, ... \end{cases}$$

$$N^{(0)}(x, x_k) = N(x, x_k)$$

where c_m^k is the combination of m terms taken k at a time. Equation 3.18 is a general form, which covers Equations 3.14-3.17. In the following, we will use Equations 3.14-3.16 to derive explicit formulations for a_{ij} and b_{ij}. Applying Equations 3.14-3.16 at the grid points and using $N(x_i, x_i) = p(x_i)$, we obtain

$$N^{(1)}(x_i, x_j) = \frac{C^{(1)}(x_i)}{\cos x_i - \cos x_j}, \quad \text{when } j \neq i \tag{3.19a}$$

$$N^{(1)}(x_i, x_i) = \frac{-C^{(2)}(x_i) - p(x_i)\cdot\cos x_i}{2\sin x_i} \tag{3.19b}$$

$$N^{(2)}(x_i, x_j) = \frac{C^{(2)}(x_i) + 2N^{(1)}(x_i, x_j)\cdot\sin x_i}{\cos x_i - \cos x_j}, \quad \text{when} \quad j \neq i \tag{3.20a}$$

$$N^{(2)}(x_i, x_i) = \frac{-C^{(3)}(x_i) - 3N^{(1)}(x_i, x_i)\cdot\cos x_i + p(x_i)\cdot\sin x_i}{3\sin x_i} \tag{3.20b}$$

Equation 3.14 also gives

$$C^{(1)}(x_i) = -N(x_i, x_i)\sin x_i = -p(x_i)\sin x_i \tag{3.21}$$

Thus, Equation 3.19*a* can be further reduced to

$$N^{(1)}(x_i, x_j) = \frac{-p(x_i)\sin x_i}{\cos x_i - \cos x_j}, \quad \text{when} \quad j \neq i \tag{3.22}$$

Substituting Equations 3.22 and 3.19*b* into Equation 3.11 leads to

$$a_{ij} = \frac{-p(x_i)\sin x_i}{(\cos x_i - \cos x_j)\cdot p(x_j)}, \quad \text{when} \quad j \neq i \tag{3.23}$$

$$a_{ii} = \frac{-C^{(2)}(x_i) - p(x_i)\cos x_i}{2p(x_i)\sin x_i} \tag{3.24}$$

Equation 3.23 is a simple algebraic formulation to compute the weighting coefficients a_{ij} with $i \neq j$. However, the computation of a_{ii} involves the second order derivative $C^{(2)}(x_i)$ which is not easy to calculate. This difficulty can be overcome by the properties of a linear vector space, which is applied in PDQ. In the linear vector space V_{N+1}, the two sets of base vectors given by Equations 3.5 and 3.6 satisfy the same linear Equation 3.1. In other words, a_{ij} should also satisfy the equation system derived by the set of base vectors 1, cos*x*, cos2*x*, ..., cos*Nx* given by Equation 3.5. Among this set of base vectors, we let the vector 1 satisfy Equation 3.1 and we then obtain

$$\sum_{j=0}^{N} a_{ij} = 0 \quad \text{or} \quad a_{ii} = -\sum_{j=0, j\neq i}^{N} a_{ij} \tag{3.25}$$

On the other hand, Equation 3.19*b* can be rewritten as

$$C^{(2)}(x_i) = -2\sin x_i \cdot N^{(1)}(x_i, x_i) - p(x_i)\cdot \cos x_i \tag{3.26}$$

Substituting Equation 3.26 into Equation 3.20*a* leads to

$$N^{(2)}(x_i, x_j) = \frac{2\sin x_i [N^{(1)}(x_i, x_j) - N^{(1)}(x_i, x_i)] - p(x_i)\cos x_i}{\cos x_i - \cos x_j}, \text{ for } j \neq i$$

From Equation 3.11, we have

$$N^{(1)}(x_i, x_j) = a_{ij} \cdot p(x_j)$$

So the expression of $N^{(2)}(x_i, x_j)$ can be further reduced to

$$N^{(2)}(x_i, x_j) = \frac{2\sin x_i [a_{ij} p(x_j) - a_{ii} p(x_i)] - p(x_i)\cos x_i}{\cos x_i - \cos x_j}, \text{ for } j \neq i \tag{3.27}$$

Finally, substituting Equation 3.27 into Equation 3.12, we obtain

$$b_{ij} = \frac{2\sin x_i [a_{ij} p(x_j) - a_{ii} p(x_i)] - p(x_i)\cos x_i}{(\cos x_i - \cos x_j) p(x_j)}, \text{ for } j \neq i \tag{3.28}$$

Using Equation 3.23, Equation 3.28 can be further simplified to

$$b_{ij} = a_{ij}\left(\frac{2\sin x_i}{\cos x_i - \cos x_j} + 2a_{ii} + \cot x_i\right), \text{ for } j \neq i \tag{3.29}$$

Equation 3.29 is a simple algebraic formulation to compute b_{ij} for $i \neq j$. Similarly, the weighting coefficient b_{ii} can be obtained by substituting Equation 3.20*b* into Equation 3.12, thus giving

$$b_{ii} = \frac{-C^{(3)}(x_i) + p(x_i)(\sin x_i - 3a_{ii} \cdot \cos x_i)}{3p(x_i)\sin x_i} \tag{3.30}$$

Like Equation 3.24, the computation of b_{ii} from Equation 3.30 is very difficult since it involves the third order derivative $C^{(3)}(x_i)$. Again, this difficulty can be eliminated by using the properties of a linear vector space. In the linear vector space V_{N+1}, the equation system for b_{ij} derived from one set of base vectors given by Equation 3.6 should be equivalent to the system derived from another set of base vectors given by Equation 3.5. Among the set of base vectors 1, cos*x*, cos2*x*, ..., cos*Nx*, we let the vector 1 satisfy Equation 3.2 and we obtain

$$\sum_{j=0}^{N} b_{ij} = 0 \quad \text{or} \quad b_{ii} = -\sum_{j=0,\, j\neq i}^{N} b_{ij} \tag{3.31}$$

From Equation 3.31, b_{ii} can be easily calculated from b_{ij} $(i \neq j)$.

Theoretically, using Equation 3.18, a recurrence relationship for $w_{ij}^{(m)}$ can be derived. However, the expression is very complicated. One can note that Equation 3.29 is more complicated than the corresponding equation in PDQ. For the third order derivative, the expression for its weighting coefficients has more terms than Equation 3.29. Applying Equations 3.16, 3.17 at the grid points, we can obtain

$$N^{(3)}(x_i, x_j) = \frac{C^{(3)}(x_i) + 3N^{(2)}(x_i, x_j) \cdot \sin x_i + 3N^{(1)}(x_i, x_j) \cdot \cos x_i}{\cos x_i - \cos x_j} \tag{3.32a}$$

$$N^{(3)}(x_i, x_i) = \frac{-C^{(4)}(x_i) - 6N^{(2)}(x_i, x_i) \cdot \cos x_i}{4 \sin x_i} + \frac{4N^{(1)}(x_i, x_i) \cdot \sin x_i + p(x_i) \cos x_i}{4 \sin x_i} \tag{3.32b}$$

Equation 3.32*a* is valid for $i \neq j$. On the other hand, $C^{(3)}(x_i)$ in Equation 3.32*a* can be expressed using Equation 3.20*b* as

$$C^{(3)}(x_i) = [p(x_i) - 3N^{(2)}(x_i, x_i)]\sin x_i - 3N^{(1)}(x_i, x_i) \cdot \cos x_i \tag{3.33}$$

Using Equation 3.33, Equation 3.32*a* can be written as

$$N^{(3)}(x_i,x_j) = \frac{3[N^{(2)}(x_i,x_j) - N^{(2)}(x_i,x_i)] \cdot \sin x_i + p(x_i)\sin x_i}{\cos x_i - \cos x_j} + \frac{3[N^{(1)}(x_i,x_j) - N^{(1)}(x_i,x_i)] \cdot \cos x_i}{\cos x_i - \cos x_j}, \text{ for } i \neq j \tag{3.34}$$

Note that $N^{(1)}(x_i,x_j) = a_{ij} \cdot p(x_j)$ and $N^{(2)}(x_i,x_j) = b_{ij} \cdot p(x_j)$. So, when Equation 3.34 is substituted into Equation 3.13, we obtain

$$w_{ij}^{(3)} = \frac{3\sin x_i \cdot w_{ij}^{(2)}}{\cos x_i - \cos x_j} + 3w_{ij}^{(1)}\left(w_{ii}^{(2)} - \frac{1}{3} + w_{ii}^{(1)} \cot x_i + \frac{\cos x_i}{\cos x_i - \cos x_j} \right) \tag{3.35}$$

for $i, j = 0, 1, ..., N$, but $i \neq j$

where $w_{ij}^{(3)}$ is the weighting coefficient for the third order derivative, and $w_{ij}^{(2)} = b_{ij}$, $w_{ij}^{(1)} = a_{ij}$. Equation 3.32*b* can be used to derive the expression to compute $w_{ii}^{(3)}$, but that expression involves the fourth order derivative $C^{(4)}(x_i)$ which is difficult to evaluate. So, for $w_{ii}^{(3)}$, we use the second set of base vectors 1, cos*x*, cos2*x*, ..., cos*Nx*. Let the vector 1 satisfy Equation 3.3, we thus obtain

$$\sum_{j=0}^{N} w_{ij}^{(3)} = 0 \quad \text{or} \quad w_{ii}^{(3)} = -\sum_{j=0, j\neq i}^{N} w_{ij}^{(3)} \tag{3.36}$$

Obviously, Equation 3.35 is much more complicated than Equation 3.29. If we consider a high order derivative, the explicit expression for the weighting coefficients would have many terms involving the weighting coefficients of all the lower order derivatives. This will cause its application to be inconvenient. Therefore, for this case, we suggest the use of the matrix multiplication method introduced in Chapter 2 to compute the weighting coefficients of the high order derivatives. However, the following formulation, derived from the set of base vectors 3.5, can still be used to compute the diagonal weighting coefficients

$$\sum_{j=0}^{N} w_{ij}^{(m)} = 0 \quad \text{or} \quad w_{ii}^{(m)} = -\sum_{j=0, j\neq i}^{N} w_{ij}^{(m)}, \quad m = 1, 2, ..., N \tag{3.37}$$

As shown in Chapter 2, the procedure to derive the explicit expressions for the weighting coefficients depends on the choice of the base vectors. In the following, we will show that Equations 3.23, 3.29, and 3.35 can be easily obtained from the

PDQ formulations when we choose another set of base vectors. It was shown in Chapter 1 that in the linear vector space V_{N+1},

$$c_k(x) = \cos^k x, \quad k = 0, 1, \ldots, N \tag{3.38}$$

is also a set of base vectors. Accordingly, $f(x)$ can be approximated by

$$f(x) = \sum_{k=0}^{N} \bar{d}_k \cos^k x \tag{3.39}$$

If we further set $\xi = \cos x$, then Equation 3.39 can be written as

$$f(x) = \sum_{k=0}^{N} \bar{d}_k \xi^k \tag{3.40}$$

Equation 3.40 indicates that the function can be approximated by a polynomial of degree N in terms of ξ. With this transformation, the interval $[0, \pi]$ for x is transformed to $[1, -1]$ for ξ. Accordingly, the coordinate x_i of a grid point in the x domain is transformed to ξ_i in the ξ domain. Since the transformation is only carried out for the coordinates, the functional and derivative values at x_i in the x domain should be the same as those at ξ_i in the ξ domain.

The DQ approximation can be applied in the x domain and can also be applied in the ξ domain. Let $f_x^{(m)}$ and $f_\xi^{(m)}$ be the mth order derivative of the function f with respect to x and ξ respectively. In addition, let $w_{ij}^{(m)}$ and $\overline{w}_{ij}^{(m)}$ be the DQ weighting coefficients in the x and ξ domains respectively. Then, we have

$$f_x^{(m)}(x_i) = \sum_{j=0}^{N} w_{ij}^{(m)} f(x_j), \quad i = 0, 1, \ldots N; m = 1, 2, \ldots, N \tag{3.41a}$$

$$f_\xi^{(m)}(\xi_i) = \sum_{j=0}^{N} \overline{w}_{ij}^{(m)} f(\xi_j), \quad i = 0, 1, \ldots, N; m = 1, 2, \ldots, N \tag{3.41b}$$

On the other hand, using the chain rule for differentiation, we have

$$f_x^{(1)} = \frac{\partial f}{\partial x} = \frac{\partial f}{\partial \xi}\frac{\partial \xi}{\partial x} = -\sin x \frac{\partial f}{\partial \xi} = -\sin x \cdot f_\xi^{(1)} \tag{3.42}$$

Application of Equation 3.41 to both sides of Equation 3.42 leads to

$$w_{ij}^{(1)} = -\sin x_i \cdot \overline{w}_{ij}^{(1)} \tag{3.43}$$

Similarly, applying the chain rule for differentiation, we obtain

$$f_x^{(2)} = -\cos x \cdot f_\xi^{(1)} + \sin^2 x \cdot f_\xi^{(2)} \tag{3.44}$$

$$f_x^{(3)} = \sin x \cdot f_\xi^{(1)} + 3\sin x \cos x \cdot f_\xi^{(2)} - \sin^3 x \cdot f_\xi^{(3)} \tag{3.45}$$

Applying Equation 3.41 to both sides of Equations 3.44 and 3.45, we get

$$w_{ij}^{(2)} = -\cos x_i \cdot \overline{w}_{ij}^{(1)} + \sin^2 x_i \cdot \overline{w}_{ij}^{(2)} \tag{3.46}$$

$$w_{ij}^{(3)} = \sin x_i \cdot \overline{w}_{ij}^{(1)} + 3\sin x_i \cos x_i \cdot \overline{w}_{ij}^{(2)} - \sin^3 x_i \cdot \overline{w}_{ij}^{(3)} \tag{3.47}$$

Using the PDQ formulations shown in Chapter 2, $\overline{w}_{ij}^{(1)}, \overline{w}_{ij}^{(2)}, \overline{w}_{ij}^{(3)}$ can be given as follows

$$\overline{w}_{ij}^{(1)} = \frac{M^{(1)}(\xi_i)}{(\xi_i - \xi_j)M^{(1)}(\xi_j)}, \quad \text{for } i \neq j \tag{3.48a}$$

$$\overline{w}_{ij}^{(2)} = 2\overline{w}_{ij}^{(1)}\left(\overline{w}_{ii}^{(1)} - \frac{1}{\xi_i - \xi_j}\right), \quad \text{for } i \neq j \tag{3.48b}$$

$$\overline{w}_{ij}^{(3)} = 3\left(\overline{w}_{ij}^{(1)} \cdot \overline{w}_{ii}^{(2)} - \frac{\overline{w}_{ij}^{(2)}}{\xi_i - \xi_j}\right), \quad \text{for } i \neq j \tag{3.48c}$$

where

$$M^{(1)}(\xi_i) = \prod_{k=0,\, k\neq i}^{N} (\xi_i - \xi_k)$$

Since $\xi = \cos x$, then

$$(\xi_i - \xi_j) = \cos x_i - \cos x_j$$

$$M^{(1)}(\xi_i) = \prod_{k=0,\, k\neq i}^{N} (\xi_i - \xi_k) = \prod_{k=0,\, k\neq i}^{N} (\cos x_i - \cos x_k) = p(x_i)$$

Therefore, Equation 3.48*a* is reduced to

$$\overline{w}_{ij}^{(1)} = \frac{p(x_i)}{(\cos x_i - \cos x_j)p(x_j)}, \quad \text{for } i \neq j \tag{3.49}$$

and furthermore, Equation 3.43 becomes

$$w_{ij}^{(1)} = \frac{-\sin x_i \cdot p(x_i)}{(\cos x_i - \cos x_j)p(x_j)}, \quad \text{for } i \neq j \tag{3.50}$$

Note that $w_{ij}^{(1)} = a_{ij}$. Equation 3.50 is exactly the same as Equation 3.23. Substituting Equations 3.43, 3.48*b* into Equation 3.46, we obtain

$$\begin{aligned} w_{ij}^{(2)} &= \cot x_i \cdot w_{ij}^{(1)} + 2w_{ij}^{(1)}\left(w_{ii}^{(1)} + \frac{\sin x_i}{\cos x_i - \cos x_j} \right) \\ &= w_{ij}^{(1)}\left(2w_{ii}^{(1)} + \frac{2\sin x_i}{\cos x_i - \cos x_j} + \cot x_i \right), \quad \text{for } i \neq j \end{aligned} \tag{3.51}$$

With $w_{ij}^{(1)} = a_{ij}$ and $w_{ij}^{(2)} = b_{ij}$, it was found that Equation 3.51 is the same as Equation 3.29. Similarly, substituting Equations 3.43, 3.46 and 3.48*c* into Equation 3.47 gives

$$\begin{aligned} w_{ij}^{(3)} &= -w_{ij}^{(1)} + 3\cot x_i \left(w_{ij}^{(2)} - \cot x_i \cdot w_{ij}^{(1)}\right) + 3w_{ij}^{(1)}\left(w_{ii}^{(2)} - \cot x_i \cdot w_{ii}^{(1)}\right) \\ &+ \frac{3\sin x_i}{\cos x_i - \cos x_j}\left(w_{ij}^{(2)} - \cot x_i \cdot w_{ij}^{(1)}\right), \quad \text{for } i \neq j \end{aligned} \tag{3.52}$$

Using Equation 3.51, we have

$$w_{ij}^{(2)} - \cot x_i \cdot w_{ij}^{(1)} = 2w_{ij}^{(1)}\left(w_{ii}^{(1)} + \frac{\sin x_i}{\cos x_i - \cos x_j} \right) \tag{3.53}$$

Finally, substituting Equation 3.53 into Equation 3.52 gives

$$w_{ij}^{(3)} = 3w_{ij}^{(1)}\left(w_{ii}^{(2)} - \frac{1}{3} + w_{ii}^{(1)} \cot x_i + \frac{\cos x_i}{\cos x_i - \cos x_j} \right) + \frac{3\sin x_i \cdot w_{ij}^{(2)}}{\cos x_i - \cos x_j} \tag{3.54}$$

$$\text{for } i, j = 0, 1, \ldots, N, \text{ but } i \neq j$$

Equation 3.54 is the same as Equation 3.35.

In the above description, an even function is defined on the interval [0, π]. For practical applications, the interval may not be [0, π] but [a, b]. In this case, we can prescribe a transformation

$$\xi = \frac{x-a}{b-a}\pi \tag{3.55}$$

to map the interval [a, b] in the x domain to the interval [0, π] in the ξ domain. Basically, there are two ways to solve a problem on the interval [a, b]. One way is to transform all the derivatives with respect to x into the derivatives with respect to ξ. Then the discretization is made in the ξ domain. For this case, the above weighting coefficient formulations 3.50, 3.51, 3.54, and 3.37 can be applied directly. Another way is to make the discretization directly in the x domain, but for this case, the formulations for the weighting coefficients should be modified accordingly. Using the transformation Equation 3.55, we have

$$f_x^{(1)} = \frac{\partial f}{\partial x} = \frac{\partial f}{\partial \xi}\frac{\partial \xi}{\partial x} = \frac{\pi}{b-a}\frac{\partial f}{\partial \xi} = \alpha f_\xi^{(1)} \tag{3.56}$$

where $\alpha = \pi / (b\text{-}a)$. Similarly, for the mth order derivative, we have

$$f_x^{(m)} = \alpha^m f_\xi^{(m)}, \quad m = 1, 2, ..., N \tag{3.57}$$

Since the interval in the ξ domain is [0, π], Equations 3.50, 3.51 and 3.54 can be used to discretize the derivative $f_\xi^{(m)}$. Then using Equation 3.57, the formulations for the weighting coefficients in the x domain can be modified to give

$$w_{ij}^{(1)} = \frac{-\alpha \cdot \sin\xi_i \cdot p(\xi_i)}{(\cos\xi_i - \cos\xi_j)p(\xi_j)}, \quad \text{for } i \neq j \tag{3.58}$$

$$w_{ij}^{(2)} = w_{ij}^{(1)}\left(2w_{ii}^{(1)} + \frac{2\alpha \cdot \sin\xi_i}{\cos\xi_i - \cos\xi_j} + \alpha \cdot \cot\xi_i\right), \quad \text{for } i \neq j \tag{3.59}$$

$$\begin{aligned} w_{ij}^{(3)} = {} & 3w_{ij}^{(1)}\left(w_{ii}^{(2)} - \frac{\alpha^2}{3} + \alpha \cdot w_{ii}^{(1)} \cot\xi_i + \frac{\alpha^2 \cdot \cos\xi_i}{\cos\xi_i - \cos\xi_j}\right) \\ & + \frac{3\alpha \cdot \sin\xi_i \cdot w_{ij}^{(2)}}{\cos\xi_i - \cos\xi_j}, \quad \text{for } i \neq j \end{aligned} \tag{3.60}$$

3.3 Sine Expansion-based Differential Quadrature (SDQ) for Odd Functions

For an odd function $f(x)$ defined on the open interval (0, π), the function can be approximated from a Fourier series expansion as

$$f(x) = \sum_{k=1}^{N} d_k \sin kx \tag{3.61}$$

It was shown in Chapter 1 that $f(x)$ in Equation 3.61 constitutes a N-dimensional linear vector space V_N, which has the following two typical sets of base vectors

$$s_k(x) = \sin kx, \ k = 1, 2, ..., N \tag{3.62a}$$

$$s_k(x) = \frac{\sin x}{\sin x_k} c_k(x), \quad k = 1, 2, ..., N \tag{3.62b}$$

where $c_k(x)$ has the same form as Equation 3.6. In Section 3.2, the problem was defined on the closed interval. Thus, the grid points include the boundary points, and there are (N+1) grid points with coordinates $0 = x_0, x_1, ..., x_N = \pi$. For the present case, the problem is defined on an open interval. So, the grid points do not include the boundary points. From Equation 3.61, it can be seen that the functional values at the boundary points $x = 0$ and $x = \pi$ are inherently zero. Thus, no approximation is needed at the boundary points. For the present problem, it is assumed that there are N grid points in the open interval with coordinates $x_1, ..., x_N$. Therefore, when Equations 3.7 and 3.8 are used for $c_k(x)$, the index for multiplication is from 1 to N.

In the following, the above two sets of base vectors will be used to derive the explicit formulations to compute the weighting coefficients for the first, second and third order derivatives. For the higher order derivatives, we suggest the use of the matrix multiplication method introduced in Chapter 2 to compute the weighting coefficients.

For the set of base vectors given by Equation 3.62a, we only use one base vector, $\sin x$. Let $\sin x$ satisfy Equations 3.1-3.3, we obtain,

$$\sum_{j=1}^{N} a_{ij} \sin x_j = \cos x_i , \ \text{for } i = 1, 2, ..., N \tag{3.63a}$$

$$\sum_{j=1}^{N} b_{ij} \sin x_j = -\sin x_i , \ \text{for } i = 1, 2, ..., N \tag{3.63b}$$

$$\sum_{j=1}^{N} w_{ij}^{(3)} \sin x_j = -\cos x_i, \text{ for } i = 1, 2, ..., N \tag{3.63c}$$

When $a_{ij}, b_{ij}, w_{ij}^{(3)}$ are obtained for $i \neq j$, we can use Equation 3.63 to compute $a_{ii}, b_{ii}, w_{ii}^{(3)}$ as

$$a_{ii} = \left[\cos x_i - \sum_{j=1, j\neq i}^{N} a_{ij} \sin x_j\right] / \sin x_i, \text{ for } i = 1, 2, ..., N \tag{3.64a}$$

$$b_{ii} = \left[-\sin x_i - \sum_{j=1, j\neq i}^{N} b_{ij} \sin x_j\right] / \sin x_i, \text{ for } i = 1, 2, ..., N \tag{3.64b}$$

$$w_{ii}^{(3)} = \left[-\cos x_i - \sum_{j=1, j\neq i}^{N} w_{ij}^{(3)} \sin x_j\right] / \sin x_i, \text{ for } i = 1, 2, ..., N \tag{3.64c}$$

The determination of $a_{ij}, b_{ij}, w_{ij}^{(3)}$ $(i \neq j)$ will be given from the second set of base vectors given by Equation 3.62*b*. To apply Equation 3.62*b*, we note that both $s_k(x)$ and $c_k(x)$ have the following properties

$$s_k(x_i) = \begin{cases} 1, & \text{when } k = i \\ 0, & \text{others} \end{cases} \tag{3.65a}$$

$$c_k(x_i) = \begin{cases} 1, & \text{when } k = i \\ 0, & \text{others} \end{cases} \tag{3.65b}$$

Substituting Equation 3.62*b* into Equations 3.1-3.3 and using the property of Equation 3.65*a*, we obtain

$$a_{ij} = s_j^{(1)}(x_i) \tag{3.66a}$$

$$b_{ij} = s_j^{(2)}(x_i) \tag{3.66b}$$

$$w_{ij}^{(3)} = s_j^{(3)}(x_i) \tag{3.66c}$$

where $s_j^{(1)}(x_i), s_j^{(2)}(x_i)$ and $s_j^{(3)}(x_i)$ are respectively the first, second, and third order derivatives of the function $s_k(x)$ at x_i. To evaluate these derivatives, we differentiate Equation 3.62*b* successively, and obtain

$$s_k^{(1)}(x) = \frac{\sin x}{\sin x_k} c_k^{(1)}(x) + \frac{\cos x}{\sin x_k} c_k(x) \tag{3.67a}$$

$$s_k^{(2)}(x) = \frac{\sin x}{\sin x_k} c_k^{(2)}(x) + \frac{2\cos x}{\sin x_k} c_k^{(1)}(x) - \frac{\sin x}{\sin x_k} c_k(x) \tag{3.67b}$$

$$s_k^{(3)}(x) = \frac{\sin x}{\sin x_k} c_k^{(3)}(x) + \frac{3\cos x}{\sin x_k} c_k^{(2)}(x) - \frac{3\sin x}{\sin x_k} c_k^{(1)}(x) - \frac{\cos x}{\sin x_k} c_k(x) \tag{3.67c}$$

Using the property of Equation 3.65*b* and applying Equation 3.67 at the grid points, we get

$$s_j^{(1)}(x_i) = \frac{\sin x_i}{\sin x_j} c_j^{(1)}(x_i) + \frac{\cos x_i}{\sin x_j} \delta_{ij} \tag{3.68a}$$

$$s_j^{(2)}(x_i) = \frac{\sin x_i}{\sin x_j} c_j^{(2)}(x_i) + \frac{2\cos x_i}{\sin x_j} c_j^{(1)}(x_i) - \frac{\sin x_i}{\sin x_j} \delta_{ij} \tag{3.68b}$$

$$s_j^{(3)}(x_i) = \frac{\sin x_i}{\sin x_j} c_j^{(3)}(x_i) + \frac{3\cos x_i}{\sin x_j} c_j^{(2)}(x_i) - \frac{3\sin x_i}{\sin x_j} c_j^{(1)}(x_i) - \frac{\cos x_i}{\sin x_j} \delta_{ij} \tag{3.68c}$$

Note that $c_k(x)$ is actually a set of base vectors in the cosine expansion-based differential quadrature (CDQ). When $c_k(x)$ is substituted into Equations 3.1-3.3, we find that $c_j^{(1)}(x_i), c_j^{(2)}(x_i), c_j^{(3)}(x_i)$ are the weighting coefficients of the first, second, and third order derivatives in CDQ. So, using Equations 3.23, 3.29 and 3.35 in CDQ, we have

$$c_j^{(1)}(x_i) = \frac{-p(x_i)\sin x_i}{(\cos x_i - \cos x_j)\cdot p(x_j)}, \quad \text{for } j \neq i \tag{3.69a}$$

$$c_j^{(2)}(x_i) = c_j^{(1)}(x_i)\left(\frac{2\sin x_i}{(\cos x_i - \cos x_j)} + 2c_i^{(1)}(x_i) + \cot x_i\right), \quad \text{for } j \neq i \tag{3.69b}$$

$$\begin{aligned} c_j^{(3)}(x_i) = {} & 3c_j^{(1)}(x_i)\left(c_i^{(2)}(x_i) - \frac{1}{3} + c_i^{(1)}(x_i)\cot x_i + \frac{\cos x_i}{\cos x_i - \cos x_j}\right) \\ & + \frac{3\sin x_i \cdot c_j^{(2)}(x_i)}{\cos x_i - \cos x_j}, \quad \text{for } j \neq i \end{aligned} \tag{3.69c}$$

Substituting Equations 3.69*a* and 3.68*a* into Equation 3.66*a* gives

$$a_{ij} = \frac{\sin x_i}{\sin x_j} c_j^{(1)}(x_i) = \frac{-p(x_i)\sin^2 x_i}{(\cos x_i - \cos x_j)\sin x_j \cdot p(x_j)}, \quad \text{for } j \neq i \quad (3.70)$$

From Equations 3.66*a* and 3.68*a*, we also find that

$$a_{ii} = c_i^{(1)}(x_i) + \cot x_i \quad (3.71)$$

Using Equations 3.70, 3.71 and 3.69*b*, we obtain

$$\frac{\sin x_i}{\sin x_j} c_j^{(2)}(x_i) = a_{ij}\left(\frac{2\sin x_i}{(\cos x_i - \cos x_j)} + 2a_{ii} - \cot x_i\right), \text{ for } j \neq i \quad (3.72)$$

Substituting Equations 3.70, 3.72 into Equation 3.68*b* and using Equation 3.66*b* yields

$$\begin{aligned} b_{ij} &= \frac{\sin x_i}{\sin x_j} c_j^{(2)}(x_i) + 2\cot x_i \cdot a_{ij} \\ &= a_{ij}\left(\frac{2\sin x_i}{(\cos x_i - \cos x_j)} + 2a_{ii} + \cot x_i\right), \text{ for } j \neq i \end{aligned} \quad (3.73)$$

Equations 3.68*b* and 3.66*b* also result in

$$b_{ii} = c_i^{(2)}(x_i) + 2\cot x_i \cdot c_i^{(1)}(x_i) - 1 \quad (3.74)$$

Substituting Equation 3.68*c* into Equation 3.66*c* leads to

$$w_{ij}^{(3)} = \frac{\sin x_i}{\sin x_j} c_j^{(3)}(x_i) + \frac{3\cos x_i}{\sin x_j} c_j^{(2)}(x_i) - \frac{3\sin x_i}{\sin x_j} c_j^{(1)}(x_i), \text{ for } j \neq i \quad (3.75)$$

Using Equations 3.70 and 3.73, we have

$$\frac{\sin x_i}{\sin x_j} c_j^{(1)}(x_i) = a_{ij} \quad (3.76)$$

$$\frac{\sin x_i}{\sin x_j} c_j^{(2)}(x_i) = b_{ij} - 2\cot x_i \cdot a_{ij} \quad (3.77)$$

Similarly, using Equation 3.69*c*, we get

$$\frac{\sin x_i}{\sin x_j} c_j^{(3)}(x_i) = 3a_{ij}\left(b_{ii} - \cot x_i (a_{ii} - \cot x_i) + \frac{2}{3} + \frac{\cos x_i}{\cos x_i - \cos x_j} \right) + \frac{3\sin x_i}{\cos x_i - \cos x_j}(b_{ij} - 2\cot x_i \cdot a_{ij}) \tag{3.78}$$

Substituting Equations 3.76-3.78 into Equation 3.75 and using

$$b_{ij} - \cot x_i \cdot a_{ij} = a_{ij}\left(\frac{2\sin x_i}{(\cos x_i - \cos x_j)} + 2a_{ii} \right) \quad \text{(from Equation 3.73)}$$

we finally obtain

$$w_{ij}^{(3)} = 3a_{ij}\left(b_{ii} - \frac{1}{3} + a_{ii} \cdot \cot x_i + \frac{\cos x_i}{\cos x_i - \cos x_j} \right) + \frac{3\sin x_i \cdot b_{ij}}{\cos x_i - \cos x_j}, \quad \text{for } j \neq i \tag{3.79}$$

Note that Equation 3.70 is slightly different from Equation 3.23, but Equations 3.73 and 3.79 are exactly the same as Equations 3.29 and 3.35. This indicates that for the second and higher order derivatives, both cosine expansion-based differential quadrature (CDQ) and sine expansion-based differential quadrature (SDQ) methods use the same formulation to compute the weighting coefficients for $j \neq i$. The computation of the diagonal weighting coefficients such as $a_{ii}, b_{ii}, w_{ii}^{(3)}$ is different for both methods.

For practical applications, the problem may be defined on an open interval (a, b) instead of $(0, \pi)$. Then for this case, we can use a similar procedure to perform a coordinate transformation using Equation 3.55. As a consequence, the formulations for the weighting coefficients of the first order derivative is modified to

$$a_{ij} = \frac{-\alpha \cdot \sin^2 \xi_i \cdot p(\xi_i)}{(\cos\xi_i - \cos\xi_j)\sin\xi_j \cdot p(\xi_j)}, \quad \text{for } i \neq j \tag{3.80}$$

Equations 3.59, 3.60 can still be used to compute the weighting coefficients of the second and third order derivatives.

3.4 Fourier Expansion-based Differential Quadrature (FDQ) for Any General Function

Now, it is supposed that $f(x)$ is approximated by a Fourier series expansion of the form

$$f(x) = c_0 + \sum_{k=1}^{N/2} (c_k \cos kx + d_k \sin kx) \tag{3.81}$$

It is easy to show that $f(x)$ in Equation 3.81 constitutes a (N+1) dimensional linear vector space V_{N+1}. As shown in Chapter 1, two typical sets of base vectors in V_{N+1} are

$$1, \cos x, \sin x, \cos 2x, \sin 2x, ..., \cos(Nx/2), \sin(Nx/2) \tag{3.82a}$$

$$s_k(x) = \frac{S(x)}{q(x_k) \cdot \sin \dfrac{x - x_k}{2}}, \quad k = 0, 1, ..., N \tag{3.82b}$$

where

$$S(x) = \prod_{k=0}^{N} \sin \frac{x - x_k}{2} \tag{3.83a}$$

$$q(x_i) = \prod_{k=0, k \neq i}^{N} \sin \frac{x_i - x_k}{2} \tag{3.83b}$$

In the following, we will use the above two sets of base vectors to derive explicit formulations to compute the weighting coefficients of the first, second, and third order derivatives. For higher order derivatives, we suggest the use of the matrix multiplication method to compute the weighting coefficients. Among the set of base vectors $1, \cos x, \sin x, ..., \cos(Nx/2), \sin(Nx/2)$, we only apply the vector 1. Let the vector 1 satisfy Equations 3.1, 3.2, and 3.3, we obtain

$$\sum_{j=0}^{N} a_{ij} = 0, \quad \text{or} \quad a_{ii} = -\sum_{j=0, j \neq i}^{N} a_{ij} \tag{3.84a}$$

$$\sum_{j=0}^{N} b_{ij} = 0, \quad \text{or} \quad b_{ii} = -\sum_{j=0, j \neq i}^{N} b_{ij} \tag{3.84b}$$

$$\sum_{j=0}^{N} w_{ij}^{(3)} = 0, \quad \text{or} \quad w_{ii}^{(3)} = -\sum_{j=0, j \neq i}^{N} w_{ij}^{(3)} \tag{3.84c}$$

Equation 3.84 will be used to compute the diagonal weighting coefficients. For the non-diagonal weighting coefficients, we need to use the second set of base vectors. For simplicity, we set

$$S(x)=\prod_{k=0}^{N}\sin\frac{x-x_k}{2}=N(x,x_k)\cdot\sin\frac{x-x_k}{2} \tag{3.85}$$

where

$$N(x_i,x_i)=\prod_{k=0,k\neq i}^{N}\sin\frac{x_i-x_k}{2}=q(x_i)$$

$N(x_i,x_j)=N(x_i,x_i)\cdot\delta_{ij}$, δ_{ij} is the Kronecker operator.

Then, Equation 3.82*b* can be reduced to

$$s_k(x)=\frac{N(x,x_k)}{q(x_k)} \tag{3.86}$$

Using the same approach as in CDQ, we let all the base vectors given by Equation 3.86 satisfy linear Equations 3.1, 3.2 and 3.3, and obtain

$$a_{ij}=\frac{N^{(1)}(x_i,x_j)}{q(x_j)} \tag{3.87a}$$

$$b_{ij}=\frac{N^{(2)}(x_i,x_j)}{q(x_j)} \tag{3.87b}$$

$$w_{ij}^{(3)}=\frac{N^{(3)}(x_i,x_j)}{q(x_j)} \tag{3.87c}$$

where $N^{(1)}(x,x_k)$, $N^{(2)}(x,x_k)$ and $N^{(3)}(x,x_k)$ are respectively the first, second and third order derivatives of $N(x,x_k)$. It is observed from Equation 3.87 that the computation of $a_{ij}, b_{ij}, w_{ij}^{(3)}$ is equivalent to the evaluation of $N^{(1)}(x,x_k)$, $N^{(2)}(x,x_k)$ and $N^{(3)}(x,x_k)$, since $q(x_j)$ can be easily determined from Equation 3.83*b*. To evaluate $N^{(1)}(x,x_k)$, $N^{(2)}(x,x_k)$ and $N^{(3)}(x,x_k)$, we successively differentiate Equation 3.85 to obtain

$$S^{(1)}(x) = N^{(1)}(x,x_k)\cdot\sin\frac{x-x_k}{2} + \frac{1}{2}N(x,x_k)\cdot\cos\frac{x-x_k}{2} \tag{3.88}$$

$$\begin{aligned} S^{(2)}(x) &= N^{(2)}(x,x_k)\cdot\sin\frac{x-x_k}{2} + N^{(1)}(x,x_k)\cdot\cos\frac{x-x_k}{2} \\ &-\frac{1}{4}N(x,x_k)\cdot\sin\frac{x-x_k}{2} \end{aligned} \tag{3.89}$$

$$\begin{aligned} S^{(3)}(x) &= N^{(3)}(x,x_k)\cdot\sin\frac{x-x_k}{2} + \frac{3}{2}N^{(2)}(x,x_k)\cdot\cos\frac{x-x_k}{2} - \\ &\frac{3}{4}N^{(1)}(x,x_k)\cdot\sin\frac{x-x_k}{2} - \frac{1}{8}N(x,x_k)\cdot\cos\frac{x-x_k}{2} \end{aligned} \tag{3.90}$$

Similar to Equation 3.19, we can also obtain a recurrence relationship for the mth order derivative of $S(x)$ as

$$S^{(m)}(x) = N^{(m)}(x,x_k)\sin\frac{x-x_k}{2} + \sum_{k=1}^{m} B_{m,k}, \; m = 1, 2, ..., N \tag{3.91}$$

where

$$B_{m,k} = \begin{cases} c_m^k \cdot (-1)^{\frac{k-1}{2}} \cdot \dfrac{1}{2^k} \cdot N^{(m-k)}(x,x_k)\cdot\cos\dfrac{x-x_k}{2}, & \text{when } k = 1, 3, 5,... \\ c_m^k \cdot (-1)^{\frac{k}{2}} \cdot \dfrac{1}{2^k} \cdot N^{(m-k)}(x,x_k)\cdot\sin\dfrac{x-x_k}{2}, & \text{when } k = 2, 4, 6,... \end{cases}$$

$$N^{(0)}(x,x_k) = N(x,x_k)$$

where c_m^k is the combination of m terms taken k at a time. Clearly, Equation 3.91 includes Equations 3.88-3.90. Applying Equations 3.88-3.90 at the grid points, we obtain

$$S^{(1)}(x_i) = \frac{1}{2}N(x_i,x_i) = \frac{1}{2}q(x_i) \tag{3.92}$$

$$N^{(1)}(x_i,x_j) = \frac{S^{(1)}(x_i)}{\sin\dfrac{x_i-x_j}{2}}, \quad \text{for } j \neq i \tag{3.93a}$$

$$N^{(1)}(x_i,x_i) = S^{(2)}(x_i) \tag{3.93b}$$

$$N^{(2)}(x_i, x_j) = \frac{S^{(2)}(x_i) - N^{(1)}(x_i, x_j) \cdot \cos\frac{x_i - x_j}{2}}{\sin\frac{x_i - x_j}{2}}, \quad \text{for } j \neq i \tag{3.94a}$$

$$N^{(2)}(x_i, x_i) = \frac{2}{3}\left[S^{(3)}(x_i) + \frac{1}{8}q(x_i)\right] \tag{3.94b}$$

$$N^{(3)}(x_i, x_j) = \left(S^{(3)}(x_i) - \frac{3}{2}N^{(2)}(x_i, x_j)\cos\frac{x_i - x_j}{2}\right) / \sin\frac{x_i - x_j}{2} + \frac{3}{4}N^{(1)}(x_i, x_j), \quad \text{for } j \neq i \tag{3.95}$$

Substituting Equation 3.92 into Equation 3.93*a* gives

$$N^{(1)}(x_i, x_j) = \frac{q(x_i)}{2\sin\frac{x_i - x_j}{2}}, \quad \text{for } j \neq i \tag{3.96}$$

and substituting Equation 3.93*b* into Equation 3.94*a* leads to

$$N^{(2)}(x_i, x_j) = \frac{N^{(1)}(x_i, x_i) - N^{(1)}(x_i, x_j) \cdot \cos\frac{x_i - x_j}{2}}{\sin\frac{x_i - x_j}{2}}, \quad \text{for } j \neq i \tag{3.97}$$

Similarly, substituting Equation 3.94*b* into Equation 3.95 gives

$$N^{(3)}(x_i, x_j) = \left(\frac{3}{2}N^{(2)}(x_i, x_i) - \frac{3}{2}N^{(2)}(x_i, x_j)\cos\frac{x_i - x_j}{2}\right) / \sin\frac{x_i - x_j}{2} + \left(\frac{3}{4}N^{(1)}(x_i, x_j)\sin\frac{x_i - x_j}{2} - \frac{1}{8}q(x_i)\right) / \sin\frac{x_i - x_j}{2}, \quad \text{for } j \neq i \tag{3.98}$$

Finally, substituting Equations 3.96-3.98 into Equation 3.87, we obtain

$$a_{ij} = \frac{q(x_i)}{2\sin\frac{x_i - x_j}{2}q(x_j)}, \quad \text{when } j \neq i \tag{3.99}$$

$$b_{ij} = a_{ij}\left[2a_{ii} - \cot\frac{x_i - x_j}{2}\right], \quad \text{when } j \neq i \tag{3.100}$$

$$w_{ij}^{(3)} = a_{ij}\left(\frac{1}{2} + 3b_{ii}\right) - \frac{3}{2}b_{ij}\cdot\cot\frac{x_i - x_j}{2}, \quad \text{when } j \neq i \tag{3.101}$$

Equations 3.99-3.101 are used to compute the non-diagonal weighting coefficients.

It is interesting to realize that Equations 3.99-3.101 can be applied to periodic problems and non-periodic problems. For non-periodic problems, the range of x in the computational domain should be $0 \leq x \leq \pi$, while for periodic problems, the range of x in the computational domain must be $0 \leq x < 2\pi$. For practical applications, the physical domain is not $[0, \pi]$ nor $[0, 2\pi)$, but rather $[a, b]$ or $[a, b)$, then for this case, we can perform a coordinate transformation from the x domain to the ξ domain. The interval in the ξ domain will either be $[0, \pi]$ or $[0, 2\pi)$. The weighting coefficients are firstly computed in the ξ domain, and then transformed back into the x domain. As a result, Equations 3.99-3.101 become

$$a_{ij} = \frac{\alpha\cdot q(\xi_i)}{2\sin\dfrac{\xi_i - \xi_j}{2}q(\xi_j)}, \quad \text{when } j \neq i \tag{3.102}$$

$$b_{ij} = a_{ij}\left[2a_{ii} - \alpha\cdot\cot\frac{\xi_i - \xi_j}{2}\right], \quad \text{when } j \neq i \tag{3.103}$$

$$w_{ij}^{(3)} = a_{ij}\left(\frac{\alpha^2}{2} + 3b_{ii}\right) - \frac{3}{2}\alpha\cdot b_{ij}\cdot\cot\frac{\xi_i - \xi_j}{2}, \quad \text{when } j \neq i \tag{3.104}$$

where ξ and α depend on the problem. For periodic problems, ξ and α are given by

$$\xi = 2\pi\frac{x-a}{b-a}, \qquad \alpha = 2\pi/(b-a) \tag{3.105}$$

while for non-periodic problems, ξ and α are given by

$$\xi = \pi\frac{x-a}{b-a}, \qquad \alpha = \pi/(b-a) \tag{3.106}$$

It should be indicated that Equation 3.82a was first used by Striz, Wang and Bert (1995) as test functions to determine the weighting coefficients in the DQ approximation. Using the same approach as Bellman's first method, they determined the weighting coefficients by solving a set of algebraic equations. Later, under the analyses of Fourier series expansion and linear vector space, Shu and Chew (1997), and Shu and Xue (1997) developed explicit formulations for computing the weighting coefficients of the first and second order derivatives.

3.5 Some Properties of Fourier Expansion-based Differential Quadrature

As shown previously, the FDQ method is based on Fourier series expansion. Thus, it has most of the properties of Fourier series expansion.

Convergence: The convergence of a truncated Fourier series expansion is guaranteed by Weierstrass' second theorem. The convergence rate of the truncated Fourier series depends on how fast the Fourier coefficients (c_k and d_k in Equation 3.81) decay to zero. This problem has been discussed in many books. For details, the reader is encouraged to consult the books by Gottlieb and Orszag (1977), Canuto et al. (1988). Here, we just cite some of the relevant conclusions. If $f(x)$ and all its derivatives are infinitely smooth and periodic, the error between $f(x)$ and its *N*th order truncated Fourier series decays faster than algebraically in 1/*N*. This kind of property is commonly termed as spectral accuracy or exponential convergence. Fourier sine and cosine series have convergence properties very similar to the general Fourier series described previously. On the other hand, the error estimate for the PDQ approach can be used for the Fourier cosine expansion-based differential quadrature. From Equation 3.40, it can be seen that the Fourier cosine series can be expressed by a polynomial of degree *N* in terms of ξ. We recall from Equation 2.69 that the error of the *m*th order derivative approximation in the polynomial-based differential quadrature (PDQ) can be estimated as

$$E_D^{(m)} \le C \frac{h^{N-m}}{(N-m)!}, \quad \text{for} \quad m = 0, 1, 2, ..., N-1 \tag{3.107}$$

where C is a constant and h could be the largest distance of two grid points. Note that when m is taken as zero, Equation 3.107 gives the error estimate for the function approximation. For Equation 3.40, the interval in the ξ domain is [1, -1]. So, for this case, $h = 2$, and accordingly, Equation 3.107 reduces to

$$E_D^{(m)} \le C \frac{2^{N-m}}{(N-m)!}, \quad \text{for} \quad m = 0, 1, 2, ..., N-1 \tag{3.108}$$

Equation 3.108 can be applied to estimate the error for Fourier cosine expansion-based differential quadrature. As for the Fourier sine expansion-based differential quadrature, a similar form can be used estimate the error since from Chapter 1, Equation 3.61 can also be expressed as

$$f(x) = \sin x \sum_{k=1}^{N} \overline{d}_k \cdot \cos^{k-1} x = \sin x \sum_{k=1}^{N} \overline{d}_k \xi^{k-1} \tag{3.109}$$

The summation in Equation 3.109 actually results in a polynomial of degree (N-1).

The Gibbs phenomenon: For a smooth function, the convergence of the truncated Fourier series is very fast. However, for a discontinuous problem, the truncated Fourier series exhibits the Gibbs phenomenon. The Gibbs phenomenon describes the characteristic oscillatory behavior of the truncated Fourier series or the discrete Fourier series of a function of bounded variation in the neighborhood of a point of discontinuity. A detailed discussion on the Gibbs phenomenon can be found in the books by Gottlieb and Orszag (1977), and Canuto et al. (1988).

Extension to the multi-dimensional case: Like the PDQ method, when the physical domain is regular, the one-dimensional FDQ formulations can be directly extended to the multi-dimensional case. In other words, each direction can be treated using the same manner as in the one-dimensional case. For the irregular physical domain, one has to first perform a coordinate transformation to map the irregular physical domain into a regular computational domain. Subsequently, in the computational domain, the FDQ formulations are applied separately in each direction. Like the PDQ method, once the functional values at the grid points are obtained from the FDQ method, the functional value at any location can be accurately computed by using the interpolation technique. Using a rectangular domain as an example, we have the following interpolation forms

$$f(x, y_j) = \sum_{i=1}^{N} f(x_i, y_j) \cdot r_i(x) \tag{3.110a}$$

$$f(x_i, y) = \sum_{j=1}^{M} f(x_i, y_j) \cdot s_j(y) \tag{3.110b}$$

$$f(x, y) = \sum_{i=1}^{N} \sum_{j=1}^{M} f(x_i, y_j) \cdot r_i(x) \cdot s_j(y) \tag{3.110c}$$

Here $r_i(x)$ and $s_j(y)$ are the Lagrange interpolation polynomials given by

$$r_i(x) = \prod_{k=0,k\neq i}^{N} \frac{\cos x - \cos x_k}{\cos x_i - \cos x_k}$$

$$s_j(y) = \prod_{k=0,k\neq j}^{M} \frac{\cos y - \cos y_k}{\cos y_j - \cos y_k}$$

for the cosine expansion-based differential quadrature, and

$$r_i(x) = \frac{\sin x}{\sin x_i} \prod_{k=1,k\neq i}^{N} \frac{\cos x - \cos x_k}{\cos x_i - \cos x_k}$$

$$s_j(y) = \frac{\sin y}{\sin y_j} \prod_{k=1,k\neq j}^{M} \frac{\cos y - \cos y_k}{\cos y_j - \cos y_k}$$

for the sine expansion-based differential quadrature, and

$$r_i(x) = \prod_{k=0,k\neq i}^{N} \frac{\sin \dfrac{x - x_k}{2}}{\sin \dfrac{x_i - x_k}{2}}$$

$$s_j(y) = \prod_{k=0,k\neq j}^{M} \frac{\sin \dfrac{y - y_k}{2}}{\sin \dfrac{y_j - y_k}{2}}$$

for the Fourier expansion-based differential quadrature. It has to be mentioned that when Equation 3.110 is applied, the x and y coordinates should be taken in the interval $[0, \pi]$ for the cosine expansion-based differential quadrature, and $(0, \pi)$ for the sine expansion-based differential quadrature. For the general Fourier expansion-based differential quadrature, the x and y coordinates should be taken in the interval $[0, \pi]$ for non-periodic problems, and $[0, 2\pi)$ for periodic problems.

For application purposes, Equation 3.110*a* is applied along $y = \text{constant}$ lines while Equation 3.110*b* is used along $x = \text{constant}$ lines. Equation 3.110*c* is utilized at any location in a regular domain.

Number of grid points used in the FDQ method: The cosine expansion-based differential quadrature is based on the expansion of cosine functions while the sine expansion-based differential quadrature is based on the expansion of sine functions. They have no restriction as to the choice of the number of grid points. In other words, the number of grid points used can be even or odd. On the other hand, for the general Fourier expansion-based differential quadrature, it is based on the expansion of both cosine and sine functions. Equation 3.81 has (N+1) terms, which constitutes a (N+1) dimensional linear vector space. So, to form a complete set of (N+1) base vectors in the linear vector space, it is advisable for the number of grid points to be chosen as (N+1), which is an odd number. Due to this fact, it may be difficult to apply the Fast Fourier Transform (FFT) technique in the FDQ method because the FFT technique usually requires the number of grid points to be on integral power of 2. For practical applications, it was found that the number of grid points in the FDQ application can also be chosen as an even number, but the accuracy of the numerical results is diminished.

Chapter 4

Some Properties of DQ Weighting Coefficient Matrices

4.1 Introduction

It has been shown in Chapters 2 and 3 that the DQ approximation for a derivative of any order has a similar form. The difference in the approximation for the respective derivatives lies only in the weighting coefficients. Consider a one-dimensional problem. It is supposed that there are N grid points in the whole domain with coordinates $x_1, x_2, ..., x_N$. At any location x_i, the nth order derivative of a function $u(x,t)$ with respect to x can be approximated by the DQ method as

$$u_x^{(n)}(x_i,t) = \sum_{j=1}^{N} w_{ij}^{(n)} u(x_j,t), \quad \text{for } i = 1, 2, ..., N \tag{4.1}$$

The index n in Equation 4.1 can be taken from 1 to N-1. The determination of the weighting coefficients $w_{ij}^{(n)}$ depends on how the function $u(x,t)$ is approximated. The explicit formulations for $w_{ij}^{(n)}$ were shown in Chapter 2, when the function is approximated by a polynomial of high degree, and in Chapter 3, when the function is approximated by a truncated Fourier series. When Equation 4.1 is applied at all the grid points, we can obtain an equation system which, when written in the matrix form, becomes

$$\{u_x^{(n)}\} = [W^{(n)}] \cdot \{u\} \tag{4.2}$$

where

$$\left\{u_x^{(n)}\right\} = \begin{pmatrix} u_x^{(n)}(x_1,t) \\ u_x^{(n)}(x_2,t) \\ \cdot \\ \cdot \\ \cdot \\ u_x^{(n)}(x_N,t) \end{pmatrix}, \quad \left[W^{(n)}\right] = \begin{bmatrix} w_{1,1}^{(n)} & w_{1,2}^{(n)} & \cdot\ \cdot\ \cdot & w_{1,N}^{(n)} \\ w_{2,1}^{(n)} & w_{2,2}^{(n)} & \cdot\ \cdot\ \cdot & w_{2,N}^{(n)} \\ \cdot & \cdot & \cdot\ \cdot\ \cdot & \cdot \\ \cdot & \cdot & \cdot\ \cdot\ \cdot & \cdot \\ \cdot & \cdot & \cdot\ \cdot\ \cdot & \cdot \\ w_{N,1}^{(n)} & w_{N,2}^{(n)} & \cdot\ \cdot\ \cdot & w_{N,N}^{(n)} \end{bmatrix}, \quad \{u\} = \begin{pmatrix} u(x_1,t) \\ u(x_2,t) \\ \cdot \\ \cdot \\ \cdot \\ u(x_N,t) \end{pmatrix}$$

In order to apply the DQ method more proficiently and meaningfully, it is necessary to understand the basic properties of the DQ weighting coefficient matrices, $[W^{(n)}], n = 1, 2, ..., N-1$. For example, if the determinant of $[W^{(n)}]$ is zero, then its rank should be less than N. We need to know the rank of $[W^{(n)}]$. If the rank of $[W^{(n)}]$ is (N-n), then Equation 4.1 can only be applied at (N-n) grid points in order to ensure that the resultant DQ discretization matrix is not singular. On the other hand, if the structure of $[W^{(n)}]$ is centrosymmetric or skew centrosymmetric, some efficient approaches can be used to simplify the solution process for the DQ algebraic equations or the DQ ordinary differential equations. The determinant, rank, and the structure of $[W^{(n)}]$ will be discussed in this chapter.

For application of the DQ method, it was found that the grid point distribution has a great effect on the convergence and accuracy of the numerical results. In this chapter, the effect of the grid point distribution is studied using the eigenvalues of the DQ discretization matrix. In addition, we will show the effect of the grid point distribution on the magnitude and distribution of the DQ weighting coefficients.

4.2 Determinant and Rank of DQ Weighting Coefficient Matrices

In this section, we will discuss the rank of the DQ weighting coefficient matrices from the computation of their determinants.

4.2.1 Definition and Properties of Determinant and Rank

A determinant is a scalar function of a square matrix. It is conventional to indicate a determinant with the same symbol as the matrix, but with a bar '|' on both sides. Thus, $|A|$ is the determinant of the matrix [A], which must be square. The determinant of a N by N matrix [A] can be computed by the following recursive formula

$$|A| = \sum_{i=1}^{N} (-1)^{i+j} a_{i,j} |A_{i,j}| \tag{4.3}$$

where $a_{i,j}$ is the element of [A] at the ith row and jth column, $|A_{i,j}|$ is the determinant of the (N-1) by (N-1) matrix, which is formed by removing the ith row and jth column from [A]. We have the following four basic properties for matrix [A].

Property 4.2.1.1: If a scalar multiple of one column of a square matrix is added to another column, then the determinant of the resulting matrix is the same as the determinant of the original matrix.

Property 4.2.1.2: If a column of a matrix consists entirely of zeros, then the determinant of that matrix is zero.

Property 4.2.1.2 can be easily obtained from Equation 4.3. If we repeatedly apply property 4.2.1.1, we can get the following property,

Property 4.2.1.3: When the scalar multiples of all columns of a square matrix are used to replace a column, the determinant of the resulting matrix is the same as the determinant of the original matrix. Mathematically, it is written as

$$\begin{vmatrix} a_{1,1} & \cdots & a_{1,k} & \cdots & a_{1,N} \\ a_{2,1} & \cdots & a_{2,k} & \cdots & a_{2,N} \\ \cdot & \cdots & \cdot & \cdots & \cdot \\ \cdot & \cdots & \cdot & \cdots & \cdot \\ \cdot & \cdots & \cdot & \cdots & \cdot \\ a_{N,1} & \cdots & a_{N,k} & \cdots & a_{N,N} \end{vmatrix} = \begin{vmatrix} a_{1,1} & \cdots & \sum_{j=1}^{N} \beta_j a_{1,j} & \cdots & a_{1,N} \\ a_{2,1} & \cdots & \sum_{j=1}^{N} \beta_j a_{2,j} & \cdots & a_{2,N} \\ \cdot & \cdots & \cdot & \cdots & \cdot \\ \cdot & \cdots & \cdot & \cdots & \cdot \\ \cdot & \cdots & \cdot & \cdots & \cdot \\ a_{N,1} & \cdots & \sum_{j=1}^{N} \beta_j a_{N,j} & \cdots & a_{N,N} \end{vmatrix} \tag{4.4}$$

Using Equation 4.4 and property 4.2.1.1, we can arrive at the following property, which will be used to analyze the DQ weighting coefficient matrices.

Property 4.2.1.4: Let $\beta_j, j = 1, 2, ..., N$ be scalars, which have at least one component with non-zero value. If the following condition

$$\sum_{j=1}^{N} \beta_j a_{i,j} = 0 \tag{4.5}$$

is satisfied for all $i = 1, 2, ..., N$, then the determinant of $[A]$ is zero.

The determinant is a very useful parameter for matrix analysis. If the determinant of a N by N matrix is zero, then the inversion of the matrix does not exist, and its rank is less than N.

Definition of rank: The rank of a matrix is the order of the highest order square submatrix with non-zero determinant. The submatrix is formed from the original matrix by removing certain rows and columns.

From the above definition, it is known that for a N by N matrix, it has an inverse if and only if its rank is N. Now, we can study the determinant and the rank of the DQ weighting coefficient matrices by using the above properties and definition.

4.2.2 Determinant and Rank of DQ Weighting Coefficient Matrices

First, we consider the PDQ weighting coefficient matrices. For simplicity, we let $a_{i,j}$, $b_{i,j}$ be the weighting coefficients of the first and second order derivatives, that is, $a_{i,j} = w_{i,j}^{(1)}, b_{i,j} = w_{i,j}^{(2)}$, and $[A]$, $[B]$ be the weighting coefficient matrices of the first and second order derivatives. It is easy to show that the determinants of $[A]$ and $[B]$ are zero. In fact, Equations 2.18 and 2.25 give

$$\sum_{j=1}^{N} a_{i,j} = 0 \tag{4.6a}$$

$$\sum_{j=1}^{N} b_{i,j} = 0 \tag{4.6b}$$

By selecting $\beta_j = 1, j = 1, 2, ..., N$ in Equation 4.5, we can conclude that $|A|$ and $|B|$ are zero. Furthermore, we can prove that the determinant of the (N-1) by (N-1) submatrix of $[B]$ is also zero. Let $[B_{N-1}]$ represent the (N-1) by (N-1) submatrix of $[B]$, which is formed by removing the kth row and the jth column from $[B]$. $[B_{N-1}]$ can also be formed from the application of Equation 4.1 with n = 2 at (N-1) grid points, x_1, x_2, ..., $x_{k-1}, x_{k+1}, ..., x_N$. $[B_{N-1}]$ can be written as

$$[B_{N-1}] = \begin{bmatrix} b_{1,1} & \cdots & b_{1,j-1} & b_{1,j+1} & \cdots & b_{1,N} \\ \cdot & \cdots & \cdot & \cdot & \cdots & \cdot \\ b_{k-1,1} & \cdots & b_{k-1,j-1} & b_{k-1,j+1} & \cdots & b_{k-1,N} \\ b_{k+1,1} & \cdots & b_{k+1,j-1} & b_{k+1,j+1} & \cdots & b_{k+1,N} \\ \cdot & \cdots & \cdot & \cdot & \cdots & \cdot \\ b_{N,1} & \cdots & b_{N,j-1} & b_{N,j+1} & \cdots & b_{N,N} \end{bmatrix} \tag{4.7}$$

Using Equation 4.5, we know that if

$$\sum_{k1=1,k1\neq j}^{N} \beta_{k1} b_{i,k1} = 0 \tag{4.8}$$

is satisfied for $i = 1, 2, ..., k-1, k+1, ..., N$, then $|B_{N-1}| = 0$. From the results of Chapter 2 obtained by using $x^k, k = 0, 1, ..., N-1$ as a set of base vectors for Equation 2.19, we have

$$\sum_{k1=1}^{N} b_{i,k1} x_{k1} = 0 \tag{4.9}$$

Using Equations 4.6*b* and 4.9, we can get

$$\sum_{k1=1,k1\neq j}^{N} b_{i,k1}(x_{k1} - x_j) = \sum_{k1=1}^{N} b_{i,k1}(x_{k1} - x_j) = \sum_{k1=1}^{N} b_{i,k1} x_{k1} - x_j \sum_{k1=1}^{N} b_{i,k1} = 0 \tag{4.10}$$

Clearly, Equation 4.10 shows that Equation 4.8 is satisfied by selecting $\beta_{k1} = x_{k1} - x_j$. Therefore, $|B_{N-1}| = 0$. Through the numerical experiments, we find that the determinants of the (*N*-1) by (*N*-1) submatrix of [*A*], and the (*N*-2) by (*N*-2) submatrix of [*B*] are non-zero. This indicates that the rank of [*A*] is (*N*-1) while the rank of [*B*] is (*N*-2). In a similar manner, we can show that the rank of the weighting coefficient matrix for the *n*th order derivative, $[W^{(n)}]$, is (*N*-*n*). This result implies that the DQ approximation should only be applied at (*N*-*n*) grid points. Otherwise, the resulting discretization matrix will be singular. This is a very interesting result, which is in well agreement with the definition of a well-posed problem. As we know, for a well-posed problem, the *n*th order differential equation requires *n* initial or boundary conditions, which provide *n* equations. When we pursue the numerical solution of a PDE, the functional values at the grid points are considered as unknowns, and the application of the discretized governing equation at a grid point gives one equation for the solution of the unknowns. Suppose that there are *N* grid points in the computational domain. So, the total number of unknowns is *N*. To solve for *N* unknowns, we need *N* equations. Since the initial or boundary conditions already provide *n* equations, the discretized governing equation can only be applied at (*N*-*n*) grid points, giving (*N*-*n*) equations. If the discretized governing equation is applied at more than (*N*-*n*) grid points, the problem is not well-posed since the number of equations exceeds the number of unknowns.

For purposes of brevity in the following discussion, the cosine expansion-based differential quadrature is termed as CDQ, the sine expansion-based differential quadrature is named as SDQ, and the general Fourier expansion-based differential quadrature is called FDQ. It was found that the weighting coefficient matrices of FDQ, CDQ, and SDQ do not exhibit the same features as those of PDQ. From the results of Chapter 3, the FDQ and CDQ weighting coefficients satisfy the following condition

$$\sum_{j=1}^{N} w_{i,j}^{(n)} = 0, \quad \text{for } i = 1, 2, \ldots, N; \; n = 1, 2, \ldots, N\text{-}1 \tag{4.11}$$

Therefore, the determinant of the FDQ and CDQ weighting coefficient matrices $[W^{(n)}], n = 1, 2, \ldots, N-1$, is zero. On the other hand, it was found from numerical experiments that for any odd number N, the determinant of the (N-1) by (N-1) submatrix of $[W^{(n)}]$ is not zero. This indicates that the rank of $[W^{(n)}]$ in the FDQ and CDQ methods is (N-1) instead of (N-n) for the PDQ method. For the SDQ method applied to the open interval (0, π), it can be observed from Equation 3.61 that the approximated functional values by SDQ are zero at $x = 0$ and π. Hence, there are implicitly two boundary conditions at $x = 0$ and π for the SDQ method. From this point of view, the SDQ method is suitable for the solution of the second order differential equations. In a similar manner, it was found from Equation 3.7 that the first order derivatives at $x = 0$ and π are always zero when the CDQ approximation is employed. This can also be confirmed from the CDQ weighting coefficients of the first order derivative. Through numerical experiments, we also found that for any grid point distribution, the CDQ weighting coefficients, $w_{1,k}^{(1)}$ and $w_{N,k}^{(1)}, k = 1, 2, \ldots, N$, are always zero, leading to zero values for the first order derivative at $x = 0$ and π. Therefore, the CDQ method is also suitable for the solution of second order differential equations. The SDQ discretization of a differential equation is applied at all grid points since the boundary points are not included in the SDQ formulations. It is confirmed from numerical experiments that the determinant of the second order weighting coefficient matrix for SDQ is not zero, and so its rank is N.

4.3 Structures and Properties of DQ Weighting Coefficient Matrices

The structures of the DQ weighting coefficient matrices for the first and second order derivatives were first reported by Quan and Chang (1989*a*, *b*). They indicated that the properties of these matrices can be used to reduce the DQ ordinary differential equation (ODE) system into two smaller ODE systems. The comprehensive study on the structures and properties of the DQ weighting coefficient matrices was given by Chen and Yu (1993*a*, *b*), Chen, Wang and Zhong (1996), and Chen (1996). In the following, we will show that the DQ weighting coefficient matrices have either centrosymmetric or skew centrosymmetric properties.

4.3.1 Definition of Centrosymmetric and Skew Centrosymmetric Matrices

Definition of centrosymmetric matrix: A N by N matrix $[Q]=[q_{i,j}]$ is said to be centrosymmetric if its elements satisfy

$$q_{i,j} = q_{N-i+1,N-j+1}, \quad \text{for } i, j = 1, 2, ..., N \tag{4.12}$$

With the contra-identity matrix $[J]$ defined by

$$[J] = \begin{bmatrix} 0 & \cdots & 0 & 1 \\ 0 & \cdots & 1 & 0 \\ \vdots & \iddots & \vdots & \vdots \\ 1 & \cdots & 0 & 0 \end{bmatrix} \tag{4.13}$$

the centrosymmetric matrix $[Q]$ can be written as

$$[Q] = [J][Q][J] \tag{4.14}$$

Note that the matrix $[J]$ has the following features:

- It has unit elements along the secondary diagonal and zeros elsewhere;
- $[J]^T = [J]$ and $[J]^2 = [I]$, where $[I]$ is the identity matrix;
- Pre-multiply any matrix by $[J]$ reverses the order of its rows;
- Post-multiply any matrix by $[J]$ reverses the order of its columns.

Definition of skew centrosymmetric matrix: A N by N matrix $[R]=[r_{i,j}]$ is said to be skew centrosymmetric if its elements satisfy

$$r_{i,j} = -r_{N-i+1,N-j+1}, \quad \text{for } i, j = 1, 2, ..., N \tag{4.15}$$

The skew centrosymmetric matrix $[R]$ can also be expressed by $[J]$ as

$$[R] = -[J][R][J] \tag{4.16}$$

In addition to the definition of centrosymmetric and skew centrosymmetric matrices, a vector can also be defined as symmetric or skew symmetric.

Definition of symmetric vector: A N dimensional vector γ is said to be symmetric if it satisfies

$$[J]\gamma = \gamma \tag{4.17}$$

Definition of skew symmetric vector: A N dimensional vector γ is said to be skew symmetric if it satisfies

$$[J]\gamma = -\gamma \tag{4.18}$$

4.3.2 Properties of Centrosymmetric and Skew Centrosymmetric Matrices

For brevity, we let $[C]_{N\times N}$ denote the set of N by N centrosymmetric matrices, and $[S]_{N\times N}$ represent the set of skew centrosymmetric matrices.

4.3.2.1 Properties of Centrosymmetric Matrices

Property 4.3.2.1.1: If $[Q_1],[Q_2]\in [C]_{N\times N}$ and $[Q_1]^{-1}$ exists, then

$$\begin{aligned} &[Q] = [Q_1][Q_2] \in [C]_{N\times N} \\ &[Q] = [Q_1] + [Q_2] \in [C]_{N\times N} \\ &[Q_1]^{-1} \in [C]_{N\times N} \end{aligned} \tag{4.19}$$

Property 4.3.2.1.2: If $[Q]\in [C]_{N\times N}$ and $N = 2M$, $[Q]$ can be written as

$$[Q] = \begin{bmatrix} A & JCJ \\ C & JAJ \end{bmatrix} \tag{4.20}$$

where $[A]$ and $[C]$ are arbitrary $M \times M$ matrices. The determinant of matrix $[Q]$ can then be computed from

$$|Q| = |A + JC| \cdot |A - JC| \tag{4.21}$$

and the inverse of the matrix [*Q*] is given by

$$[Q]^{-1} = \begin{bmatrix} B & JDJ \\ D & JBJ \end{bmatrix} \tag{4.22}$$

where

$$2[B] = [A + JC]^{-1} + [A - JC]^{-1} \text{ and } 2[D] = [A + JC]^{-1} - [A - JC]^{-1}$$

Using Equations 4.21 and 4.22, the computational effort to evaluate the inverse and determinant of a centrosymmetric matrix can be greatly reduced. Chen (1996) showed that the reduction in computational effort can be as high as 75%.

Property 4.3.2.1.3: If $[Q] \in [C]_{N\times N}$ and $N = 2M$, then

$$[Q] = \begin{bmatrix} A & JCJ \\ C & JAJ \end{bmatrix} \quad \text{and} \quad \begin{bmatrix} A - JC & 0 \\ 0 & A + JC \end{bmatrix} \tag{4.23}$$

are orthogonally similar. Thus, the calculation of the eigenvalues and eigenvectors of [*Q*] is equivalent to the evaluation of the eigenvalues and eigenvectors of two *M* by *M* matrices [*A-JC*] and [*A+JC*]. The eigenvalues are then given from the following two equations

$$|\lambda I - A + JC| = 0 \tag{4.24a}$$

$$|\lambda I - A - JC| = 0 \tag{4.24b}$$

Note that the dimension of the matrices in Equation 4.24 is *M* by *M*. So, both Equation 4.24*a* and 4.24*b* provide *M* eigenvalues, and in total, Equation 4.24 gives 2*M* eigenvalues. The eigenvectors of [*Q*] are formed from the eigenvectors of [*A-JC*] and [*A+JC*]. Let $\xi_i, i = 1,2,...,M$, and $\eta_i, i = M+1, M+2,...,2M$ be the eigenvectors of [*A-JC*] and [*A+JC*] respectively. Then ξ_i , and η_i are given by

$$[A - JC]\xi_i = \lambda_i \xi_i, \qquad \text{for } i = 1, 2, ..., M \tag{4.25a}$$

$$[A + JC]\eta_i = \lambda_i \eta_i, \qquad \text{for } i = M+1, M+2, ..., 2M \tag{4.25b}$$

Using ξ_i and η_i, the eigenvector of [*Q*], U_i can be formed as

$$U_i = \frac{1}{\sqrt{2}}[(\xi_i)^T, -(J\xi_i)^T]^T, \quad \text{for } i = 1, 2, ..., M \tag{4.26a}$$

$$U_i = \frac{1}{\sqrt{2}}[(\eta_i)^T, (J\eta_i)^T]^T, \text{for } i = M+1, M+2, ..., 2M \tag{4.26b}$$

Since the dimension of ξ_i and η_i is M, Equation 4.26 shows that the dimension of U_i is $2M$.

Property 4.3.2.1.4: If $[Q] \in [C]_{N\times N}$ and $N = 2M+1$, then $[Q]$ can be written as

$$[Q] = \begin{bmatrix} A & J\alpha & JCJ \\ \beta & \gamma & \beta J \\ C & \alpha & JAJ \end{bmatrix} \tag{4.27}$$

where $[A]$ and $[C]$ are arbitrary M by M matrices, $[\alpha]$ is a M by 1 matrix, $[\beta]$ is a 1 by M matrix, and γ is a scalar. It can be easily proven that

$$[Q] = \begin{bmatrix} A & J\alpha & JCJ \\ \beta & \gamma & \beta J \\ C & \alpha & JAJ \end{bmatrix} \quad \text{and} \quad \begin{bmatrix} A - JC & 0 & 0 \\ 0 & \gamma & 2\beta \\ 0 & 2\alpha & A + JC \end{bmatrix} \tag{4.28}$$

are orthogonally similar. So, for this case, the determinant, eigenvectors and eigenvalues of $[Q]$ can be computed from those of a M by M matrix and a $(M+1)$ by $(M+1)$ matrix. The determinant of $[Q]$ can be evaluated from

$$|Q| = |A - JC| \cdot \begin{vmatrix} \gamma & 2\beta \\ 2\alpha & A + JC \end{vmatrix} \tag{4.29}$$

And the eigenvalues of $[Q]$ are given from

$$|\lambda I - A + JC| = 0 \tag{4.30a}$$

$$\begin{vmatrix} \lambda - \gamma & -2\beta \\ -2\alpha & \lambda I - A - JC \end{vmatrix} = 0 \tag{4.30b}$$

Equation 4.30*a* provides M eigenvalues while Equation 4.30*b* gives $(M+1)$ eigenvalues. Thus, in total, we have $(2M+1)$ eigenvalues. The corresponding eigenvectors of Equation 4.30 can be obtained by solving the following equations

$$[A - JC]\xi_i = \lambda_i \xi_i, \qquad \text{for } i = 1, 2, ..., M \qquad (4.31a)$$

$$\begin{bmatrix} \gamma & 2\beta \\ 2\alpha & A + JC \end{bmatrix} \cdot \begin{bmatrix} \sigma_i \\ \eta_i \end{bmatrix} = \lambda_i \begin{bmatrix} \sigma_i \\ \eta_i \end{bmatrix}, \quad \text{for } i = M+1, M+2, ..., 2M+1 \qquad (4.31b)$$

Note that ξ_i and η_i in Equation 4.31 are M-dimensional vectors, while σ_i is a scalar. Using ξ_i, η_i and σ_i, the eigenvectors of [Q], V_i, can be formed as

$$V_i = \frac{1}{\sqrt{2}}[(\xi_i)^T, 0, -(J\xi_i)^T]^T, \qquad \text{for } i = 1, 2, ..., M \qquad (4.32a)$$

$$V_i = \frac{1}{\sqrt{2}}[(\eta_i)^T, 2\sigma_i, (J\eta_i)^T]^T, \qquad \text{for } i = M+1, M+2, ..., 2M+1 \qquad (4.32b)$$

Since ξ_i and η_i are M-dimensional vectors, Equation 4.32 indicates that the dimension of V_i is $2M+1$.

For proof and more details of the above properties, the reader may refer to the work of Good (1970), Datta and Morgera (1989).

4.3.2.2 Properties of Skew Centrosymmetric Matrices

The properties of skew centrosymmetric matrices are very similar to those of centrosymmetric matrices. These properties are listed as follows.

Property 4.3.2.2.1: If $[Q_1], [Q_2] \in [S]_{N\times N}$ and $[Q_1]^{-1}$ exists, then

$$\begin{aligned} &[Q] = [Q_1][Q_2] \in [C]_{N\times N} \\ &[Q] = [Q_1] + [Q_2] \in [S]_{N\times N} \\ &[Q_1]^{-1} \in [S]_{N\times N} \end{aligned} \qquad (4.33)$$

Property 4.3.2.2.2: If $[Q] \in [S]_{N\times N}$ and $N = 2M$, [Q] can be written as

$$[Q] = \begin{bmatrix} A & -JCJ \\ C & -JAJ \end{bmatrix} \qquad (4.34)$$

where [A] and [C] are arbitrary $M\times M$ matrices. The determinant of matrix [Q] can be calculated from

$$|Q| = |A + JC| \cdot |A - JC| \tag{4.35}$$

and the inverse of the matrix $[Q]$ is given by

$$[Q]^{-1} = \begin{bmatrix} B & -JDJ \\ D & -JBJ \end{bmatrix} \tag{4.36}$$

where

$$2[B] = [A + JC]^{-1} + [A - JC]^{-1}$$

and

$$2[D] = [A + JC]^{-1} - [A - JC]^{-1}$$

Property 4.3.2.2.3: If $[Q] \in [S]_{N\times N}$ and $N = 2M$, then

$$[Q] = \begin{bmatrix} A & -JCJ \\ C & -JAJ \end{bmatrix} \quad \text{and} \quad \begin{bmatrix} 0 & A - JC \\ A + JC & 0 \end{bmatrix} \tag{4.37}$$

are orthogonally similar. Thus, the determinant, eigenvalues and eigenvectors of $[Q]$ can be easily computed from those of the M by M matrices $[A-JC]$ and $[A+JC]$.

Property 4.3.2.2.4: If $[Q] \in [S]_{N\times N}$ and $N = 2M + 1$, then $[Q]$ can be written as

$$[Q] = \begin{bmatrix} A & -J\alpha & -JCJ \\ \beta & 0 & -\beta J \\ C & \alpha & -JAJ \end{bmatrix} \tag{4.38}$$

where $[A]$ and $[C]$ are arbitrary M by M matrices, $[\alpha]$ is a M by 1 matrix, and $[\beta]$ is a 1 by M matrix. It is found that

$$[Q] = \begin{bmatrix} A & -J\alpha & -JCJ \\ \beta & 0 & -\beta J \\ C & \alpha & -JAJ \end{bmatrix} \quad \text{and} \quad \begin{bmatrix} 0 & -2\alpha & A - JC \\ 2\beta & 0 & 0 \\ A + JC & 0 & 0 \end{bmatrix} \tag{4.39}$$

are orthogonally similar. Obviously, the determinant of $[Q]$ is zero for this case. The eigenvectors and eigenvalues of $[Q]$ can be computed from those of a $(M+1)$ by $(M+1)$ matrix.

4.3.3 Structures of DQ Weighting Coefficient Matrices

In this section, we will show that when the grid distribution is symmetric with respect to the center point, the DQ weighting coefficient matrices are either centrosymmetric or skew centrosymmetric, depending on the order of the approximated derivatives.

4.3.3.1 Structure of the First Order DQ Weighting Coefficient Matrix

For purposes of brevity in the following discussion, we let $a_{i,j}$ be the DQ weighting coefficients of the first order derivative, *N* be the number of grid points in the whole computational domain. The computational domain is set to be [0, 1] for the PDQ formulation, [0, π] for the CDQ formulation and FDQ formulation with non-periodic conditions, (0, π) for the SDQ formulation, and [0, 2π) for the FDQ formulation with periodic conditions. For the PDQ formulation, we recall from Equation 2.17*a* that

$$a_{i,j} = \frac{M^{(1)}(x_i)}{(x_i - x_j)\cdot M^{(1)}(x_j)}, \quad \text{for } j \neq i \tag{4.40}$$

where

$$M^{(1)}(x_i) = \prod_{k=1,k\neq i}^{N}(x_i - x_k) \tag{4.41}$$

If the grid points are symmetrically distributed with respect to the center point in the interval [0, 1], then we have

$$x_{N-i+1} = 1 - x_i \tag{4.42}$$

Using Equation 4.42, Equation 4.41 gives

$$\begin{aligned} M^{(1)}(x_{N-i+1}) &= \prod_{k=1,k\neq N-i+1}^{N}(x_{N-i+1} - x_k) = \prod_{k=1,k\neq N-i+1}^{N}[(1-x_i)-(1-x_{N-k+1})] \\ &= -\prod_{k=1,k\neq N-i+1}^{N}(x_i - x_{N-k+1}) = -\prod_{k1=N,k1\neq i}^{1}(x_i - x_{k1}) = -M^{(1)}(x_i) \end{aligned} \tag{4.43}$$

Similarly, we have

$$M^{(1)}(x_{N-j+1}) = -M^{(1)}(x_j) \tag{4.44}$$

Since $x_{N-i+1} - x_{N-j+1} = -(x_i - x_j)$, and using Equations 4.43 and 4.44, we can simplify Equation 4.40 as

$$a_{N-i+1,N-j+1} = -a_{i,j} \tag{4.45}$$

According to the definition in Section 4.3.1, the PDQ weighting coefficient matrix of the first order derivative is skew centrosymmetric if the grid distribution is symmetric. Similarly, using the results in Chapter 3, $a_{i,j}$ can be given by

$$a_{i,j} = \frac{-c(x_i)\sin x_i}{(\cos x_i - \cos x_j)\cdot c(x_j)}, \quad \text{when } j \neq i \tag{4.46a}$$

where

$$c(x_i) = \prod_{k=0,k\neq i}^{N} (\cos x_i - \cos x_k) \tag{4.46b}$$

for the cosine expansion-based differential quadrature (CDQ), and

$$a_{ij} = \frac{-s(x_i)\sin^2 x_i}{(\cos x_i - \cos x_j)\sin x_j \cdot s(x_j)}, \quad \text{when } j \neq i \tag{4.47a}$$

where

$$s(x_i) = \prod_{k=1,k\neq i}^{N} (\cos x_i - \cos x_k) \tag{4.47b}$$

for the sine expansion-based differential quadrature (SDQ), and

$$a_{ij} = \frac{1}{2}\cdot\frac{g(x_i)}{\sin\frac{x_i - x_j}{2}\cdot g(x_j)}, \quad \text{when } j \neq i \tag{4.48a}$$

where

$$g(x_i) = \prod_{k=0,k\neq i}^{N} \sin\frac{x_i - x_k}{2} \tag{4.48b}$$

for the general Fourier expansion-based differential quadrature (FDQ). If the grid distribution is symmetric with respect to the center point in the interval $[0, \pi]$ or $(0, \pi)$, then we have

$$x_{N-i+1} = \pi - x_i \tag{4.49}$$

Therefore,

$$c(x_{N-i+1}) = \prod_{k=0,k\neq N-i+1}^{N} (\cos x_{N-i+1} - \cos x_k) = -c(x_i) \tag{4.50a}$$

$$s(x_{N-i+1}) = \prod_{k=1,k\neq N-i+1}^{N} (\cos x_{N-i+1} - \cos x_k) = -s(x_i) \tag{4.50b}$$

$$g(x_{N-i+1}) = \prod_{k=0,k\neq N-i+1}^{N} \sin\frac{x_{N-i+1} - x_k}{2} = -g(x_i) \tag{4.50c}$$

On the other hand, using Equation 4.49, we obtain

$$\begin{aligned} &\sin x_{N-i+1} = \sin x_i, \quad \cos x_{N-i+1} - \cos x_{N-j+1} = -(\cos x_i - \cos x_j) \\ &\sin\frac{x_{N-i+1} - x_{N-j+1}}{2} = -\sin\frac{x_i - x_j}{2} \end{aligned} \tag{4.51}$$

Using Equations 4.50 and 4.51, we can conclude from Equations 4.46, 4.47, 4.48 that Equation 4.45 is also true for the CDQ, SDQ, and FDQ weighting coefficient matrices. In a similar manner, we can show that if the interval is $[0, 2\pi)$, the FDQ weighting coefficient matrix also satisfies Equation 4.45. This indicates that all the DQ weighting coefficient matrices of the first order derivative are skew centrosymmetric.

4.3.3.2 Structures of Higher Order DQ Weighting Coefficient Matrices

For the second and higher order derivatives, the PDQ weighting coefficients can be computed from the following recurrence relationship

$$w_{i,j}^{(m)} = m\left(a_{i,j} \cdot w_{i,i}^{(m-1)} - \frac{w_{i,j}^{(m-1)}}{x_i - x_j}\right), \text{ for } j \neq i \tag{4.52}$$

where $w_{i,j}^{(m)}$ denotes the PDQ weighting coefficients of the mth order derivative. As shown in the previous section, when the grid distribution is symmetric with respect to the center point, we have

$$a_{N-i+1,N-j+1} = -a_{i,j}, \quad x_{N-i+1} - x_{N-j+1} = -(x_i - x_j) \tag{4.53}$$

Using Equation 4.53, Equation 4.52 can be written as

$$w^{(m)}_{N-i+1,N-j+1} = -m\left(a_{i,j} \cdot w^{(m-1)}_{N-i+1,N-i+1} - \frac{w^{(m-1)}_{N-i+1,N-j+1}}{x_i - x_j} \right), \text{ for } j \neq i \quad (4.54)$$

Equation 4.54 is a recurrence relationship. For simplicity, let $[W^{(m-1)}],[W^{(m)}]$ be the PDQ weighting coefficient matrices of the (m-1)th and mth order derivatives. Equation 4.54 shows that if $[W^{(m-1)}]$ is skew centrosymmetric, then $[W^{(m)}]$ is centrosymmetric, and if $[W^{(m-1)}]$ is centrosymmetric, then $[W^{(m)}]$ is skew centrosymmetric. This conclusion can also be obtained from the matrix multiplication method. Taking $[A]$ to be the DQ weighting coefficient matrix of the first order derivative, the matrix multiplication method gives

$$[W^{(m)}] = [W^{(m-1)}][A] \quad (4.55a)$$

or

$$w^{(m)}_{i,j} = \sum_{k=1}^{N} w^{(m-1)}_{i,k} a_{k,j} = \sum_{k1=N}^{1} w^{(m-1)}_{i,N-k1+1} a_{N-k1+1,j} \quad (4.55b)$$

Equation 4.55 is valid for PDQ, CDQ, SDQ and FDQ formulations. Since the weighting coefficient matrix of the first order derivative is skew centrosymmetric, Equation 4.55 shows that the DQ weighting coefficient matrices are skew centrosymmetric for odd order derivatives (m is odd), and centrosymmetric for even order derivatives (m is even) when the grid distribution is symmetric with respect to the center point. The above conclusion is true for both uniform and non-uniform grids.

It has been demonstrated in the work of Chen (1996) that for the flexural vibration analysis of beams and plates with simply supported and clamped conditions, the centrosymmetric and skew centrosymmetric properties of the DQ weighting coefficient matrices can be used to factorize the resultant eigenvalue system to yield two smaller systems. As a consequence, the computational effort is reduced by 75%.

4.4 Effect of Grid Point Distribution on Eigenvalues of DQ Discretization Matrices

Consider a time-dependent problem

$$\frac{\partial u}{\partial t} = \ell(u) \quad (4.56)$$

with proper initial and boundary conditions, where ℓ is a spatial differential operator and is in general a non-linear operator. After discretization using the DQ method and linearization of the non-linear terms, Equation 4.56 can be reduced into a set of ordinary differential equations in time

$$\frac{d\{u\}}{dt} = [A]\{u\} + \{s\} \tag{4.57}$$

where $\{u\}$ is an unknown vector of the functional values at the interior grid points, $\{s\}$ is a known vector which is made up of the non-homogeneous part and the boundary conditions, and $[A]$ is a matrix. Before we apply a numerical scheme for the temporal discretization to obtain the converged numerical solution of Equation 4.57, we need to ensure that Equation 4.57 satisfies the stability condition for ordinary differential equations. If Equation 4.57 is unstable, then the stable numerical scheme for the temporal discretization may not generate the converged solution. The stability of Equation 4.57 is related to the eigenvalues of the matrix $[A]$, which is called the DQ discretization matrix. In the following, we will first discuss the stability condition for ordinary differential equations, and then study the effect of the grid point distribution on the stability of Equation 4.57 through the eigenvalue distribution of some specific DQ discretization matrices.

4.4.1 Stability of Ordinary Differential Equations

The stability analysis of Equation 4.57 is based on the eigenvalue distribution of the matrix $[A]$, since its exact solution is directly determined by the eigenvalues and eigenvectors of $[A]$. Let $\lambda_i, i = 1, 2, ..., N$ be the N eigenvalues of $[A]$, ξ_i be the associated eigenvectors, and $[P]$ be the matrix formed from the N eigenvectors, the diagonal matrix $[D]$ be formed from the eigenvalues. We then obtain

$$[D] = [P]^{-1}[A][P] \tag{4.58}$$

Pre-multiplying by the matrix $[P]^{-1}$ on both sides of Equation 4.57 and setting

$$\{U\} = [P]^{-1}\{u\}, \ \{S\} = [P]^{-1}\{s\} \tag{4.59}$$

we thus obtain

$$\frac{d\{U\}}{dt} = [D]\{U\} + \{S\} \tag{4.60}$$

Since $[D]$ is a diagonal matrix, Equation System 4.60 is an uncoupled set of ordinary differential equations. In other words, each equation in Equation System 4.60 can be solved independently. Consider the ith equation of Equation System 4.60

$$\frac{dU_i}{dt} = \lambda_i U_i + S_i \tag{4.61}$$

The solution of Equation 4.61 consists of two parts: the complementary solution and the particular solution. If S_i is time-independent, then the solution of Equation 4.61 can be written as

$$U_i = [U_i(0) + S_i / \lambda_i] e^{\lambda_i t} - S_i / \lambda_i \tag{4.62}$$

For this case, using Equation 4.59, the solution $\{u\}$ can be obtained as

$$\{u\} = [P]\{U\} = \sum_{i=1}^{N} U_i \xi_i = \sum_{i=1}^{N} \left[U_i(0) e^{\lambda_i t} + \frac{S_i}{\lambda_i} \left(e^{\lambda_i t} - 1 \right) \right] \xi_i \tag{4.63}$$

Clearly, the stable solution of $\{u\}$ when $t \to \infty$ requires

$$\mathrm{Re}(\lambda_i) \le 0 \quad \text{for all } i \tag{4.64}$$

where $\mathrm{Re}(\lambda_i)$ denotes the real part of λ_i. Equation 4.64 is the stability condition for the ordinary differential Equation 4.57.

4.4.2 Eigenvalues of Some Specific DQ Discretization Matrices

From the previous analysis, the stability condition for ordinary differential equations is related to the eigenvalues of the spatial discretization matrix $[A]$. We will investigate, in this section, the eigenvalue distribution of some typical matrices obtained using the DQ discretization technique for three model problems, and the effect of boundary conditions and grid point distribution on the eigenvalue distribution.

4.4.2.1 The Convection Operator

For this case, the differential operator $\ell(u)$ is taken to be

$$\ell(u) = -\frac{\partial u}{\partial x} \quad \text{on } [0, \pi] \tag{4.65}$$

with the Dirichlet boundary condition

$$u(0) = g(x) \tag{4.66}$$

As discussed in Section 4.2.2, the CDQ and SDQ formulations are not applicable to this problem. Thus, we will only apply the PDQ and FDQ methods to discretize $\ell(u)$, and the resultant discretization matrices are noted as PDQ and FDQ matrices. Let N be the number of grid points in the interval [0, π] with coordinates of the grid points given by $0 = x_1, x_2, ..., x_N = \pi$. Since the boundary condition is implemented at x_1, the DQ discretization of $\ell(u)$ should only be applied at (N-1) grid points, $x_2,..., x_N$. Hence, the dimension of matrix [A] is (N-1) by (N-1). To study the effects of grid point distribution on the eigenvalue distribution of [A], we consider the three typical grids,

Grid I: Chebyshev-Gauss-Lobatto grid (basic grid is given by $|T_N(x)| = 1$)

Grid II: Chebyshev grid (basic grid is generated by $T_N(x) = 0$)

Grid III: Uniform grid

The simplified PDQ formulations of the first order derivative for the above three grids have been shown in Chapter 2.

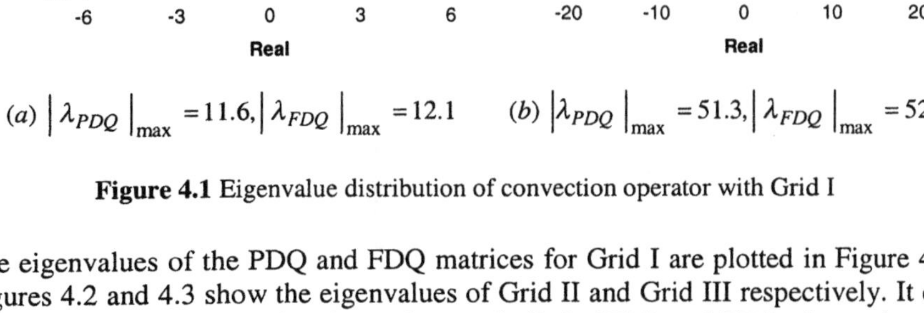

(a) $|\lambda_{PDQ}|_{\max} = 11.6, |\lambda_{FDQ}|_{\max} = 12.1$ (b) $|\lambda_{PDQ}|_{\max} = 51.3, |\lambda_{FDQ}|_{\max} = 52.5$

Figure 4.1 Eigenvalue distribution of convection operator with Grid I

The eigenvalues of the PDQ and FDQ matrices for Grid I are plotted in Figure 4.1. Figures 4.2 and 4.3 show the eigenvalues of Grid II and Grid III respectively. It can be seen from Figure 4.1 that the real part of all the PDQ and FDQ eigenvalues of Grid I are strictly negative. The modulus of most FDQ eigenvalues is less than that of the PDQ eigenvalues. However, the maximum modulus of the FDQ eigenvalues is larger than that of the PDQ eigenvalues. Since the maximum time step size employed for satisfying the stability condition is usually determined by the maximum modulus of the eigenvalues, the eigenvalue distribution in Figure 4.1 indicates that the PDQ method allows the use of a slightly larger time step size for Grid I.

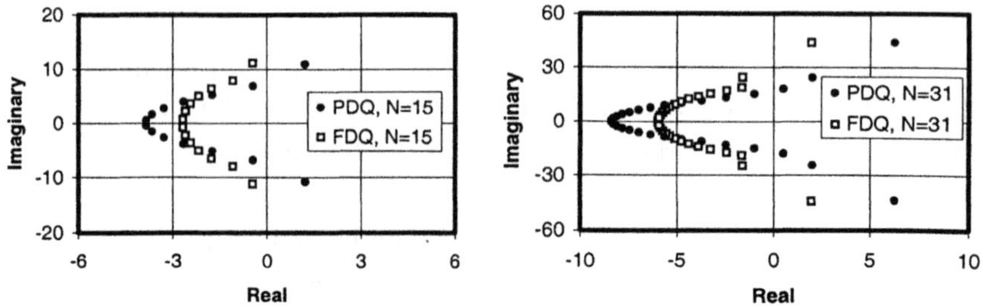

(*a*) $\left|\lambda_{PDQ}\right|_{\max} = 10.9, \left|\lambda_{FDQ}\right|_{\max} = 11.2$ (*b*) $\left|\lambda_{PDQ}\right|_{\max} = 44.4, \left|\lambda_{FDQ}\right|_{\max} = 44.1$

Figure 4.2 Eigenvalue distribution of convection operator with Grid II

Unlike Grid I, when the Grid II is employed, the real part of the maximum PDQ eigenvalue is always positive, which corresponds to the unstable region. This behavior is independent of the number of the grid points used. For the FDQ eigenvalues, when the number of grid points is small, for example, $N = 15$ as shown in Figure 4.2 (*a*), the real parts of all the eigenvalues are negative, which thus lead to a stable solution. However, when N is large, for example, for $N = 31$ as shown in Figure 4.2 (*b*), the real part of the maximum FDQ eigenvalue is positive, which can lead to an unstable solution. In general, the modulus of the FDQ eigenvalues is less than that of the PDQ eigenvalues for Grid II. For Grid III, the real parts of the eigenvalues of large amplitudes for both the PDQ and FDQ methods are always positive, which are shown in Figure 4.3. This implies that the uniform grid (Grid III) is not suitable for the PDQ and FDQ solution of convection problems.

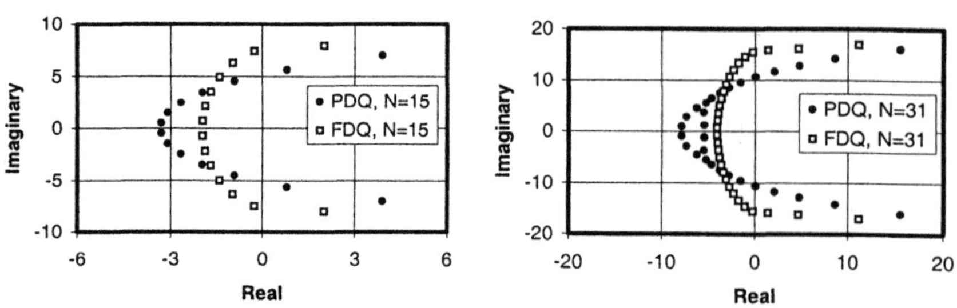

(*a*) $\left|\lambda_{PDQ}\right|_{\max} = 8.0, \left|\lambda_{FDQ}\right|_{\max} = 8.2$ (*b*) $\left|\lambda_{PDQ}\right|_{\max} = 22.2, \left|\lambda_{FDQ}\right|_{\max} = 20.3$

Figure 4.3 Eigenvalue distribution of convection operator with Grid III

By comparing the results in Figures 4.1-4.3, it seems that the distribution of the grid points can greatly affect the stability behavior of a global method such as the DQ approach. Here, one may wonder what is the suitable grid point distribution for a stable solution. To study this, we investigate the details of the above three grids, and find that Grids I and II which are non-uniform, are stretched towards the boundaries, while Grid III is uniform. We also found that the minimum step size $|\Delta x|_{\min}$ for the Grid I is the smallest among the three grids. For example, when $N = 15$, the values of $|\Delta x|_{\min}$ are 0.03938, 0.06865 and 0.2244 respectively for Grids I, II and III, and when $N = 31$, the values of $|\Delta x|_{\min}$ are reduced to 0.008605, 0.01612, 0.1047 respectively for Grids I, II, III. It seems that for a stable solution, $|\Delta x|_{\min}$ should be sufficiently small. To further study the effect of grid distribution on the stability, we introduce the following transformation

$$\bar{x} = (1-\alpha)(3\pi x^2 - 2x^3)/\pi^2 + \alpha x, \quad \alpha \geq 0 \tag{4.67}$$

where $\bar{x}$ is the transformed coordinate still within the interval $[0, \pi]$. When $\alpha \leq 1$, the transformed grid is stretched near the boundary (the grid points are more clustered near the boundary), when $\alpha > 1$, the transformed grid is stretched and clustered near the center of the interval. Using Equation 4.67, we can get the following two new grids,

Grid IV: Transformed from Grid II with $\alpha \leq 1$

Grid V: Transformed from Grid III with $\alpha \leq 1$

The eigenvalue distributions of Grids IV and V are displayed in Figures 4.4-4.5. For these cases, the real parts of all the eigenvalues are strictly negative when the proper choice of the stretching parameter α is used. When $N = 15$, α is taken to be 0.73 for Grid IV and 0.03 for Grid V. Although the α values are quite different for Grids IV and V, the values of $|\Delta x|_{\min}$ for these grids are almost the same. For Grid IV, $|\Delta x|_{\min}$ is reduced from 0.06865 of Grid II to 0.05131, while for the Grid V, $|\Delta x|_{\min}$ is reduced from 0.2244 of Grid III to 0.05115. When N is increased, the stretching parameter α for stability should be decreased. As shown in Figures 4.4-4.5, when N=31, α should be reduced to 0.66 for Grid IV and 0.005 for Grid V in order to obtain a stable solution. For this case, $|\Delta x|_{\min}$ is reduced from 0.01612 for Grid II to 0.01072 for Grid IV, and from 0.1047 for Grid III to 0.01071 for Grid V. By comparing the PDQ and FDQ methods, it was found that the FDQ method is less sensitive to the grid point distribution. For Grids IV and V, the FDQ method allows a larger value of α to be used than PDQ method to ensure that the real parts of all the eigenvalues are negative.

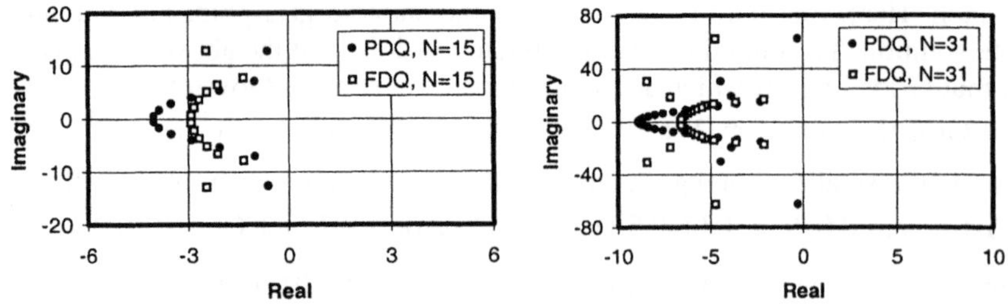

(a) $\left|\lambda_{PDQ}\right|_{\max}=12.8, \left|\lambda_{FDQ}\right|_{\max}=13.1, \alpha=0.73$ (b) $\left|\lambda_{PDQ}\right|_{\max}=62.6, \left|\lambda_{FDQ}\right|_{\max}=62.8, \alpha=0.66$

Figure 4.4 Eigenvalue distribution of convection operator with Grid IV

Under the condition that the real parts of all the eigenvalues are strictly negative, when $|\Delta x|_{\min}$ is relaxed near the boundary, the modulus of the maximum eigenvalue is reduced. Thus, the time step size for stability is relaxed. The reduction in maximum magnitude of the eigenvalues for larger $|\Delta x|_{\min}$ can be found by comparing Figure 4.6 and Figure 4.1. Figure 4.6 corresponds to Grid VI defined by

Grid VI: Transformed from Grid I with $\alpha > 1$

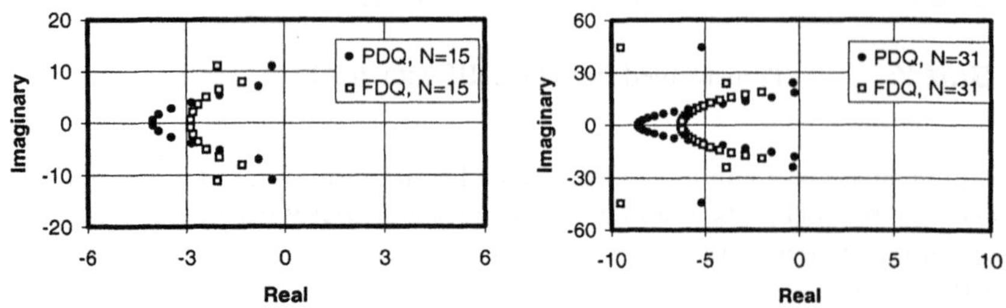

(a) $\left|\lambda_{PDQ}\right|_{\max}=11.0, \left|\lambda_{FDQ}\right|_{\max}=11.2, \alpha=0.03$ (b) $\left|\lambda_{PDQ}\right|_{\max}=44.7, \left|\lambda_{FDQ}\right|_{\max}=45.4, \alpha=0.005$

Figure 4.5 Eigenvalue distribution of convection operator with Grid V

For Grid VI, we select α ($\alpha > 1$) to increase $|\Delta x|_{\min}$. It was found that when $|\Delta x|_{\min}$ is increased to the value in Grids IV and V, the real parts of all the eigenvalues are still negative. When α is further increased, the real parts of some of the eigenvalues will become positive. This study shows that in order to obtain a stable solution, $|\Delta x|_{\min}$ should be less than a certain value. Figures 4.6 and 4.1 demonstrate that when $|\Delta x|_{\min}$ is increased, the maximum modulus of the eigenvalue will be decreased.

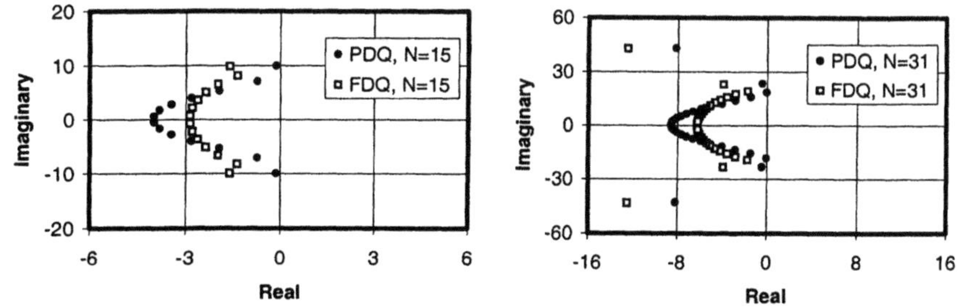

(*a*) $|\lambda_{PDQ}|_{max} = 9.9,\ |\lambda_{FDQ}|_{max} = 10.0, \alpha = 1.31$ (*b*) $|\lambda_{PDQ}|_{max} = 43.9,\ |\lambda_{FDQ}|_{max} = 44.9, \alpha = 1.17$

Figure 4.6 Eigenvalue distribution of convection operator with Grid VI

In the above study, the grid is stretched near the boundary. In other words, $|\Delta x|_{min}$ is the grid spacing adjacent to the boundary point. Here, one may wonder what the behavior will be like if the grid is stretched at other points. To study this, we introduce Grid VII

Grid VII: Transformed from Grid III with $\alpha > 1$

For the Grid VII, when $\alpha > 1$, the grid points are clustered near the center point. Figure 4.7 shows the eigenvalue distribution of Grid VII. Obviously, the grid stretched towards the center point does not improve the stability behavior. Actually, it worsens eigenvalue distribution (more eigenvalues have positive real parts).

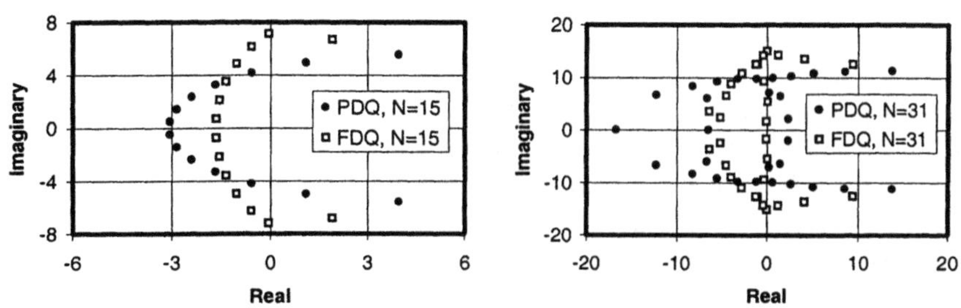

(*a*) $|\lambda_{PDQ}|_{max} = 6.8,\ |\lambda_{FDQ}|_{max} = 7.0, \alpha = 2.0$ (*b*) $|\lambda_{PDQ}|_{max} = 17.8,\ |\lambda_{FDQ}|_{max} = 15.7, \alpha = 2.0$

Figure 4.7 Eigenvalue distribution of convection operator with Grid VII

4.4.2.2 The Diffusion Operator

The diffusion operator $\ell(u)$ is chosen as

$$\ell(u) = \frac{\partial^2 u}{\partial x^2} \quad \text{on } [0, \pi] \tag{4.68}$$

The boundary conditions imposed is either of the Dirichlet type

$$u(0) = u(1) = 0 \tag{4.69}$$

or of the Neumann type

$$u_x(0) = u_x(1) = 0 \tag{4.70}$$

where the subscript represents the first order derivative with respect to x. We will use Grids I, II and III to study the eigenvalue distribution of the diffusion operator. It was found that when Grid I is used, the PDQ and FDQ eigenvalues for both the Dirichlet and Neumann boundary conditions are real and negative. But the Neumann boundary condition gives smaller absolute eigenvalues than the Dirichlet boundary condition. Thus, the former may allow a larger time step size to be used. For example, for the PDQ method, the maximum absolute eigenvalue for N = 31 is 1.5647×10^4 for the Dirichlet condition, and 4.7282×10^3 for the Neumann condition. For the same boundary condition, the maximum absolute eigenvalue of the FDQ method is slightly less than that of the PDQ method. For example, when N = 31 and the Dirichlet condition is implemented, the maximum absolute eigenvalue is 1.5647×10^4 for the PDQ method, and 1.5298×10^4 for the FDQ method. Similarly, when the Neumann condition is considered and N = 31, the maximum absolute eigenvalue is 4.7282×10^3 for the PDQ method, and 4.7697×10^3 for the FDQ method. It may be concluded from the above results that the FDQ method permits the use of a slightly larger time step size for a stable solution to be obtained than the PDQ method when Grid I is used. For Grid II, it was found that the PDQ and FDQ eigenvalues are real for the Dirichlet condition, and complex for the Neumann condition. For all cases, the real parts of the eigenvalues are negative, which thus lead to a stable solution. Unlike Grid I, the maximum modulus of the FDQ eigenvalues for Grid II is slightly larger than that for the PDQ eigenvalues. For example, when N = 31 and the Dirichlet condition is used, the maximum absolute eigenvalue is 5.6121×10^3 for the PDQ method, and 5.7883×10^3 for the FDQ method. So, for Grid II, the PDQ method allows the use of a slightly larger time step size than the FDQ method. When Grid III is applied, the PDQ and FDQ eigenvalues are complex for both the Dirichlet and Neumann conditions. By taking N as 31, it was found that the real parts of all the PDQ and FDQ eigenvalues are negative for the Dirichlet condition. However, for the Neumann condition, the real part of the PDQ eigenvalue of largest magnitude is positive, which may cause the computation to be unstable. The real part of the FDQ eigenvalue of largest magnitude still remains negative for the Neumann condition. The eigenvalue distribution of the diffusion operator for Grid III is displayed in Figure 4.8. It can be observed from the figure that for both the Dirichlet and Neumann conditions, the maximum modulus of the FDQ eigenvalues is less than that of the PDQ eigenvalues.

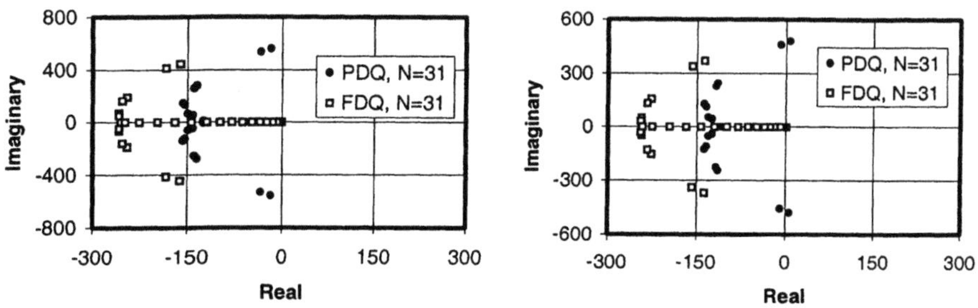

Dirichlet Condition Neumann Condition

(*a*) $\left|\lambda_{PDQ}\right|_{\max} = 556.6, \left|\lambda_{FDQ}\right|_{\max} = 473.6$ (*b*) $\left|\lambda_{PDQ}\right|_{\max} = 479.7, \left|\lambda_{FDQ}\right|_{\max} = 394.2$

Figure 4.8 Eigenvalue distribution of diffusion operator with Grid III

For the diffusion operator, the CDQ and SDQ formulations can also be applied. However, the CDQ formulations can only be applied to the Neumann condition while the SDQ formulations can only be applied to the Dirichlet condition. Unlike the application of the PDQ and FDQ formulations, when the CDQ and SDQ formulations are used, the discretization of the differential operator has to be applied at all the grid points. Thus, the dimension of the discretization matrix for the CDQ and SDQ formulations is N by N rather than (N-2) by (N-2) for the PDQ and FDQ formulations. It was found that only when the uniform mesh (Grid III) is used, the CDQ and SDQ eigenvalues are real and negative. When Grids I and II are applied, some of the CDQ and SDQ eigenvalues are complex, and their real parts may be positive. Therefore, for the CDQ and SDQ formulations, it may be better to use a uniform mesh. This behavior is different from that of the PDQ and FDQ formulations. As demonstrated above, for the PDQ and FDQ formulations, it is better to use a non-uniform mesh with grid points clustered near the boundaries.

4.4.2.3 The Convection-Diffusion Operator

We consider the convection-diffusion operator

$$\ell(u) = \frac{1}{\text{Re}}\frac{\partial^2 u}{\partial x^2} - \frac{\partial u}{\partial x} \quad \text{on } [0,\pi] \tag{4.71}$$

with a Dirichlet type of boundary condition. Equation 4.71 is a model equation for the one-dimensional Navier-Stokes equation. When Re is in the order of 1, the problem is dominated by both the convection and diffusion terms. When $\text{Re} >> 1$, the problem is mainly dominated by convection.

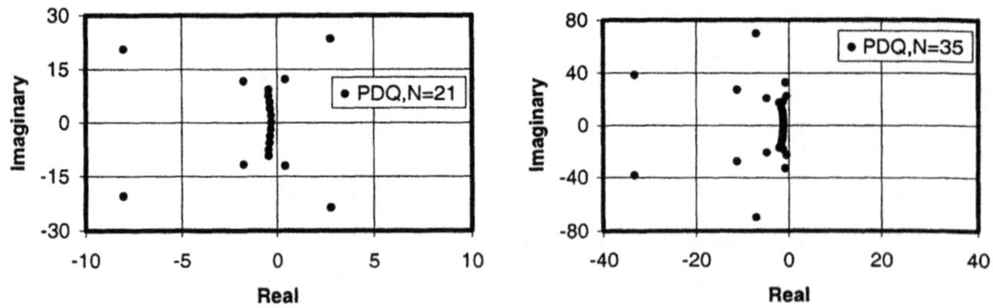

(*a*) $\mathrm{Re} = 500, \left|\lambda_{PDQ}\right|_{\max} = 23.8$ (*b*) $\mathrm{Re} = 500, \left|\lambda_{PDQ}\right|_{\max} = 70.2$

Figure 4.9 Eigenvalue distribution of convection-diffusion operator with Grid I

The convection-diffusion operator can be discretized by the PDQ or FDQ method. The PDQ and FDQ eigenvalue distribution for this problem is very similar. Here, for purposes of demonstration, we only consider the PDQ discretization of the convection-diffusion operator. It was found that when Re is small, the real parts of all the eigenvalues for all the grid types are strictly negative, but when Re is large, the real parts of some eigenvalues may be positive, leading to instability. It was found that the maximum Re for maintaining stability is greatly affected by the grid point distribution. For example, to obtain a stable solution, the maximum value of Re is around 6 for Grid III, 245 for Grid II, and 250 for Grid I when $N = 21$. Thus, for very large Re, a uniform mesh is not recommended. The stability condition can be improved by increasing the number of grid points for all the cases when Re is very large. It was also found that the stability condition can be enhanced if the grid is stretched towards the boundary. Figure 4.9 shows the eigenvalue distribution for the convection-diffusion operator with Re = 500 for Grid I. It is observed from the figure that, when $N = 21$, the real part of the eigenvalue with maximum magnitude is positive, but when $N = 35$, the real parts of all the eigenvalues are strictly negative. It may be concluded from this result that for the convection-diffusion operator, the stability condition can be improved by increasing the number of grid points for a larger value of Re.

4.5 Effect of Grid Point Distribution on Magnitude of DQ Weighting Coefficients

In the above section, it has been shown that when a uniform grid is used, the PDQ and FDQ methods may encounter instability problems. To obtain a converged solution, one should use a non-uniform grid with grid points clustered near the boundary. In this section, we will confirm this conclusion from the magnitude of the weighting coefficients. At first, we consider the three basic grids of Grids I, II, and III. It was found that for each grid, the distributions of the weighting coefficients for the first and higher order derivatives are very similar. The distribution of the PDQ weighting coefficients is also similar to the distribution of the FDQ weighting coefficients. Thus, we only study the distribution of the PDQ weighting coefficients for the first order derivative, $a_{i,j}$, over the indices i and j.

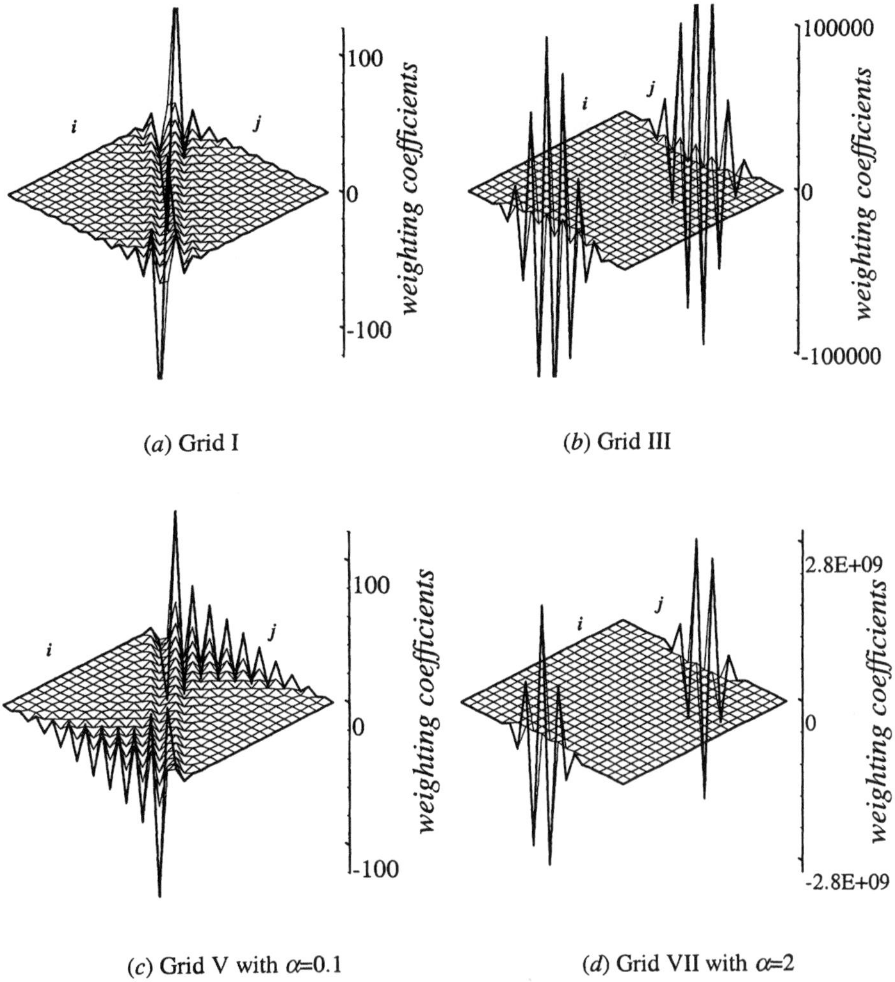

Figure 4.10 Distribution of PDQ weighting coefficients for the first order derivative, N=21

Grids I and II have the same feature, where the grid points are clustered near the boundary. We found that the distributions of the weighting coefficients for these two grids are very similar. Thus, only the distribution of the weighting coefficients for the Grid I is displayed in Figure 4.10(a). The distribution of the corresponding weighting coefficients for the uniform grid (Grid III) is displayed in Figure 4.10(b).

By comparing Figure 4.10(*a*) with Figure 4.10(*b*), we can find two major differences. First, the magnitudes of the weighting coefficients are different. For Grid I, the maximum weighting coefficient is in the order of 10^2 (N=21), but for Grid III, the maximum weighting coefficient is in the order of 10^5 (N=21). Such large values of the weighting coefficients would introduce significant round-off errors into the computation. The other difference pertains to the distribution of the large weighting coefficients. For Grid I, the large weighting coefficients are located in the vicinity of the diagonal line. However, for Grid III, the large weighting coefficients are distributed along the first and last rows ($i = 1, N$).

Using Grid III as the basic grid, we found that when the grid points are clustered towards the boundary (Grid V with $\alpha \leq 1$), the values of the largest weighting coefficients are greatly reduced, and they move towards the diagonal line. This can be observed from Figure 4.10(*c*), in which Grid V with $\alpha = 0.1$ is used. When the grid points are clustered near the center point, the distribution of the weighting coefficients gets worse. This can be seen from Figure 4.10(*d*), in which Grid VII with $\alpha = 2$ is used. Obviously, for this case, the large weighting coefficients still remain in the first and last rows, but the maximum weighting coefficient is increased to the order of 10^9 (N=21). So, for practical applications of the DQ method, we should avoid the use of a uniform mesh or a mesh with grid points clustered near the center point.

Chapter 5

Solution Techniques for DQ Resultant Equations

5.1 Introduction

In most applications of the DQ method to engineering problems, which are governed by time-dependent partial differential equations (PDEs), the spatial derivatives are discretized by the DQ method whereas the time derivatives are discretized by low order finite difference schemes. For the general case, we consider a time-dependent PDE as follows

$$\frac{\partial w}{\partial t} + \ell(w) = g \tag{5.1}$$

where $\ell(w)$ is a differential operator containing all the spatial derivatives and g is a given function. Equation 5.1 should be specified with proper initial and boundary conditions for the solution to a specific problem. By using the DQ method to discretize the spatial derivatives in the differential operator $\ell(w)$ and applying Equation 5.1 at all the interior grid points, we can obtain a set of ordinary differential equations (ODEs)

$$\frac{d\{W\}}{dt} + L\{W\} = \{G\} \tag{5.2}$$

where $\{W\}$ is a vector representing a set of unknown functional values at all the interior points, $L\{W\}$ is a vector resulting from the DQ discretization, $\{G\}$ is a known vector arising from the function g and the given initial and boundary conditions.

For a time-dependent problem, Equation 5.2 constitutes a standard form of ODEs. The time derivative in Equation 5.2 can be discretized explicitly or implicitly using low order finite difference schemes. In the following, explicit schemes will be considered for the solution of the DQ ordinary differential Equation 5.2. We will introduce some standard explicit schemes in Section 5.2 to solve Equation 5.2.

If $\ell(w)$ is a linear differential operator, then the vector $L\{W\}$ can be directly written as the product of a matrix $[H]$ and the vector $\{W\}$, that is,

$$L\{W\} = [H]\{W\} \tag{5.3}$$

If $\ell(w)$ is a nonlinear differential operator, then after linearization of the nonlinear terms, $L\{W\}$ can also be expressed by Equation 5.3. For a time-independent problem, when Equation 5.3 is substituted into Equation 5.2, we obtain

$$[H]\{W\} = \{G\} \tag{5.4}$$

Equation 5.4 is a set of algebraic equations. For a time-dependent problem, when implicit schemes such as the Euler backward scheme are applied to discretize the time derivative in Equation 5.2, Equation 5.2 can also be reduced to a set of algebraic equations similar to Equation 5.4. The algebraic equations can be solved using direct methods or iterative methods. These solution techniques will be introduced in Section 5.3. Apart from the solution techniques, the implementation of the boundary conditions is also important. The DQ implementation of the boundary conditions will be addressed in Section 5.4.

As shown in Chapters 2 and 3, both the PDQ and FDQ methods use the same form to approximate the derivatives. The only difference lies in the way the weighting coefficients are computed. So, the resultant PDQ and FDQ equations have the same form. In the following, these equations are termed as the DQ resultant equations for brevity.

5.2 Solution Techniques for DQ Ordinary Differential Equations

When explicit schemes are used to solve Equation 5.2, we must ensure that the schemes are stable in order to arrive at a converged solution. In the following, we will introduce some conventional explicit schemes to solve Equation 5.2. The stability region of each explicit scheme is different, and is independent of the spatial discretization. However, the required time step size for a stable solution of Equation 5.2 is related to the spatial discretization. To illustrate this, we linearize $L\{W\}$ without loss of generality, and write Equation 5.2 as

$$\frac{d\{W\}}{dt} = [A]\{W\} + \{G\} \tag{5.5}$$

where $[A]$ is a matrix resulting from the linearization. Definitely, $[A]$ depends on the spatial discretization using the DQ method. As shown in Chapter 4, both the grid point distribution and boundary conditions affect the structure of the matrix $[A]$. It is also shown in Chapter 4 that for a stable solution of Equation 5.5 to be obtained, the real part of the eigenvalues of $[A]$ should be negative or zero. Let $[E]$ be the right eigenvector matrix of $[A]$, and $[D]$ be the diagonal matrix with the eigenvalues of $[A]$ as the elements. Then, as shown in Chapter 4, Equation 5.5 can be decoupled using $[E]$ and $[D]$ as

$$\frac{d\{\overline{W}\}}{dt} = [D]\{\overline{W}\} + \{\overline{G}\} \tag{5.6}$$

where $\{\overline{W}\} = [E]^{-1}\{W\}, \{\overline{G}\} = [E]^{-1}\{G\}$. Since $[D]$ is a diagonal matrix, each equation in Equation System 5.6 is independent. So, we can select a model scalar equation

$$\frac{d\overline{w}}{dt} = \lambda\overline{w} + \overline{g} \tag{5.7}$$

where λ denotes the representative eigenvalue of $[A]$, to study the stability of an explicit scheme used for the temporal discretization. For a stable solution of Equation 5.7 to be obtained using an explicit scheme, $\lambda\Delta t$, where Δt is the time step size, should be within the region of absolute stability. Clearly, the effect of DQ discretization on stability depends on the eigenvalues of the DQ discretization matrix $[A]$.

In the following, we will use the following representative ordinary differential equation

$$\frac{dw}{dt} = f(w,t), \quad \text{with } w(0) \text{ given} \tag{5.8}$$

to illustrate the different explicit schemes employed for temporal discretization.

Euler forward (explicit) scheme: This is a first order scheme given by

$$w^{n+1} = w^n + \Delta t \cdot f^n \tag{5.9}$$

where the superscripts denote the time-level at which the term is evaluated. Euler forward scheme is the simplest explicit scheme for the solution of ODEs. The stability condition of the Euler forward scheme is given by

$$\left|1 + \lambda\Delta t\right| \leq 1 \tag{5.10}$$

Equation 5.10 represents a unit circle with the origin at (-1, 0) in the complex $\lambda\Delta t$ - plane.

Adams-Bashforth schemes: This is a class of explicit schemes. Its first order version is actually the Euler forward scheme as introduced above. Among this class of the schemes, the second order Adams-Bashforth (AB2) scheme

$$w^{n+1} = w^n + \Delta t\left(\frac{3}{2}f^n - \frac{1}{2}f^{n-1}\right) \tag{5.11}$$

and the third order Adams-Bashforth (AB3) scheme

$$w^{n+1} = w^n + \frac{\Delta t}{12}\left(23f^n - 16f^{n-1} + 5f^{n-2}\right) \tag{5.12}$$

are the most popular versions. It is noted that the size of the stability region for this class of schemes decreases as the order of the scheme increases. As shown in Equations 5.11 and 5.12, AB2 involves three time levels while AB3 involves four time levels. For Equation 5.8, the initial condition is only given at one time level ($n = 0$). This indicates that AB2 and AB3 are not self-starting schemes. Before we apply these schemes, we need to use some lower order schemes to provide additional condition at one time level for the AB2 scheme, and two time levels for the AB3 scheme.

Runge-Kutta schemes: Runge-Kutta schemes are the most popular explicit schemes for the solution of ODEs. The general second order Runge-Kutta (RK2) scheme can be written as

$$\begin{cases} w = w^n \\ g = f(w, t_n) \end{cases} \tag{5.13a}$$

$$\begin{cases} w = w + \alpha\Delta t \cdot g \\ g = ag + f(w, t_n + \alpha\Delta t) \end{cases} \tag{5.13b}$$

$$w^{n+1} = w + \frac{1}{2\alpha}\Delta t \cdot g \tag{5.13c}$$

where $a = -1 + 2\alpha - 2\alpha^2$, and α is a parameter selected by the user. Note that only two levels of storage are required for Equation 5.13. The RK2 scheme is also called the two-stage Runge-Kutta scheme. Similarly, there are many versions of the third and fourth order Runge-Kutta (RK3 and RK4) schemes. To reduce the requirements for storage, we are interested in the low storage versions of Runge-Kutta schemes. A three-stage RK3 scheme, which requires only two levels of storage, is given by Williamson (1980)

$$\begin{cases} w = w^n \\ g = f(w, t_n) \end{cases} \tag{5.14a}$$

$$\begin{cases} w = w + \dfrac{\Delta t}{3} \cdot g \\ g = -\dfrac{5}{9} g + f(w, t_n + \dfrac{\Delta t}{3}) \end{cases} \tag{5.14b}$$

$$\begin{cases} w = w + \dfrac{15\Delta t}{16} \cdot g \\ g = -\dfrac{153}{128} g + f(w, t_n + \dfrac{3\Delta t}{4}) \end{cases} \tag{5.14c}$$

$$w^{n+1} = w + \frac{8}{15} \Delta t \cdot g \tag{5.14d}$$

Pike and Roe (1985) presented the following four-stage RK4 scheme for the event that f contains no explicit dependence on t,

$$\begin{cases} w = w^n \\ g = f(w) \end{cases} \tag{5.15a}$$

$$\begin{cases} w = w + \dfrac{\Delta t}{4} \cdot g \\ g = f(w) \end{cases} \tag{5.15b}$$

$$\begin{cases} w = w + \dfrac{\Delta t}{3} \cdot g \\ g = f(w) \end{cases} \tag{5.15c}$$

$$\begin{cases} w = w + \dfrac{\Delta t}{2} \cdot g \\ g = f(w) \end{cases} \tag{5.15d}$$

$$w^{n+1} = w + \Delta t \cdot g \tag{5.15e}$$

The above RK4 scheme also requires two levels of storage. All the Runge-Kutta schemes of a given order have the same stability properties. The stability regions for this class of schemes expand as the order increases.

There are many other explicit schemes such as the Leap Frog scheme, Adams-Moulton schemes, Predictor-Corrector schemes, etc.. For details of these schemes, the reader may consult the books by Canuto et al. (1988), Fornberg (1996).

5.3 Solution Techniques for DQ Algebraic Equations

To demonstrate the solution techniques for the DQ algebraic equations, we consider the solution of a two-dimensional Poisson equation

$$\frac{\partial^2 w}{\partial x^2}+\frac{\partial^2 w}{\partial y^2}=f(x,y) \tag{5.16}$$

on a rectangular domain, $0\le x\le L, 0\le y\le H$, where $f(x,y)$ is a given function. Suppose that Equation 5.16 is subjected to a Dirichlet boundary condition, that is, w is given at the boundary points. Let N and M be the number of grid points in the x and y direction respectively, $b_{i,j}$ and $\bar{b}_{i,j}$ be the DQ weighting coefficients of the second order derivatives of w with respect to x and y respectively. At any point (x_i, y_j), Equation 5.16 can be discretized as

$$\sum_{k=1}^{N} b_{i,k} w_{k,j} + \sum_{k=1}^{M} \bar{b}_{j,k} w_{i,k} = f_{i,j} \tag{5.17}$$

where $w_{i,j}$ and $f_{i,j}$ respectively represent the values of w and f at the grid point (x_i, y_j).

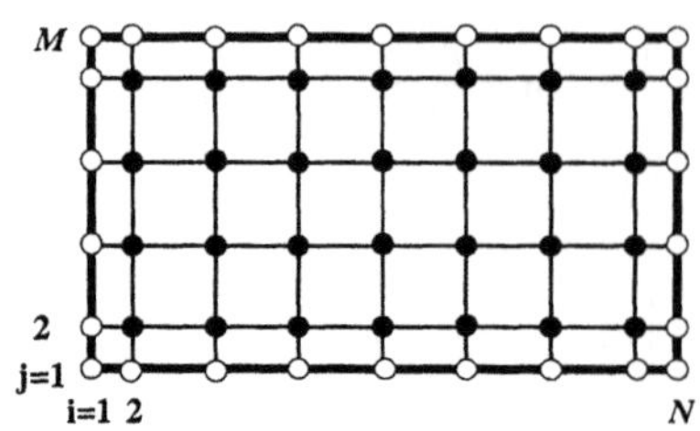

Figure 5.1 A two-dimensional grid

As shown in Figure 5.1, for the Dirichlet condition, the functional values at the boundary points, denoted by the open circles, are known. So, Equation 5.17 should only be applied at the interior points denoted by the solid circles. All the interior points can be represented by $2\le i\le N-1, 2\le j\le M-1$. By shifting the functional values at the boundary points within the summation sign to the right hand side, Equation 5.17 can be rewritten as

$$\sum_{k=2}^{N-1} b_{i,k} w_{k,j} + \sum_{k=2}^{M-1} \bar{b}_{j,k} w_{i,k} = s_{i,j} \tag{5.18}$$
$$\text{for } 2 \le i \le N-1, 2 \le j \le M-1$$

where

$$s_{i,j} = f_{i,j} - (b_{i,1} w_{1,j} + b_{i,N} w_{N,j} + \bar{b}_{j,1} w_{i,1} + \bar{b}_{j,M} w_{i,M})$$

Equation 5.18 is a set of DQ algebraic equations, which can be written in matrix form

$$[A]\{w\} = \{s\} \tag{5.19}$$

where $\{w\}$ is a vector of unknown functional values at all the interior points given by

$$\{w\} = (w_{2,2}, w_{2,3}, \ldots, w_{2,M-1}, w_{3,2}, \ldots, w_{3,M-1}, \ldots, w_{N-1,2}, \ldots, w_{N-1,M-1})^T$$

and $\{s\}$ is a known vector given by

$$\{s\} = (s_{2,2}, s_{2,3}, \ldots, s_{2,M-1}, s_{3,2}, \ldots, s_{3,M-1}, \ldots, s_{N-1,2}, \ldots, s_{N-1,M-1})^T$$

The dimension of the matrix $[A]$ is $(N-2)(M-2)$ by $(N-2)(M-2)$. Equation 5.19 can be solved by direct methods or iterative methods. Some conventional approaches for the solution of Equation 5.19 will be introduced in this section.

For the multi-dimensional case, the matrix $[A]$ contains many zero elements, which are irregularly distributed in the matrix. Thus, the matrix $[A]$ is still considered as a full matrix. In other words, we need to store all the elements of $[A]$ although many of them are zero. There are two major disadvantages in the use of Equation 5.19. The first pertains to the virtual storage. When the mesh size is increased by a small amount, the virtual storage required is increased considerably. For example, when the mesh size is increased from 11×11 to 21×21, the dimension of matrix $[A]$ will be increased from 121×121 to 441×441. Clearly, the problem of virtual storage will become a critical one when the mesh size is very large. The other disadvantage is the construction of $[A]$. The structure of $[A]$ should follow the ordering of $\{w\}$. One should carefully select the ordering of $\{w\}$ to get a well-conditioned matrix $[A]$. In general, this is not an easy task. On the other hand, we find that the DQ algebraic Equation 5.18 can be alternatively written in other matrix forms. For example, it can be expressed as the Lyapunov matrix form

$$[C][W]+[W][D]=[S] \tag{5.20}$$

where [*C*], [*D*], [*W*], and [*S*] are matrices, given by

$$[W]=\begin{bmatrix} w_{2,2} & w_{2,3} & \cdots & w_{2,M-1} \\ w_{3,2} & w_{3,3} & \cdots & w_{3,M-1} \\ \cdot & \cdot & \cdots & \cdots \\ \cdot & \cdot & \cdots & \cdots \\ \cdot & \cdot & \cdots & \cdots \\ w_{N-1,2} & w_{N-1,3} & \cdots & w_{N-1,M-1} \end{bmatrix}, \quad [S]=\begin{bmatrix} s_{2,2} & s_{2,3} & \cdots & s_{2,M-1} \\ s_{3,2} & s_{3,3} & \cdots & s_{3,M-1} \\ \cdot & \cdot & \cdots & \cdots \\ \cdot & \cdot & \cdots & \cdots \\ \cdot & \cdot & \cdots & \cdots \\ s_{N-1,2} & s_{N-1,3} & \cdots & s_{N-1,M-1} \end{bmatrix}$$

$$[C]=\begin{bmatrix} b_{2,2} & b_{2,3} & \cdots & b_{2,N-1} \\ b_{3,2} & b_{3,3} & \cdots & b_{3,N-1} \\ \cdot & \cdot & \cdots & \cdots \\ \cdot & \cdot & \cdots & \cdots \\ \cdot & \cdot & \cdots & \cdots \\ b_{N-1,2} & b_{N-1,3} & \cdots & b_{N-1,N-1} \end{bmatrix}, \quad [D]=\begin{bmatrix} \bar{b}_{2,2} & \bar{b}_{2,3} & \cdots & \bar{b}_{2,M-1} \\ \bar{b}_{3,2} & \bar{b}_{3,3} & \cdots & \bar{b}_{3,M-1} \\ \cdot & \cdot & \cdots & \cdots \\ \cdot & \cdot & \cdots & \cdots \\ \cdot & \cdot & \cdots & \cdots \\ \bar{b}_{M-1,2} & \bar{b}_{M-1,3} & \cdots & \bar{b}_{M-1,M-1} \end{bmatrix}$$

The dimension of the matrices [*W*], [*S*], [*C*] and [*D*] is (N-2)×(M-2), (N-2)×(M-2), (N-2)×(N-2), (M-2)×(M-2) respectively. It is obvious that [*C*] and [*D*] are full matrices, which can be easily constructed from the DQ weighting coefficients. In addition, the dimension of [*C*] and [*D*] are very small as compared to the dimension of [*A*]. So, the problem of virtual storage can be greatly diminished by using Equation 5.20. In this section, we will also introduce some efficient approaches to solve Equation 5.20.

5.3.1 Direct Methods

There are many standard methods to solve Equation 5.19. Among them, the Gaussian elimination method and the *LU* decomposition approach are used extensively. The details of these methods can be found in textbooks on numerical analysis. In the following, we just list the basic procedures of these methods.

Gaussian elimination method: The Gaussian elimination method is used to reduce Equation 5.19 to an upper triangular equation system by a sequence of elimination steps. Its solution procedure for Equation 5.19 can be summarized as

(1) Augment the N by N matrix [*A*] with the vector on the right hand side, $\{s\}$, to form an N by (N+1) matrix.

(2) Interchange rows if necessary to make the value of $a_{1,1}$ the largest magnitude of any coefficient in the first column.

(3) Create zeros in the second through *N*th rows in the first column by subtracting $a_{i,1}/a_{1,1}$ times the first row from the *i*th row. Store the values of $a_{i,1}/a_{1,1}$ in $a_{i,1}$, $i=1,2,...,N$.

(4) Repeat steps (2) and (3) for the second through the (*N*-1)th rows, putting the largest magnitude coefficient on the diagonal by interchanging rows (considering only rows *j* to *N*), and then subtracting $a_{i,j}/a_{j,j}$ times the *j*th row from the *i*th row so as to create zeros in all positions of the *j*th column below the diagonal. Store the values of $a_{i,j}/a_{j,j}$ in $a_{i,j}, i=j+1,...,N$. Upon the completion of this step, the system is upper triangular.

(5) Solve for w_N from the *N*th equation by

$$w_N = a_{N,N+1}/a_{N,N} \tag{5.21}$$

(6) Solve for $w_{N-1}, w_{N-2},..., w_1$ from the (*N*-1)th through the first equation in turn, by

$$w_i = \left(a_{i,N+1} - \sum_{j=i+1}^{N} a_{i,j} w_j \right) / a_{i,i}, \quad i=N-1, N-2,...,1 \tag{5.22}$$

LU Decomposition: The *LU* decomposition method is a modification of the elimination method. In this method, the matrix [*A*] is decomposed into the product of two matrices [*L*] and [*U*], where [*L*] is a lower triangular matrix with 1's as its diagonal, and [*U*] is an upper triangular matrix. Upon *LU* decomposition, Equation 5.19 can be written as

$$[L][U]\{w\} = \{s\} \tag{5.23}$$

Introducing an intermediate vector $\{z\}$, and setting

$$[U]\{w\} = \{z\}, \tag{5.24}$$

Equation 5.23 can be reduced to

$$[L]\{z\} = \{s\} \tag{5.25}$$

Since [*L*] and [*U*] are lower and upper triangular matrices, the solution of Equations 5.25 and 5.24 can be easily given by

$$z_i = s_i - \sum_{j=1}^{i-1} l_{i,j} z_j , \quad i = 1, 2, ..., N \tag{5.26a}$$

$$w_i = \left(z_i - \sum_{j=i+1}^{N} u_{i,j} w_j \right) / u_{i,i} , \quad i = N, N-1, ..., 1 \tag{5.26b}$$

where $u_{i,j}$ and $l_{i,j}$ are respectively the elements of $[U]$ and $[L]$ given by

$$u_{i,j} = a_{i,j} - \sum_{k=1}^{i-1} l_{i,k} u_{k,j} , \quad j \ge i, \; j = 1, 2, ..., N \tag{5.27a}$$

$$l_{i,j} = \left(a_{i,j} - \sum_{k=1}^{j-1} l_{i,k} u_{k,j} \right) / u_{j,j} , \quad i \ge j, \; i = 2, 3, ..., N \tag{5.27b}$$

$a_{i,j}$ are elements of the matrix $[A]$.

Schur-Decomposition: The Schur-decomposition presented by Bartels and Stewart (1972) can be used to solve Equation 5.20. This method uses orthogonal transformations to reduce $[C]$ into a block-lower-triangular matrix $[\overline{C}]$, and $[D]$ into a block-upper-triangular matrix $[\overline{D}]$. If $[P]$ and $[Q]$ denote the orthogonal transformations, then we have

$$[\overline{C}] = [P]^T [C][P] \tag{5.28a}$$

$$[\overline{D}] = [Q]^T [D][Q] \tag{5.28b}$$

$$[I] = [P][P]^T = [Q][Q]^T \tag{5.28c}$$

where the superscript T represents the transpose of a matrix. By pre-multiplying the matrix $[P]^T$ and post-multiplying the matrix $[Q]$ to both sides of Equation 5.20, we obtain

$$[\overline{C}][\overline{W}] + [\overline{W}][\overline{D}] = [\overline{S}] \tag{5.29}$$

where

$$[\overline{W}] = [P]^T [W][Q] \tag{5.30a}$$

$$[\overline{S}] = [P]^T [S][Q] \tag{5.30b}$$

Since $[\overline{C}]$ and $[\overline{D}]$ are block-triangular matrices, the solution of Equation 5.29 can be easily obtained. Basically, the solution procedure involves the following four steps,

(1) Reduce $[C]$ and $[D]$ into real Schur form by using the *QR* algorithm and determine the orthogonal transformation matrices $[P]$ and $[Q]$.

(2) Use the matrices $[P]$ and $[Q]$ to construct the matrix $[\overline{S}]$ *via* Equation 5.30*b*.

(3) Solve Equation 5.29 and obtain the solution for $[\overline{W}]$.

(4) Transform $[\overline{W}]$ to $[W]$ *via* Equation 5.30*a*.

The details of the Schur-decomposition method are given in Bartels and Stewart (1972), who also supplied a complete Fortran program for this solution technique.

Matrix-diagonalization: The matrix-diagonalization method is similar to the Schur-decomposition approach. The difference is that the matrices $[C]$ and $[D]$ are diagonalized rather than merely reduced to block-triangular form. Let $[P]$ and $[Q]$ be right eigenvector matrices of $[C]$ and $[D]$, $[\Lambda_C]$ and $[\Lambda_D]$ be the diagonal matrices with the eigenvalues of $[C]$ and $[D]$ along the diagonal. Then, we have

$$[\Lambda_C] = [P]^{-1}[C][P] \tag{5.31a}$$

$$[\Lambda_D] = [Q]^{-1}[D][Q] \tag{5.31b}$$

Similar to the Schur-decomposition method, we pre-multiply the matrix $[P]^{-1}$ and post-multiply the matrix $[Q]$ to both sides of Equation 5.20 and obtain

$$[\Lambda_C][\overline{W}] + [\overline{W}][\Lambda_D] = [\overline{S}] \tag{5.32}$$

where

$$[\overline{W}] = [P]^{-1}[W][Q] \tag{5.33a}$$

$$[\overline{S}] = [P]^{-1}[S][Q] \tag{5.33b}$$

Since $[\Lambda_C]$ and $[\Lambda_D]$ are diagonal matrices, the solution of Equation 5.32 is much easier than the solution of Equation 5.29. The solution procedure of the matrix-diagonalization method also involves four steps, namely

(1) Diagonalize $[C]$ and $[D]$ by computing the eigenvalues and eigenvectors of $[C]$ and $[D]$ so that $[\Lambda_C]$, $[\Lambda_D]$, $[P]$ and $[Q]$ are obtained.

(2) Use the matrices $[P]^{-1}$ and $[Q]$ to construct the matrix $[\overline{S}]$ via Equation 5.33*b*.

(3) Solve Equation 5.32 and obtain the solution for $[\overline{W}]$.

(4) Transform $[\overline{W}]$ to $[W]$ *via* Equation 5.33*a*.

The Schur-decomposition and matrix-diagonalization methods have been efficiently applied by Chen, Zhong and Shu (1998) to solve the Lyapunov equation system which results from the DQ discretization of the Poisson and convection-diffusion equations.

In general, direct methods involve a lot of operations when applied to the *LU* decomposition, orthogonal transformation, and diagonalization. When the same problem has to be solved repeatedly, this work can be done once, and the relevant matrices are stored and applied repeatedly. An example of such an application is the solution of the stream function formulation in the computation of the incompressible Navier-Stokes equations, which will be described in detail in Chapter 6.

5.3.2 Iterative Methods

Direct methods are very efficient when the dimension of the matrix is not large. However, as mentioned above, as the number of grid points is increased, the dimension of the matrix will increase accordingly. Hence the problem of virtual storage will become critical. Furthermore, the DQ discretization matrix tends to become ill-conditioned when the mesh size is large. This would lead to difficulties in obtaining the solution or even worse, reduce the accuracy of the solution. The drawbacks of direct methods can be eliminated using iterative methods. In the following, we will introduce some efficient iterative approaches for the conventional Equation 5.19 and the Lyapunov Equation 5.20.

5.3.2.1 Iterative Methods for Conventional System

The matrix $[A]$ in Equation 5.19 can be decomposed into two matrices $[H]$ and $[P]$, that is

$$[A]=[H]+[P] \tag{5.34}$$

With Equation 5.34, Equation 5.19 can be rewritten as

$$[H]\{w\}=\{s\}-[P]\{w\} \tag{5.35}$$

The iterative expression for Equation 5.35 can be written as

$$[H](\{w^{n+1}\}-\{w^n\})=\{s\}-[P]\{w^n\}-[H]\{w^n\}=\{s\}-[A]\{w^n\}=\{r^n\} \tag{5.36}$$

where n represents the iteration level and $\{r^n\}$ represents the vector of residuals. In practical applications, a relaxation factor ω is introduced on the right hand side of Equation 5.36, and the final iteration expression becomes

$$\{w^{n+1}\} = \{w^n\} + \omega[H]^{-1}\{r^n\} \tag{5.37}$$

Equation 5.37 is a general iterative expression for Equation 5.19. By using different forms for [*H*], we can obtain different iterative expressions for Equation 5.19. Before we introduce specific iterative methods, we shall briefly discuss the stability condition for Equation 5.37. Let $\{w\}$ be the exact solution for Equation 5.19. We can define the error at iteration level *n* as

$$\{e^n\} = \{w^n\} - \{w\} \tag{5.38}$$

Then Equation 5.37 can be written as

$$\{e^{n+1}\} = (I - \omega[H]^{-1}[A])\{e^n\} = [G]\{e^n\} \tag{5.39}$$

where $[G] = I - \omega[H]^{-1}[A]$ is called the iteration matrix. Clearly, to obtain a stable solution for an iterative method, the spectral radius, ρ, of $[G]$ should be less than 1. When $\omega = 1$, $[H] = [A]$, $[G]$ becomes the null matrix. For this case, the error $\{e^{n+1}\}$ is always zero no matter how large the error $\{e^n\}$ is. This indicates that the exact solution can be obtained by using only one iteration. In fact, this special case is equivalent to the direct method. It can be said that if [*H*] is a good approximation to $[A]$, the iterative method will be very efficient. On the other hand, we require that [*H*] be easily inverted or decomposed by [*L*] and [*U*].

Richardson Iteration: Richardson iteration is the simplest iteration method. For this case, [*H*] is taken as the identity matrix [*I*], and the iterative expression becomes

$$\{w^{n+1}\} = \{w^n\} + \omega\{r^n\} \tag{5.40}$$

Equation 5.40 can also be expressed by

$$w_i^{n+1} = w_i^n + \omega r_i^n \tag{5.41}$$

where $r_i^n = s_i - \sum_{j=1}^{M} a_{i,j} w_j^n$, *M* is the dimension of the matrix $[A]$.

Jacobi and J-OE Iteration: Jacobi iteration is a special case of J-OE (Jacobi over relaxation) iteration when ω is taken as 1. For purposes of simplicity in the following discussion, we split the matrix $[A]$ into the summation of three matrices $[A_L]$, $[A_U]$ and $[A_D]$,

$$[A]=[A_L]+[A_D]+[A_U] \tag{5.42}$$

where $[A_L]$ is the lower triangular matrix with diagonal elements being zero, $[A_U]$ is the upper triangular matrix with diagonal elements being zero, and $[A_D]$ is the diagonal matrix with elements being the diagonal elements of $[A]$. It is noted that the elements of $[A_L]$, $[A_U]$ and $[A_D]$ equal to those of $[A]$ at the corresponding position. For the J-OE method, $[H]$ is taken as $[A_D]$, and the iterative expression is

$$\{w^{n+1}\}=\{w^n\}+\omega[A_D]^{-1}\{r^n\} \tag{5.43}$$

Since $[A_D]$ is a diagonal matrix, Equation 5.43 can also be expressed by

$$w_i^{n+1}=w_i^n+\omega r_i^n / a_{i,i} \tag{5.44}$$

r_i^n in Equation 5.44 has the same expression as in the Richardson iteration method.

Gauss-Seidel (G-S) and SOR Iteration: Gauss-Seidel (G-S) iteration is a special case of SOR (successive over-relaxation) iteration when ω is taken as 1. SOR iteration is used to improve the convergence speed of J-OE iteration. It is noted that SOR is a point iteration method. When w_i^{n+1} is evaluated, the values of $w_k^{n+1}, k=1,2,\ldots,i-1$ have already been calculated. These new values at the iteration level (n+1) can then be used to compute the residuals. The residuals of SOR iteration are computed from

$$\{r^n\}=\{s\}-[A_L]\{w^{n+1}\}-([A_D]+[A_U])\{w^n\} \tag{5.45}$$

The iterative expression of SOR method is the same as Equation 5.44. The only difference between J-OE and SOR methods is in the computation of $\{r^n\}$. r_i^n in the SOR method can be expressed as

$$r_i^n=s_i-\sum_{j=1}^{i-1} a_{i,j} w_j^{n+1}-\sum_{j=i}^{M} a_{i,j} w_j^n \tag{5.46}$$

Conjugate Gradient (CG) Iteration: Conjugate gradient iteration is a preconditioning method. The principle of this method is to update the solution using

$$\{w^{n+1}\} = \{w^n\} + \alpha_n \{p^n\} \tag{5.47}$$

where n is the iteration level, α_n is a scalar to minimize the error, and $\{p^n\}$ is a direction vector. The method is initialized with an initial guess $\{w^0\}$. The initial direction vector is chosen to be

$$\{p^0\} = \{r^0\} \tag{5.48}$$

where $\{r^0\}$ is the vector of initial residuals. Subsequent iterations are made according to

$$\alpha_n = \frac{(\{p^n\},\{r^n\})}{(\{p^n\},[A]\{p^n\})} \tag{5.49a}$$

$$\{w^{n+1}\} = \{w^n\} + \alpha_n \{p^n\} \tag{5.49b}$$

$$\{r^{n+1}\} = \{s\} - [A]\{w^{n+1}\} \tag{5.49c}$$

$$\beta_{n+1} = -\frac{(\{r^{n+1}\},[A]\{p^n\})}{(\{p^n\},[A]\{p^n\})} \tag{5.49d}$$

$$\{p^{n+1}\} = \{r^{n+1}\} + \beta_{n+1}\{p^n\} \tag{5.49e}$$

$(\{p^n\},\{r^n\})$ in Equation 5.49 represents the inner product of two vectors $\{p^n\}$ and $\{r^n\}$. The CG method is very efficient for symmetric, positive-definite linear systems.

5.3.2.2 Iterative Methods for Lyapunov System

For Equation 5.20, it is supposed that [*S*] can be split into two matrices $[S_1]$ and $[S_2]$ such that

$$[C][W] = [S_1] \tag{5.50a}$$

$$[W][D] = [S_2] \tag{5.50b}$$

where $[S] = [S_1] + [S_2]$. Equation 5.50*b* can be rewritten as

$$[D]^T[W]^T = [S_2]^T \tag{5.51}$$

Applying the conventional iterative approach of Equation 5.36 to Equations 5.50*a* and 5.51 gives

$$[H_1]([W^{n+1}]-[W^n]) = \omega([S_1]-[C][W^n]) \tag{5.52a}$$

$$[H_2]^T([W^{n+1}]^T - [W^n]^T) = \omega([S_2]^T - [D]^T[W^n]^T) \tag{5.52b}$$

where $[H_1]$ and $[H_2]$ are two iteration matrices given by different iterative methods. Equation 5.52*b* can be rewritten as

$$([W^{n+1}]-[W^n])[H_2] = \omega([S_2]-[W^n][D]) \tag{5.53}$$

By combining Equations 5.52*a* and 5.53, we can obtain the final iterative expression for the Lyapunov equation system as

$$[H_1]([W^{n+1}]-[W^n]) + ([W^{n+1}]-[W^n])[H_2] = \omega[R^n] \tag{5.54}$$

where $[R^n]$ is the matrix of residuals given by $[R^n]=[S]-[C][W^n]-[W^n][D]$.

In a similar manner, we can split the matrices $[C]$ and $[D]$ into the summation of three matrices, that is,

$$[C] = [C_L]+[C_D]+[C_U] \tag{5.55a}$$

$$[D] = [D_L]+[D_D]+[D_U] \tag{5.55b}$$

where $[C_L],[D_L]$ are lower triangular matrices, $[C_U],[D_U]$ are upper triangular matrices, $[C_D],[D_D]$ are diagonal matrices. Taking $[H_1]$ and $[H_2]$ as different forms, we can generate the Richardson, J-OE and SOR iteration schemes for the Lyapunov equation system.

Richardson Iteration: For this case, we take $[H_1]$ and $[H_2]$ to be the identity matrix, and we then obtain

$$[W^{n+1}] = [W^n] + \frac{\omega}{2}[R^n] \tag{5.56}$$

where $[R^n]=[S]-[C][W^n]-[W^n][D]$. Equation 5.56 can also be written as

$$w_{i,j}^{n+1} = w_{i,j}^{n} + \frac{\omega}{2}\left(S_{i,j} - \sum_{k=1}^{N} C_{i,k} w_{k,j}^{n} - \sum_{k=1}^{M} D_{j,k} w_{i,k}^{n} \right) \tag{5.57}$$

J-OE Iteration: For the J-OE iteration, $[H_1]$ is taken to be $[C_D]$ while $[H_2]$ is selected to be $[D_D]$. Its iterative expression becomes

$$[C_D]([W^{n+1}]-[W^{n}]) + ([W^{n+1}]-[W^{n}])[D_D] = \omega[R^{n}] \tag{5.58}$$

where $[R^n]=[S]-[C][W^n]-[W^n][D]$. The algebraic form of Equation 5.58 can be written as

$$w_{i,j}^{n+1} = w_{i,j}^{n} + \frac{\omega}{C_{i,i}+D_{j,j}}\left(S_{i,j} - \sum_{k=1}^{N} C_{i,k} w_{k,j}^{n} - \sum_{k=1}^{M} D_{j,k} w_{i,k}^{n} \right) \tag{5.59}$$

SOR Iteration: SOR iteration uses the same form as Equation 5.58 to update the solution, but the expression for the residuals is modified to

$$\begin{aligned}[R^n] = [S] - [C_L][W^{n+1}] - ([C_D]+[C_U])[W^n] \\ - [W^{n+1}][D_L] - [W^n]([D_D]+[D_U])\end{aligned} \tag{5.60}$$

The algebraic form of the SOR iteration becomes

$$\begin{aligned} w_{i,j}^{n+1} = w_{i,j}^{n} + \frac{\omega}{C_{i,i}+D_{j,j}}\Bigg(S_{i,j} - \sum_{k=1}^{i-1} C_{i,k} w_{k,j}^{n+1} - \\ \sum_{k=i}^{N} C_{i,k} w_{k,j}^{n} - \sum_{k=1}^{j-1} D_{j,k} w_{i,k}^{n+1} - \sum_{k=j}^{M} D_{j,k} w_{i,k}^{n} \Bigg)\end{aligned} \tag{5.61}$$

Conjugate Gradient (CG) Iteration: The formulations for the conjugate gradient iteration in Section 5.3.2.1 for the conventional system can be easily extended to the Lyapunov system. The solution matrix at the iteration level $(n+1)$ can be evaluated from

$$[W^{n+1}] = [W^n] + \alpha_n [P^n] \tag{5.62}$$

where $[W]$ is the solution matrix, $[P]$ is the direction matrix, α_n is a scalar included to minimize the error. The method is initialized with an initial guess $[W^0]$. The initial direction matrix is chosen to be

$$[P^0] = [R^0] \tag{5.63}$$

where $[R^0]$ is the matrix of initial residuals. Subsequent iterations are made according to

$$[G^n] = [C][P^n] + [P^n][D] \tag{5.64a}$$

$$\alpha_n = \frac{([P^n],[R^n])}{([P^n],[G^n])} \tag{5.64b}$$

$$[W^{n+1}] = [W^n] + \alpha_n [P^n] \tag{5.64c}$$

$$[R^{n+1}] = [S] - [C][W^{n+1}] - [W^{n+1}][D] \tag{5.64d}$$

$$\beta_{n+1} = -\frac{([R^{n+1}],[G^n])}{([P^n],[G^n])} \tag{5.64e}$$

$$[P^{n+1}] = [R^{n+1}] + \beta_{n+1}[P^n] \tag{5.64f}$$

Similar to the definition of inner product of two vectors which appears in Equation 5.49, $([A],[B])$ in Equation 5.64 is defined as the inner product of the two matrices $[A]$ and $[B]$. Suppose that the dimensions of $[A]$ and $[B]$ are N by M. Then $([A],[B])$ is defined as

$$([A],[B]) = \sum_{i=1}^{N}\sum_{j=1}^{M} a_{i,j} b_{i,j} \tag{5.65}$$

where $a_{i,j}$ and $b_{i,j}$ are respectively the elements of $[A]$ and $[B]$.

5.4 Implementation of Boundary Conditions

Proper implementation of the boundary conditions is very important for the accurate numerical solution of engineering problems. We will demonstrate that the derivatives in the boundary conditions can also be approximated by the DQ method. For simplicity, we consider a problem involving a second order differential equation over a one-dimensional interval $0 \le x \le 1$. For this case, two boundary conditions have to be specified at the two ends $x = 0$ and $x = 1$. We suppose that the number of grid points is N with coordinates of grid points given as $0 = x_1, x_2, \ldots, x_N = 1$, and the weighting coefficients of the first order derivative are represented by $c_{i,j}^{(1)}$,

$i, j = 1, 2, ..., N$. The detailed implementation of different types of boundary conditions is shown as follows.

Dirichlet Condition: For the Dirichlet condition, the functional values at the two ends are given, that is,

$$w(0) = c_1, w(1) = c_2 \tag{5.66}$$

where c_1 and c_2 are constants. The implementation of Equation 5.66 is straightforward for the numerical computation. The numerical condition can be written as

$$w_1 = c_1, w_N = c_2 \tag{5.67}$$

Neumann Condition: For the Neumann condition, the first order derivatives at the two ends are specified, that is,

$$\left.\frac{dw}{dx}\right|_{x=0} = c_3, \quad \left.\frac{dw}{dx}\right|_{x=1} = c_4 \tag{5.68}$$

where c_3 and c_4 are constants. Application of the DQ method to discretize the derivatives in Equation 5.68 gives

$$\sum_{k=1}^{N} c_{1,k}^{(1)} w_k = c_3 \tag{5.69a}$$

$$\sum_{k=1}^{N} c_{N,k}^{(1)} w_k = c_4 \tag{5.69b}$$

Equation System 5.69 results in two equations, which can be used to solve for two unknowns. Let the two unknowns be w_1 and w_N. Equation 5.69 can be rewritten as

$$c_{1,1}^{(1)} w_1 + c_{1,N}^{(1)} w_N = c_3 - \sum_{k=2}^{N-1} c_{1,k}^{(1)} w_k \tag{5.70a}$$

$$c_{N,1}^{(1)} w_1 + c_{N,N}^{(1)} w_N = c_4 - \sum_{k=2}^{N-1} c_{N,k}^{(1)} w_k \tag{5.70b}$$

The solution of Equation System 5.70 is

$$w_1 = \frac{c_3 c_{N,N}^{(1)} - c_4 c_{1,N}^{(1)} + \sum_{k=2}^{N-1} (c_{1,N}^{(1)} c_{N,k}^{(1)} - c_{N,N}^{(1)} c_{1,k}^{(1)}) w_k}{c_{1,1}^{(1)} c_{N,N}^{(1)} - c_{1,N}^{(1)} c_{N,1}^{(1)}} \tag{5.71a}$$

$$w_N = \frac{c_4 c_{1,1}^{(1)} - c_3 c_{N,1}^{(1)} + \sum_{k=2}^{N-1} (c_{N,1}^{(1)} c_{1,k}^{(1)} - c_{1,1}^{(1)} c_{N,k}^{(1)}) w_k}{c_{1,1}^{(1)} c_{N,N}^{(1)} - c_{1,N}^{(1)} c_{N,1}^{(1)}} \tag{5.71b}$$

Equation 5.71 shows that for the Neumann condition, the functional values at the boundary points can be updated using the functional values at the interior points.

Mixed Condition: The mixed condition is a combination of the Dirichlet and Neumann conditions, which can be expressed as

$$d_1 w(0) + d_2 \left.\frac{dw}{dx}\right|_{x=0} = c_5 \tag{5.72a}$$

$$d_3 w(1) + d_4 \left.\frac{dw}{dx}\right|_{x=1} = c_6 \tag{5.72b}$$

where d_1, d_2, d_3, d_4, c_5 and c_6 are constants. The values of d_1 and d_2 cannot be taken as zero simultaneously. Similarly, d_3 and d_4 cannot be taken as zero at the same time. The DQ approximation for Equation 5.72 can be written as

$$d_1 w_1 + d_2 \sum_{k=1}^{N} c_{1,k}^{(1)} w_k = c_5 \tag{5.73a}$$

$$d_3 w_N + d_4 \sum_{k=1}^{N} c_{N,k}^{(1)} w_k = c_6 \tag{5.73b}$$

Equation System 5.73 can be employed to solve for two unknowns w_1 and w_N, which can be expressed as

$$w_1 = \frac{1}{AXN} \left[c_5 (d_3 + d_4 c_{N,N}^{(1)}) - c_6 d_2 c_{1,N}^{(1)} + \sum_{k=2}^{N-1} AXK1 \cdot w_k \right] \tag{5.74a}$$

$$w_N = \frac{1}{AXN} \left[c_6 (d_1 + d_2 c_{1,1}^{(1)}) - c_5 d_4 c_{N,1}^{(1)} + \sum_{k=2}^{N-1} AXKN \cdot w_k \right] \tag{5.74b}$$

where

$$AXN = (d_1 + d_2 c_{1,1}^{(1)})(d_3 + d_4 c_{N,N}^{(1)}) - d_2 d_4 c_{1,N}^{(1)} c_{N,1}^{(1)}$$

$$AXK1 = d_2 d_4 c_{1,N}^{(1)} c_{N,k}^{(1)} - (d_3 + d_4 c_{N,N}^{(1)}) d_2 c_{1,k}^{(1)}$$

$$AXKN = d_2 d_4 c_{N,1}^{(1)} c_{1,k}^{(1)} - (d_1 + d_2 c_{1,1}^{(1)}) d_4 c_{N,k}^{(1)}$$

Periodic Condition: For the periodic condition, it is suggested that the FDQ method be applied. Since the periodic condition is naturally considered in the FDQ formulation, no specific implementation of the periodic condition is needed in the FDQ computation. One just needs to apply the FDQ discretization for the governing equation at all the grid points. More details of the FDQ computation for a periodic problem will be shown in the following section.

5.5 Sample Applications of DQ Method

In this section, we will show some sample applications of the DQ method, which involve the solution of DQ ordinary differential equations, DQ algebraic equations, DQ eigenvalue problems, and also a comparison of the different DQ methods.

5.5.1 Burgers Equation

The one-dimensional Burgers equation

$$\frac{\partial u}{\partial t} + u\frac{\partial u}{\partial x} = \nu\frac{\partial^2 u}{\partial x^2} \tag{5.75}$$

is of considerable physical interest because of its role in the hierarchy of approximation to the Navier-Stokes equations. The value of ν in Equation 5.75 is assumed to be a constant. It is supposed that the physical domain is $0 \le x \le 1$, and the initial condition is

$$u(x,0) = f(x) \tag{5.76}$$

To obtain the analytical solution of Equation 5.75 for the purpose of comparison, we can adopt the following transformation

$$u(x,t) = -2\nu\frac{\partial w(x,t)}{\partial x} / w(x,t) \tag{5.77a}$$

$$f(x) = -2\nu \frac{dg(x)}{dx} / g(x) \tag{5.77b}$$

Using Equation 5.77, Equation 5.75 can be reduced to a linear heat conduction or diffusion equation as follows

$$\frac{\partial w}{\partial t} = \nu \frac{\partial^2 w}{\partial x^2} \tag{5.78}$$

with

$$w(x,0) = g(x)$$

For the test case, $f(x)$ is chosen as

$$f(x) = -2\nu \frac{b\pi \cos \pi x + 0.5c\pi \cos(\pi x/2)}{b \sin \pi x + c \sin(\pi x/2) + d} \tag{5.79}$$

where b, c and d are constants. Thus, the analytical solution of this problem can be expressed as

$$w(x,t) = b \cdot e^{-\nu\pi^2 t} \sin \pi x + c \cdot e^{-0.25\nu\pi^2 t} \sin \frac{\pi x}{2} + d \tag{5.80}$$

For the numerical computation using the DQ method, the constants b, c, d and ν are respectively taken as 0.2, 0.1, 0.3 and 0.01. Let N be the number of grid points in the whole domain, $c_{i,j}^{(1)}, c_{i,j}^{(2)}$ be the weighting coefficients of the first and second order derivatives respectively. After spatial discretization by the DQ method, Equation 5.75 can be reduced to the following ordinary differential equation

$$\frac{du_i}{dt} = -u_i \sum_{k=1}^{N} c_{i,k}^{(1)} u_k + \nu \sum_{k=1}^{N} c_{i,k}^{(2)} u_k \tag{5.81}$$

The given problem is an initial value problem. There is no boundary condition given at $x = 0$, and $x = 1$. So, Equation 5.81 should be applied at all the grid points including the boundary points at $x = 0$ and $x = 1$. The solution of Equation 5.81 can be given by the four-stage Runge-Kutta method, which is introduced in Section 5.2. In the work of Shu (1991), the weighting coefficients in Equation 5.81 were computed by the PDQ formulations, and a uniform mesh was used. Table 5.1 lists the unsteady DQ solution with different number of grid points $N = 7$, 11 and 21. The corresponding values obtained from the exact solution are also included in the table for comparison. The time step size was chosen to be 0.01. Clearly, the DQ solution is very accurate even though very few grid points are used.

Table 5.1 Unsteady DQ solution of Burgers equation (Shu 1991)

t	*x*	*PDQ Solution*			*Exact*
		$N=7$	$N=11$	$N=21$	
0.1	0.0	-0.051923	-0.051923	-0.051923	-0.051923
	0.5	-0.003897	-0.003897	-0.003897	-0.003897
	1.0	-0.031138	-0.031126	-0.031126	-0.031127
0.5	0.0	-0.050243	-0.050226	-0.050214	-0.050215
	0.5	-0.003917	-0.003917	-0.003917	-0.003917
	1.0	0.030068	0.029993	0.029995	0.029995
1.0	0.0	-0.048263	-0.048219	-0.048170	-0.048168
	0.5	-0.003939	-0.003939	-0.003939	-0.003939
	1.0	0.028819	0.028628	0.028640	0.028638

5.5.2 Two-dimensional Poisson Equation

The two-dimensional Poisson equation is employed as a sample to show the iterative process for the solution of DQ algebraic equations. The normalized Poisson equation can be written as

$$\frac{\partial^2 u}{\partial x^2}+\beta^2\frac{\partial^2 u}{\partial y^2}=g,\quad 0\le x\le 1, 0\le y\le 1 \tag{5.82}$$

where β is the aspect ratio and g is a constant. Taking g to be zero, and imposing the following boundary conditions

$$u(x,0)=0,\qquad 0\le x\le 1 \tag{5.83a}$$

$$u(x,1)=\sin(\pi x/2),\qquad 0\le x\le 1 \tag{5.83b}$$

$$u(0,y)=0,\qquad 0<y<1 \tag{5.83c}$$

$$u_x(1,y)=0,\qquad 0<y<1 \tag{5.83d}$$

the problem has an exact solution given by

$$u(x,y)=\sinh[\pi y/(2\beta)]\cdot\sin(\pi x/2)/\sinh[\pi/(2\beta)] \tag{5.84}$$

The subscript x in Equation 5.83*d* represents the first derivative of $u(x,y)$ with respect to x. The exact solution given by Equation 5.84 can be used to validate the efficiency and accuracy of the DQ results. Application of the DQ method to discretize the derivatives in Equation 5.82 leads to

$$\sum_{k=1}^{N} c_{i,k}^{(2)} u_{k,j} + \beta^2 \sum_{k=1}^{M} \bar{c}_{j,k}^{(2)} u_{i,k} = g \tag{5.85}$$

where N and M are the number of grid points in the x and y directions respectively, $c_{i,k}^{(2)}$ and $\bar{c}_{j,k}^{(2)}$ are the weighting coefficients of the second order derivatives with respect to x and y. By moving the functional values at the boundary points to the right hand side, Equation 5.85 can be rewritten as

$$\begin{aligned} &\sum_{k=2}^{N-1} c_{i,k}^{(2)} u_{k,j} + \beta^2 \sum_{k=2}^{M-1} \bar{c}_{j,k}^{(2)} u_{i,k} \\ &= g - c_{i,1}^{(2)} u_{1,j} - c_{i,N}^{(2)} u_{N,j} - \bar{c}_{j,1}^{(2)} u_{i,1} - \bar{c}_{j,M}^{(2)} u_{i,M} \end{aligned} \tag{5.86}$$

Equation 5.86 should be applied at all the interior points, $i = 2, 3, ..., N-1$, $j = 2, 3, ..., M-1$, to obtain a set of DQ algebraic equations. Equation 5.86 can be expressed as the Lyapunov form. The two iterative expressions introduced in Section 5.3.2.2, namely, J-OE and SOR, can be directly applied to solve Equation 5.86. Note that when the relaxation factor is taken as 1, the J-OE becomes the Jacobi method, and the SOR reduces to the Gauss-Seidel (G-S) method. For the DQ computation, the weighting coefficients are calculated by the PDQ formulations, and Grid I shown in Chapter 4 is adopted to generate the grid point distribution, which can be written as

$$x_i = \frac{1}{2}\left(1 - \cos\frac{i-1}{N-1}\pi\right), i = 1, 2, ..., N \tag{5.87a}$$

$$y_j = \frac{1}{2}\left(1 - \cos\frac{j-1}{M-1}\pi\right), j = 1, 2, ..., M \tag{5.87b}$$

Similar to the discretization of derivatives in Equation 5.82, the derivative in the Neumann condition is also approximated by the DQ method. The convergence criterion of the iterative process is set as

$$\left|R_{i,j}\right|_{\max} \le 10^{-10} \tag{5.88}$$

where $R_{i,j}$ is the residual given by

$$R_{i,j} = g - \sum_{k=1}^{N} c_{i,k}^{(2)} u_{k,j} - \beta^2 \sum_{k=1}^{M} \bar{c}_{j,k}^{(2)} u_{i,k} \tag{5.89}$$

Shu, Yeo and Yao (1998) have comparatively studied the performance of Jacobi, J-OE, G-S, and SOR methods for the solution of Equation 5.86, and the solution of algebraic equations derived from the second order finite difference (FD) scheme. The performance of an iterative method is measured by the ratio of iterations, *rk*, defined by

$$rk = \frac{\text{number of iterations required for an iterative method}}{\text{number of iterations required for the J - OE method}} \tag{5.90}$$

It was found that for the FD algebraic equations, the SOR method is much more efficient than the G-S and Jacobi methods. As the mesh size increases, the ratio of iterations *rk* decreases and tends to a constant. The G-S method is approximately two times faster than the Jacobi method. By comparing the SOR results with the G-S results, it was found that the relaxation factor has a great effect on the convergence rate. For the sample problem, the optimal relaxation factor ω was found to be 1.8 through a trial-and-error process. With $\omega = 1.8$, the SOR method is approximately 7.8 times faster than the G-S method when the mesh size is large. Through numerical experiments, it was found that for the SOR method, the ratio of iterations *rk* tends to 0.067, while for the G-S method, the ratio of iterations *rk* tends to 0.525.

When the above iterative methods are applied to solve Equation 5.86, an interesting question is whether the performance of these methods is the same as that for the FD algebraic equations. Through extensive numerical experiments, it was found that, for the DQ algebraic equations, the J-OE method could only result in a converged solution when the relaxation factor is less than or equal to 1. The G-S method is approximately two times faster than the J-OE method. The SOR method is still more efficient than the J-OE and G-S methods. However, as compared to the cases for the FD algebraic equations, the ratio of iterations *rk* for the SOR method is reduced by about half. This may imply that the SOR method for the DQ algebraic equations is less sensitive to the relaxation factor than the FD algebraic equations. In other words, the effect of the relaxation factor for the DQ algebraic equations is reduced, and the SOR method for the DQ algebraic equations becomes less efficient.

In the work of Shu, Yeo and Yao (1998), the DQ method is also compared with the second order FD scheme in terms of accuracy and efficiency. With the convergence criterion given by Equation 5.88, it was found that when a numerical method (FD or DQ) is chosen and a grid point distribution is given, the values of $\Delta u_{\max}$ defined by

$$\Delta u_{\max} = \max_{i,j} \left| u_{i,j}^{exact} - u_{i,j}^{\text{numerical}} \right| \tag{5.91}$$

for different iterative methods are almost the same. This means that the accuracy is mainly dependent on the numerical method (FD or DQ) and the grid points used. The values of $\Delta u_{\max}$ computed by the DQ and second order FD schemes are listed in Table 5.2. It can be observed from the table that the DQ method can generate a very accurate solution using very few grid points. In contrast, to reach the same order of accuracy, the second order FD scheme requires a large number of grid points, which costs a lot of CPU time and virtual storage. For example, if we want to obtain the solution with $\Delta u_{\max} \leq 10^{-5}$, the DQ method needs 7x7 grid points, but the second order FD scheme requires about 65x65 grid points. It can be seen from Table 5.2 that the DQ results can achieve a high degree of accuracy very quickly. However, the improvement of accuracy by the second order FD scheme is very slow. Using two extra mesh sizes of 65×65 and 95×95 for the computation by the second order FD scheme, it was found that the improvement in accuracy is not promising.

Table 5.2 Comparison of $\Delta u_{\max}$ computed by DQ and 2nd FD schemes (Shu, Yeo and Yao 1998)

Mesh	7×7	11×11	15×15	21×21	65×65	95×95
2nd *FD*	5.79×10^{-4}	2.76×10^{-4}	1.56×10^{-4}	8.34×10^{-5}	9.45×10^{-6}	4.49×10^{-6}
DQ	3.64×10^{-6}	3.79×10^{-11}	9.72×10^{-13}	3.52×10^{-13}	-	-

5.5.3 Helmholtz Eigenvalue Problem

The third sample problem involves the computation of the eigenvalues of the two-dimensional Helmholtz equation with periodic and non-periodic boundary conditions. The two-dimensional Helmholtz equation can be written as

$$\nabla^2 u + k^2 u = 0 \quad \text{in } \Omega \tag{5.92}$$

where ∇^2 is the Laplacian operator, u and k are respectively the potential and the wave number defined in the two-dimensional domain Ω surrounded by the boundary Γ. Two types of boundary conditions on Γ are considered,

$$\text{Dirichlet type: } u = 0 \tag{5.93a}$$

$$\text{Neumann type: } \frac{\partial u}{\partial n} = 0 \tag{5.93b}$$

By applying the DQ method, Equation 5.92 can be discretized in the Cartesian coordinate system as

$$\sum_{k=1}^{N} c_{i,k}^{(2)} \cdot u_{k,j} + \sum_{k=1}^{M} \bar{c}_{j,k}^{(2)} \cdot u_{i,k} = -k^2 u_{i,j} \tag{5.94}$$

where N and M are respectively the numbers of grid points in the x and y directions, $c_{i,k}^{(2)}$ and $\bar{c}_{j,k}^{(2)}$ are the weighting coefficients of the second order derivatives with respect to x and y respectively. Similarly, the derivative in the Neumann boundary condition can be discretized by the DQ method. Applying Equation 5.94 at all the interior points with the given boundary conditions gives the following eigenvalue equation system

$$[A]\cdot\{u\} = -k^2\{u\} \tag{5.95}$$

From the above equation, the k values (wavenumbers) can be obtained from the eigenvalues of matrix $[A]$. The eigenvalues of matrix $[A]$ can be sorted out by *QR* algorithm.

Non-Periodic Problems: Consider a rectangular domain of length $L_x = 1$ and width $L_y = 0.6$ with two kinds of boundary conditions at all boundaries: Dirichlet type and Neumann type. For this problem, the analytical solution can be expressed as

$$k = \pi\sqrt{(m/L_x)^2 + (n/L_y)^2}\,, \qquad m,n = 0,1,2,\ldots \tag{5.96}$$

The m, n in above equation are taken as $m,n = 1,2,3,\ldots$ for the Dirichlet type condition, and $m,n = 0,1,2,\ldots$ for the Neumann type condition. We will use this problem to study the performance of the PDQ, FDQ, CDQ and SDQ approaches. The PDQ, FDQ, and SDQ methods are applied for the Dirichlet type condition, while the PDQ, FDQ, and CDQ methods are employed for the Neumann type condition. For the PDQ application, Grid I introduced in Chapter 4 is used to generate the coordinates of the grid points, which can be written as

$$x_i = \frac{1}{2}\left(1 - \cos\frac{i-1}{N-1}\pi\right)L_x, \quad i = 1,2,\ldots,N \tag{5.97a}$$

$$y_j = \frac{1}{2}\left(1 - \cos\frac{j-1}{M-1}\pi\right)L_y, \quad j = 1,2,\ldots,M \tag{5.97b}$$

As discussed in Chapter 3, the coordinates of the grid points for CDQ, SDQ and FDQ with a non-periodic condition should be in the range of $[0, \pi]$. Therefore, the rectangular domain should be transformed into a square of $0 \le x \le \pi$, $0 \le y \le \pi$ before these approaches can be applied. Then in the computational domain, a uniform mesh is used for the FDQ, CDQ and SDQ applications.

Table 5.3 shows the absolute errors of the first ten wavenumbers for the Neumann condition, computed using the PDQ, FDQ, and CDQ approaches. It is observed that the accuracy of the computed wavenumbers by the PDQ method is different from the other approaches. The computed values for the lower wavenumbers have a better accuracy than those for the higher wavenumbers. Moreover, as the mesh size is refined, the accuracy of PDQ results is improved. When the mesh size is taken to be 13×13, the computed wavenumbers for the first 10 modes by the PDQ method are very accurate. In contrast, all the computed wavenumbers by the FDQ and CDQ methods have the same order of accuracy.

However, when the mesh size is taken as 5×5 for the FDQ, and 3×3 for the CDQ computation, there are some missing wavenumbers as denoted by a dash '-' in the table. For the FDQ computation, the discretization of the governing equation is applied at all the interior points, and therefore, the dimension of the eigenvalue system is (N-2) by (M-2), giving (N-2)×(M-2) eigenvalues. So, a mesh size of 5×5 can only generate 9 eigenvalues using the FDQ method. As $k = 0$ is not shown, only 8 numerical values are available, which are not in strict ascending order. Unlike the FDQ method, the CDQ discretization of the governing equation is applied at all the points including the boundary points. So, the dimension of the eigenvalue system is N by M, which provides $N \times M$ eigenvalues. Thus, a mesh size of 3×3 can also generate 9 eigenvalues using the CDQ method. Again, these eigenvalues are not in ascending order. As a result, there are some missing eigenvalues. Once the mesh size is increased to the one which can provide 11 eigenvalues, the computed FDQ and CDQ wavenumbers for the first ten modes are very accurate. Table 5.3 also shows that the CDQ method requires fewer mesh points to generate accurate numerical results.

Table 5.3 Absolute errors of PDQ, FDQ, CDQ results for Neumann condition

k	*PDQ*		*FDQ*		*CDQ*	
	9×9	13×13	5×5	9×9	3×3	5×5
k_1	1.31×10^{-5}	4.34×10^{-10}	4.00×10^{-15}	3.60×10^{-14}	2.22×10^{-15}	2.22×10^{-14}
k_2	2.19×10^{-5}	7.25×10^{-10}	8.88×10^{-15}	1.04×10^{-13}	7.11×10^{-15}	2.66×10^{-15}
k_3	2.55×10^{-5}	8.45×10^{-10}	6.22×10^{-15}	1.07×10^{-14}	2.66×10^{-15}	2.04×10^{-14}
k_4	1.14×10^{-3}	4.11×10^{-7}	2.66×10^{-15}	1.71×10^{-13}	0.0	1.95×10^{-14}
k_5	8.87×10^{-4}	3.16×10^{-7}	1.78×10^{-15}	1.95×10^{-14}	1.78×10^{-15}	7.11×10^{-15}
k_6	4.70×10^{-2}	1.45×10^{-4}	-	1.60×10^{-14}	-	1.42×10^{-14}
k_7	1.89×10^{-3}	6.85×10^{-7}	3.55×10^{-15}	2.84×10^{-14}	1.78×10^{-15}	3.55×10^{-15}
k_8	4.11×10^{-2}	1.27×10^{-4}	-	7.11×10^{-15}	-	7.11×10^{-15}
k_9	1.82×10^{-3}	6.56×10^{-7}	5.33×10^{-15}	8.35×10^{-14}	1.78×10^{-15}	5.33×10^{-15}
k_{10}	2.21×10^{-3}	7.99×10^{-7}	3.55×10^{-15}	1.74×10^{-13}	5.33×10^{-15}	7.11×10^{-15}

The absolute errors of the first ten wavenumbers for the Dirichlet condition, computed by the PDQ, FDQ, and SDQ approaches, are listed in Table 5.4. The behavior of the PDQ results for the Dirichlet condition is very similar to that for the Neumann condition. That is, the lower computed wavenumbers have a better accuracy than the higher computed wavenumbers. The behavior of the FDQ results for the Dirichlet condition is slightly different from that for the Neumann condition. When the mesh size is not sufficiently large (7×7 in Table 5.4), the computed values for the lower wavenumbers are very accurate while those computed for the higher wavenumbers are quite erroneous. When the mesh size is sufficiently large (11×11 in Table 5.4), the computed wavenumbers for the first ten modes have the same high order of accuracy. As compared to the Neumann condition, both the PDQ and FDQ methods require more mesh points to obtain numerical results with the same order of accuracy for the Dirichlet condition. Unlike the performance of the CDQ method for the Neumann condition, the performance of the SDQ method for the Dirichlet condition is not promising. As shown in Table 5.4, even when the mesh size is taken as 21×21, the accuracy of the SDQ results is still not good enough. The low order accuracy of the SDQ results is probably due to the computation of the weighting coefficients by the SDQ formulations. As shown in Chapter 3, the computation of the weighting coefficients by the SDQ method involves the division by $\sin x_i$. At some mesh points, the value of $\sin x_i$ is very small, which can lead to large round-off errors. As a consequence, the computation of the weighting coefficients is not very accurate.

Table 5.4 Absolute errors of PDQ, FDQ, SDQ results for Dirichlet condition

k	*PDQ*		*FDQ*		*SDQ*	
	9×9	15×15	7×7	11×11	11×11	21×21
k_1	1.40×10^{-7}	1.12×10^{-13}	1.78×10^{-15}	5.42×10^{-14}	5.15×10^{-14}	1.32×10^{-13}
k_2	1.11×10^{-4}	2.60×10^{-10}	0.0	1.42×10^{-14}	7.89×10^{-3}	1.23×10^{-3}
k_3	3.93×10^{-3}	3.75×10^{-8}	3.55×10^{-15}	1.42×10^{-14}	1.76×10^{-2}	2.78×10^{-3}
k_4	2.32×10^{-4}	5.40×10^{-10}	1.07×10^{-14}	5.51×10^{-14}	1.64×10^{-2}	2.55×10^{-3}
k_5	2.82×10^{-4}	6.58×10^{-10}	5.33×10^{-15}	1.60×10^{-14}	2.00×10^{-2}	3.10×10^{-3}
k_6	1.53×10^{-1}	5.79×10^{-6}	8.92×10^{-1}	3.55×10^{-14}	5.81×10^{-2}	8.91×10^{-3}
k_7	2.83×10^{-3}	2.91×10^{-8}	8.69×10^{-1}	5.33×10^{-15}	2.62×10^{-2}	4.10×10^{-3}
k_8	2.93×10^{-1}	7.02×10^{-8}	1.93×10^{0}	7.11×10^{-15}	3.28×10^{-2}	5.19×10^{-3}
k_9	3.46×10^{-1}	4.82×10^{-6}	7.35×10^{-1}	6.75×10^{-14}	5.93×10^{-2}	9.12×10^{-3}
k_{10}	7.22×10^{-2}	2.38×10^{-5}	5.39×10^{-1}	4.26×10^{-14}	9.63×10^{-2}	1.52×10^{-2}

Since the FDQ method performs equally well for both the Neumann and Dirichlet conditions, we recommend the use of the FDQ method for practical applications. For the sample problem considered, the exact solution has a harmonic behavior. Hence, when the same mesh size is applied, the accuracy of FDQ results is much better than the PDQ results. This may not be true for other problems, which do not behave harmonically.

Periodic Problems: As discussed in Chapter 3, the FDQ method can also be applied to periodic problems when the interval is $[0, 2\pi]$. The first such application was given in Shu and Chew (1997). They considered the following one-dimensional problem

$$d^2u/dx^2 + k^2u = 0 \tag{5.98}$$

over the interval [0, 2π] to study the performance of the FDQ method for the periodic condition. This problem has been used by Schroeder (1994) to study the spurious modes in the finite element solutions. Obviously, the exact solution for this problem is $k = 0, \pm 1, \pm 2, \pm 3, \ldots$. Both the PDQ and FDQ methods are applied to solve this problem. The PDQ method uses Equation 5.97*a* to generate the grid point distribution with $L_x = 2\pi$ while the FDQ method uses the following formulation

$$x_i = \frac{i-1}{N}2\pi, \quad i = 1, 2, \ldots, N \tag{5.99}$$

to determine the coordinates of the grid points. When Equation 5.99 is applied, the periodic condition is automatically satisfied by the FDQ method. As for the PDQ method, the periodic condition is implemented in such a way that the continuity of the function and its first order derivative is enforced at $x = 0$ and $x = 2\pi$. Table 5.5 shows the computed FDQ and PDQ non-negative eigenvalues of Equation 5.98. The FDQ results are obtained using 15 grid points and the PDQ results are obtained using 25 grid points. Also included in the table are the analytical solution and the Galerkin finite element solution of Schroeder (1994) using 50 elements. It should be indicated that by using the Galerkin finite element method, there are some spurious results, which are not shown in the table. In this study, no spurious results were found in the FDQ and PDQ solutions. It is observed from the table that the FDQ results match the analytical solution exactly and the PDQ results are more accurate than the Galerkin finite element results even though fewer grid points are used in the PDQ solution. From this example, we may conclude that the FDQ method is very suitable for periodic problems.

Table 5.5 Non-negative eigenvalues of one-dimensional periodic problem (Shu and Chew 1997)

Methods	*N*	*Non-negative eigenvalues*					
FDQ	15	1.00000	2.00000	3.00000	4.00000	5.00000	6.00000
PDQ	25	1.00000	2.00000	3.00000	4.00000	5.00001	6.00328
FEM	50	1.00000	1.99996	2.99966	3.99854	4.99546	5.98847
Exact		1	2	3	4	5	6

Chapter 6

Application of Differential Quadrature Method to Solve Incompressible Navier-Stokes Equations

6.1 Introduction

The basic theory and properties of the DQ method have been described in the previous chapters. In this chapter, we will show how the DQ method can be applied to solve the incompressible Navier-Stokes equations.

In engineering, many fluid flow problems such as those encountered in low speed aerodynamics, industrial channel flows and hydraulics can be considered as problems of incompressible flows. If a Newtonian fluid is considered, these flows are governed, in mathematical terms, by the incompressible Navier-Stokes equations. For detailed derivations and physical interpretations of these equations, the reader is advised to consult the reference books written by Streeter et al. (1998), White (1999), Anderson (1995) and Hirsch (1988).

The incompressible Navier-Stokes equations are a mixed set of elliptic-parabolic equations which can be written in several forms. One of the most popular methods for solving the two-dimensional incompressible Navier-Stokes equations is through the use of the vorticity-stream function formulation. This approach reduces the original equations to a transport equation for the vorticity ω, and a Poisson equation for the stream function ψ. However, in the vorticity-stream function formulation, the implementation of the boundary conditions for ψ is not straightforward. For example, two physical boundary conditions for the velocity u, v can give two boundary conditions for ψ, but conventional techniques normally use only one boundary condition (Dirichlet type) for ψ in the calculation. When the DQ method is applied to solve the vorticity-stream function formulation, special attention should be paid to the implementation of the boundary condition for ψ. Shu and his colleagues (1990, 1991, 1992*a, b*, 1994) have made a lot of contributions in this aspect. Some of the techniques for solving the vorticity-stream function formulation will be shown in this chapter.

Although the use of the vorticity-stream function formulation is very efficient for two-dimensional problems, it loses its attractiveness when applied to three-dimensional flows because a single scalar stream function does not exist in that case. As a consequence, the primitive variable form is usually used for three-dimensional problems. However, the primitive variable form of the incompressible Navier-Stokes equations cannot be solved in a straightforward way since there is evidently no transport or other equations for pressure. The momentum equation and the continuity equation are intricately coupled since the velocity appears in both the momentum and continuity equations but the pressure appears only in the momentum equation. To overcome these difficulties, Patankar and Spalding (1972) proposed an iterative process involving a pressure correction algorithm. In this algorithm, a guessed pressure field is used to solve the momentum equations. A pressure correction equation, deduced from the continuity equation, is solved to obtain the pressure correction field that is in turn used to update the velocity and pressure fields. The process continues until the velocity and pressure fields satisfy both the momentum and continuity equations. The pressure correction method is often applied with finite difference or finite volume discretization. The discretization is usually based on a staggered grid, in which the velocity components, pressure, density and temperature are defined at different mesh points. The use of a staggered grid ensures that the continuity equation is automatically satisfied on the boundary, but this unfortunately results in considerable large amount of programming effort. When the pressure correction method is applied with DQ discretization, there exist two major difficulties. First, since the DQ discretization is based on a non-staggered grid, the continuity equation on the boundary is not automatically satisfied. Some technique is needed to enforce this condition. Second, the boundary condition for the pressure correction has to be implemented properly. These two difficulties have to be overcome when pressure correction method is applied to the DQ computation in primitive variable form. In this chapter, we will show two approaches for solving the incompressible Navier-Stokes equations in primitive variable form.

6.2 Governing Equations

The basic equations of fluid dynamics are based on the universal laws of conservation, i.e., the conservation of mass, momentum and energy. The equation derived from applying the conservation law of mass to a fluid flow is usually called the continuity equation. The conservation law of momentum is based on Newton's second law, which yields a vector equation known as the momentum equation. The conservation law of energy is equivalent to the first law of thermodynamics and the resultant equation is called the energy equation. In the following section, we will describe the governing equations for incompressible viscous flows. These equations are actually the continuity, momentum and energy equations.

6.2.1 Dimensional Form

The incompressible flow has a feature that

$$\frac{D\rho}{Dt} = 0 \tag{6.1}$$

where $\frac{D}{Dt} = \frac{\partial}{\partial t} + u\frac{\partial}{\partial x} + v\frac{\partial}{\partial y} + w\frac{\partial}{\partial z}$. With this feature, the governing equations can be written as

continuity equation:

$$\nabla \cdot V = 0 \tag{6.2}$$

momentum equation:

$$\rho \frac{DV}{Dt} = -\nabla p + \mu \nabla^2 V + \rho f_e \tag{6.3}$$

energy equation:

$$\rho c \frac{DT}{Dt} = k\nabla^2 T + \phi + q_h \tag{6.4}$$

where $\nabla = \vec{i}\frac{\partial}{\partial x} + \vec{j}\frac{\partial}{\partial y} + \vec{k}\frac{\partial}{\partial z}$, ∇^2 is the Laplacian operator, μ is the dynamic viscosity of the fluid, V is the velocity vector, f_e is the external volume force, q_h is the external volume heat source, c is the coefficient of specific heat, k is the coefficient of thermal conductivity and ϕ is the rate of dissipation of mechanical energy defined by

$$\phi = \mu \left[2\left(\frac{\partial u}{\partial x}\right)^2 + 2\left(\frac{\partial v}{\partial y}\right)^2 + 2\left(\frac{\partial w}{\partial z}\right)^2 + \left(\frac{\partial v}{\partial x} + \frac{\partial u}{\partial y}\right)^2 + \right.$$
$$\left. \left(\frac{\partial v}{\partial z} + \frac{\partial w}{\partial y}\right)^2 + \left(\frac{\partial u}{\partial z} + \frac{\partial w}{\partial x}\right)^2 - \frac{2}{3}\left(\frac{\partial u}{\partial x} + \frac{\partial v}{\partial y} + \frac{\partial w}{\partial z}\right)^2 \right]$$

ϕ is usually small in incompressible flows, thus it can be neglected in Equation 6.4. Only cases whereby there is no external volume heat source, i.e., $q_h = 0$, will be studied. So the energy equation can be reduced to

$$\rho c \frac{DT}{Dt} = k\nabla^2 T \tag{6.5}$$

In most cases, the temperature appears only in the energy equation so that we can decouple this equation from the continuity and momentum equations. For many applications, the temperature changes are either insignificant or unimportant and it is not necessary to solve the energy equation. Therefore, for general incompressible flows, the governing equations (from Equations 6.2 and 6.3) in the Cartesian coordinate system can be written as

$$\frac{\partial u}{\partial x}+\frac{\partial v}{\partial y}+\frac{\partial w}{\partial z}=0 \tag{6.6a}$$

$$\frac{\partial u}{\partial t}+u\frac{\partial u}{\partial x}+v\frac{\partial u}{\partial y}+w\frac{\partial u}{\partial z}=-\frac{1}{\rho}\frac{\partial p}{\partial x}+\upsilon\left(\frac{\partial^2 u}{\partial x^2}+\frac{\partial^2 u}{\partial y^2}+\frac{\partial^2 u}{\partial z^2}\right) \tag{6.6b}$$

$$\frac{\partial v}{\partial t}+u\frac{\partial v}{\partial x}+v\frac{\partial v}{\partial y}+w\frac{\partial v}{\partial z}=-\frac{1}{\rho}\frac{\partial p}{\partial y}+\upsilon\left(\frac{\partial^2 v}{\partial x^2}+\frac{\partial^2 v}{\partial y^2}+\frac{\partial^2 v}{\partial z^2}\right) \tag{6.6c}$$

$$\frac{\partial w}{\partial t}+u\frac{\partial w}{\partial x}+v\frac{\partial w}{\partial y}+w\frac{\partial w}{\partial z}=-\frac{1}{\rho}\frac{\partial p}{\partial z}+\upsilon\left(\frac{\partial^2 w}{\partial x^2}+\frac{\partial^2 w}{\partial y^2}+\frac{\partial^2 w}{\partial z^2}\right) \tag{6.6d}$$

where u, v, w are the velocity components in the x, y, z directions respectively, and $\upsilon=\mu/\rho$ is the kinematic viscosity.

On the other hand, for natural convection problems, the energy equation should be directly coupled with the momentum equations, because the buoyancy force, caused by the temperature difference, is the dominant driving force for the system. For natural convection problems, the external body force is the gravitational force, given by

$$\rho\, f_e=\rho\, \mathbf{g} \tag{6.7}$$

where $\mathbf{g}$ denotes the vector of acceleration due to gravity. It is common to use the Boussinesq approximation for the study of natural convection. The variation of density ρ is, according to the Boussinesq approximation, included only in the calculation of the buoyancy term (i.e., $\rho\mathbf{g}$), and is proportional to the temperature. The buoyancy force is then written as

$$\rho\, f_e=\rho_0\left[1-\beta(T-T_0)\right]\mathbf{g} \tag{6.8}$$

where ρ_0, T_0 are respectively the reference density and temperature, and β is the thermal expansion coefficient. Using Equation 6.8, Equation 6.3 can be simplified to

$$\rho_0\frac{D\mathbf{V}}{Dt}=-\nabla p+\mu\nabla^2\mathbf{V}+\rho_0\mathbf{g}\left[1-\beta(T-T_0)\right] \tag{6.9}$$

In the Cartesian coordinate system, if the coordinate axis y is chosen to be in the direction opposite to the gravity force, then

$$\mathbf{g} = (0, -g, 0) \tag{6.10}$$

As such, the governing equations for natural convection (from Equations 6.2-6.4) in the Cartesian coordinate system are written as

$$\frac{\partial u}{\partial x} + \frac{\partial v}{\partial y} + \frac{\partial w}{\partial z} = 0 \tag{6.11a}$$

$$\frac{\partial u}{\partial t} + u\frac{\partial u}{\partial x} + v\frac{\partial u}{\partial y} + w\frac{\partial u}{\partial z} = -\frac{1}{\rho_0}\frac{\partial p}{\partial x} + \upsilon\left(\frac{\partial^2 u}{\partial x^2} + \frac{\partial^2 u}{\partial y^2} + \frac{\partial^2 u}{\partial z^2}\right) \tag{6.11b}$$

$$\frac{\partial v}{\partial t} + u\frac{\partial v}{\partial x} + v\frac{\partial v}{\partial y} + w\frac{\partial v}{\partial z} = -\frac{1}{\rho_0}\frac{\partial p}{\partial y} + \upsilon\left(\frac{\partial^2 v}{\partial x^2} + \frac{\partial^2 v}{\partial y^2} + \frac{\partial^2 v}{\partial z^2}\right) - g[1 - \beta(T - T_0)] \tag{6.11c}$$

$$\frac{\partial w}{\partial t} + u\frac{\partial w}{\partial x} + v\frac{\partial w}{\partial y} + w\frac{\partial w}{\partial z} = -\frac{1}{\rho_0}\frac{\partial p}{\partial z} + \upsilon\left(\frac{\partial^2 w}{\partial x^2} + \frac{\partial^2 w}{\partial y^2} + \frac{\partial^2 w}{\partial z^2}\right) \tag{6.11d}$$

$$\frac{\partial T}{\partial t} + u\frac{\partial T}{\partial x} + v\frac{\partial T}{\partial y} + w\frac{\partial T}{\partial z} = \frac{k}{\rho_0 c}\left(\frac{\partial^2 T}{\partial x^2} + \frac{\partial^2 T}{\partial y^2} + \frac{\partial^2 T}{\partial z^2}\right) \tag{6.11e}$$

6.2.2 Non-dimensional Form

For general incompressible flows in which only the continuity and momentum equations are solved, the flow variables and coordinates can be non-dimensionalized using

$$\begin{cases} x^* = \dfrac{x}{L}, \quad y^* = \dfrac{y}{L}, \quad z^* = \dfrac{z}{L}, \quad t^* = \dfrac{t}{L/V_\infty} \\ u^* = \dfrac{u}{V_\infty}, \quad v^* = \dfrac{v}{V_\infty}, \quad w^* = \dfrac{w}{V_\infty}, \quad p^* = \dfrac{p}{\rho_\infty V_\infty^2} \end{cases} \tag{6.12}$$

where the non-dimensional variables are denoted by an asterisk, freestream conditions are denoted by ∞ and L is the reference length used in the Reynolds number

$$\mathrm{Re} = \frac{\rho_\infty V_\infty L}{\mu} \tag{6.13}$$

Substituting the above expressions into Equation 6.6 yields

$$\frac{\partial u^*}{\partial x^*} + \frac{\partial v^*}{\partial y^*} + \frac{\partial w^*}{\partial z^*} = 0 \tag{6.14a}$$

$$\frac{\partial u^*}{\partial t^*} + u^* \frac{\partial u^*}{\partial x^*} + v^* \frac{\partial u^*}{\partial y^*} + w^* \frac{\partial u^*}{\partial z^*} = -\frac{\partial p^*}{\partial x^*} + \frac{1}{\mathrm{Re}}\left(\frac{\partial^2 u^*}{\partial x^{*2}} + \frac{\partial^2 u^*}{\partial y^{*2}} + \frac{\partial^2 u^*}{\partial z^{*2}}\right) \tag{6.14b}$$

$$\frac{\partial v^*}{\partial t^*} + u^* \frac{\partial v^*}{\partial x^*} + v^* \frac{\partial v^*}{\partial y^*} + w^* \frac{\partial v^*}{\partial z^*} = -\frac{\partial p^*}{\partial y^*} + \frac{1}{\mathrm{Re}}\left(\frac{\partial^2 v^*}{\partial x^{*2}} + \frac{\partial^2 v^*}{\partial y^{*2}} + \frac{\partial^2 v^*}{\partial z^{*2}}\right) \tag{6.14c}$$

$$\frac{\partial w^*}{\partial t^*} + u^* \frac{\partial w^*}{\partial x^*} + v^* \frac{\partial w^*}{\partial y^*} + w^* \frac{\partial w^*}{\partial z^*} = -\frac{\partial p^*}{\partial z^*} + \frac{1}{\mathrm{Re}}\left(\frac{\partial^2 w^*}{\partial x^{*2}} + \frac{\partial^2 w^*}{\partial y^{*2}} + \frac{\partial^2 w^*}{\partial z^{*2}}\right) \tag{6.14d}$$

For natural convection problems, there is no freestream velocity, but V_∞ can be taken as $V_\infty = k/(\rho_0 cL)$. Setting $\rho_\infty = \rho_0$ and $T^* = (T - T_0)/\Delta T$, Equation 6.11 can be non-dimensionalized as

$$\frac{\partial u^*}{\partial x^*} + \frac{\partial v^*}{\partial y^*} + \frac{\partial w^*}{\partial z^*} = 0 \tag{6.15a}$$

$$\frac{\partial u^*}{\partial t^*} + u^* \frac{\partial u^*}{\partial x^*} + v^* \frac{\partial u^*}{\partial y^*} + w^* \frac{\partial u^*}{\partial z^*} = -\frac{\partial p^*}{\partial x^*} + \mathrm{Pr}\left(\frac{\partial^2 u^*}{\partial x^{*2}} + \frac{\partial^2 u^*}{\partial y^{*2}} + \frac{\partial^2 u^*}{\partial z^{*2}}\right) \tag{6.15b}$$

$$\begin{aligned}\frac{\partial v^*}{\partial t^*} + u^* \frac{\partial v^*}{\partial x^*} + v^* \frac{\partial v^*}{\partial y^*} + w^* \frac{\partial v^*}{\partial z^*} &= -\frac{\partial p^*}{\partial y^*} \\ &+ \mathrm{Pr}\left(\frac{\partial^2 v^*}{\partial x^{*2}} + \frac{\partial^2 v^*}{\partial y^{*2}} + \frac{\partial^2 v^*}{\partial z^{*2}}\right) - \frac{\mathrm{Pr}\, Ra}{\beta \Delta T}(1 - \beta \Delta T \cdot T^*)\end{aligned} \tag{6.15c}$$

$$\frac{\partial w^*}{\partial t^*} + u^* \frac{\partial w^*}{\partial x^*} + v^* \frac{\partial w^*}{\partial y^*} + w^* \frac{\partial w^*}{\partial z^*} = -\frac{\partial p^*}{\partial z^*} + \mathrm{Pr}\left(\frac{\partial^2 w^*}{\partial x^{*2}} + \frac{\partial^2 w^*}{\partial y^{*2}} + \frac{\partial^2 w^*}{\partial z^{*2}}\right) \tag{6.15d}$$

$$\frac{\partial T^*}{\partial t^*} + u^* \frac{\partial T^*}{\partial x^*} + v^* \frac{\partial T^*}{\partial y^*} + w^* \frac{\partial T^*}{\partial z^*} = \frac{\partial^2 T^*}{\partial x^{*2}} + \frac{\partial^2 T^*}{\partial y^{*2}} + \frac{\partial^2 T^*}{\partial z^{*2}} \tag{6.15e}$$

where

$$\Pr = \frac{\mu c}{k}, \quad Ra = \frac{\rho_0^2 c g \beta L^3 \Delta T}{k\mu}$$

For simplicity, the asterisks in Equations 6.14 and 6.15 will be dropped hereafter. Equation 6.14 or 6.15 is called the primitive variable form of the incompressible Navier-Stokes equations.

6.2.3 Vorticity-Stream Function Formulation

For two-dimensional problems, the use of the vorticity-stream function formulation can greatly save computational effort and virtual storage. This is because in the vorticity-stream function formulation, the dependent variables are the vorticity ω and stream function ψ, while in the primitive variable form, the dependent variables are the velocity components u, v and the pressure p. Thus, only two differential equations are solved in the vorticity-stream function formulation. The vorticity and stream function are defined as

$$\omega = \frac{\partial u}{\partial y} - \frac{\partial v}{\partial x} \tag{6.16}$$

$$u = \frac{\partial \psi}{\partial y}, \quad v = -\frac{\partial \psi}{\partial x} \tag{6.17}$$

It is noted that with Equation 6.17, the continuity equation is automatically satisfied. When Equation 6.17 is substituted into Equation 6.16, we can obtain the following equation for the stream function

$$\frac{\partial^2 \psi}{\partial x^2} + \frac{\partial^2 \psi}{\partial y^2} = \omega \tag{6.18}$$

The vorticity transport equation can be derived from the two momentum equations. From Equation 6.14 for a general two-dimensional incompressible flow, the vorticity equation is given as

$$\frac{\partial \omega}{\partial t} + u\frac{\partial \omega}{\partial x} + v\frac{\partial \omega}{\partial y} = \frac{1}{\mathrm{Re}}\left(\frac{\partial^2 \omega}{\partial x^2} + \frac{\partial^2 \omega}{\partial y^2}\right) \tag{6.19}$$

In addition, from Equation 6.15 for two-dimensional problems involving natural convection, the vorticity equation can be written as

$$\frac{\partial \omega}{\partial t} + u\frac{\partial \omega}{\partial x} + v\frac{\partial \omega}{\partial y} = \Pr\left(\frac{\partial^2 \omega}{\partial x^2} + \frac{\partial^2 \omega}{\partial y^2}\right) - \Pr Ra\frac{\partial T}{\partial x} \tag{6.20}$$

The energy (temperature) equation in the vorticity-stream function formulation remains the same as the one in primitive variable form.

6.3 Solution of Vorticity-Stream Function Formulation

In this section, we will show how the DQ method can be directly applied to solve the vorticity-stream function formulation. The natural convection occurring in a square cavity is taken as an example to illustrate the solution procedure. A Fortran program using the PDQ method to solve this problem is attached in Appendix A.

6.3.1 Discretization of Governing Equations

As shown in the previous section, the governing equations for the natural convection in a square cavity are given by Equations 6.18, 6.20 and two-dimensional version of Equation 6.15*e*. The computational domain is given by $0 \le x \le 1$, $0 \le y \le 1$. For the numerical discretization, we let *N* and *M* be the total number of grid points in the *x* and *y* directions respectively. As shown in Figure 6.1, the value of a function *f* at a mesh point (x_i, y_j) is represented by $f_{i,j}$. According to the DQ discretization, all the derivatives of the function *f* with respect to *x* at the mesh point (x_i, y_j) are approximated by a linear summation of the functional values at all mesh points along the line y_j which are denoted by the symbol "O" in Figure 6.1. Similarly, all the derivatives of the function *f* with respect to *y* at the mesh point (x_i, y_j) are discretized by a linear summation of the functional values at all mesh points along the line x_i which are denoted by the symbol " " in Figure 6.1.

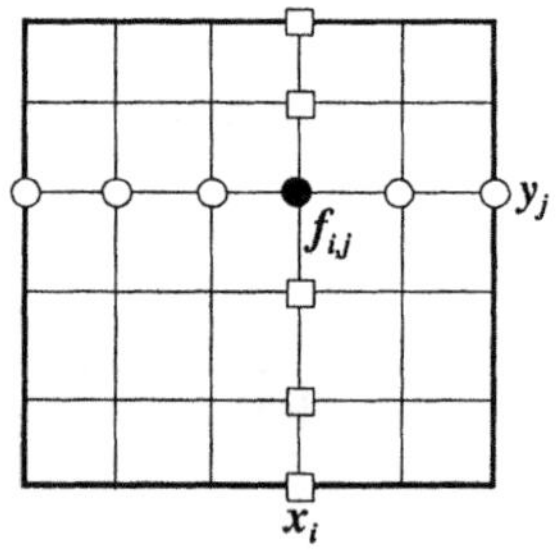

Figure 6.1 Illustration of functional value at a mesh point

Let $c_{i,j}^{(n)}$ be the weighting coefficients of the nth order derivative in the x direction and $\bar{c}_{i,j}^{(m)}$ be the weighting coefficients of the mth order derivative in the y direction. The three governing equations can be discretized by the DQ method as

$$\sum_{k=1}^{N} c_{i,k}^{(2)} \cdot \psi_{k,j} + \sum_{k=1}^{M} \bar{c}_{j,k}^{(2)} \cdot \psi_{i,k} = \omega_{i,j} \tag{6.21}$$

$$\begin{aligned} &\frac{d\omega_{i,j}}{dt} + u_{i,j} \cdot \sum_{k=1}^{N} c_{i,k}^{(1)} \cdot \omega_{k,j} + v_{i,j} \cdot \sum_{k=1}^{M} \bar{c}_{j,k}^{(1)} \cdot \omega_{i,k} \\ &\qquad = \Pr \cdot \left[\sum_{k=1}^{N} c_{i,k}^{(2)} \cdot \omega_{k,j} + \sum_{k=1}^{M} \bar{c}_{j,k}^{(2)} \cdot \omega_{i,k} \right] - Ra \cdot \Pr \cdot \sum_{k=1}^{N} c_{i,k}^{(1)} \cdot T_{k,j} \end{aligned} \tag{6.22}$$

$$\begin{aligned} &\frac{dT_{i,j}}{dt} + u_{i,j} \cdot \sum_{k=1}^{N} c_{i,k}^{(1)} \cdot T_{k,j} + v_{i,j} \cdot \sum_{k=1}^{M} \bar{c}_{j,k}^{(1)} \cdot T_{i,k} \\ &\qquad = \sum_{k=1}^{N} c_{i,k}^{(2)} \cdot T_{k,j} + \sum_{k=1}^{M} \bar{c}_{j,k}^{(2)} \cdot T_{i,k} \end{aligned} \tag{6.23}$$

6.3.2 Implementation of Boundary Conditions

The physical boundary conditions for natural convection occurring in a square cavity are shown in Figure 6.2, and are given by

$$u = v = 0,\ T = 1, \text{ at } x = 0,\ 0 \le y \le 1 \tag{6.24a}$$

$$u = v = 0,\ T = 0, \text{ at } x = 1,\ 0 \le y \le 1 \tag{6.24b}$$

$$u = v = \frac{\partial T}{\partial y} = 0, \text{ at } y = 0,\ 1,\ 0 < x < 1 \tag{6.24c}$$

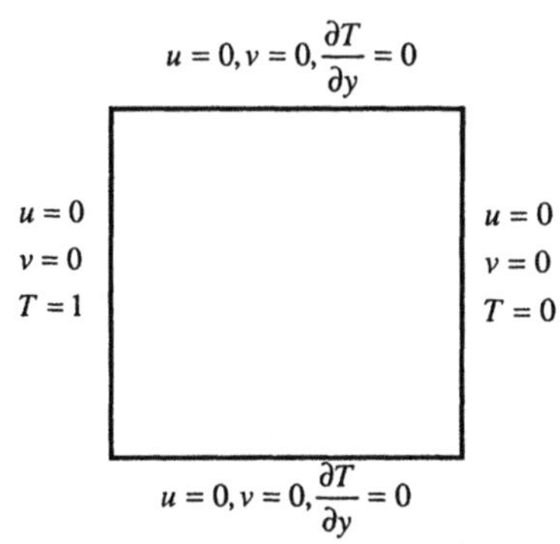

Figure 6.2 Configuration of natural convection in a square cavity

Using Equation 6.17, the boundary conditions for the two velocity components u and v can be converted to two boundary conditions for the stream function, that is

$$\psi = \frac{\partial \psi}{\partial x} = 0, \text{ at } x = 0,\ 1,\ 0 \le y \le 1 \tag{6.25a}$$

$$\psi = \frac{\partial \psi}{\partial y} = 0, \text{ at } y = 0,\ 1,\ 0 < x < 1 \tag{6.25b}$$

The derivatives which appear in the boundary conditions can also be discretized using the DQ method in a similar manner as described in Section 6.3.1.

6.3.2.1 Implementation of Boundary Condition for Vorticity

Physically, there is no boundary condition for the vorticity. But an expression can be obtained from Equation 6.18. With the boundary conditions of Equation 6.25, Equation 6.18 gives

$$\omega = \frac{\partial^2 \psi}{\partial x^2}, \text{ at } x = 0,\ 1,\ 0 \le y \le 1 \tag{6.26a}$$

$$\omega = \frac{\partial^2 \psi}{\partial y^2}, \text{ at } y = 0,\ 1,\ 0 < x < 1 \tag{6.26b}$$

After discretization of the derivatives using the DQ method, Equation 6.26 is simplified, on the four boundaries, to become

$$\omega_{1,j} = \sum_{k=1}^{N} c_{1,k}^{(2)} \cdot \psi_{k,j}, \text{ for } j = 1, 2, \ldots, M \tag{6.27a}$$

$$\omega_{N,j} = \sum_{k=1}^{N} c_{N,k}^{(2)} \cdot \psi_{k,j}, \text{ for } j = 1, 2, \ldots, M \tag{6.27b}$$

$$\omega_{i,1} = \sum_{k=1}^{M} \bar{c}_{1,k}^{(2)} \cdot \psi_{i,k}, \text{ for } i = 2, 3, \ldots, N-1 \tag{6.27c}$$

$$\omega_{i,M} = \sum_{k=1}^{M} \bar{c}_{M,k}^{(2)} \cdot \psi_{i,k}, \text{ for } i = 2, 3, \ldots, N-1 \tag{6.27d}$$

Note that $i = 1$ is the left boundary and $i = N$ is the right boundary, while $j = 1$ and $j = M$ correspond to the bottom and top boundaries respectively.

6.3.2.2 Implementation of Boundary Condition for Stream Function

In this section, two approaches for the implementation of the boundary condition for the stream function will be presented. As shown in Equation 6.25, there are two boundary conditions for ψ at each boundary. Numerically, the Dirichlet condition can be easily implemented by

$$\psi_{1,j} = 0, \quad \psi_{N,j} = 0, \quad \psi_{i,1} = 0, \quad \psi_{i,M} = 0 \tag{6.28}$$

$$\text{for } j = 1, 2, ..., M; \; i = 2, 3, ..., N-1$$

Equation 6.28 is applied by two approaches exactly at the boundary point. The Neumann condition is treated differently by two approaches which are shown as follows.

Two-layer approach: For this approach, Equation 6.28 gives the first layer condition at all boundary points, and the Neumann condition in Equation 6.25 is discretized by the DQ method and then combined to provide conditions for the second layer. Applying the DQ method to discretize the derivatives in the Neumann condition gives

$$c_{1,2}^{(1)} \cdot \psi_{2,j} + \sum_{k=1,k\neq 2,N-1}^{N} c_{1,k}^{(1)} \cdot \psi_{k,j} + c_{1,N-1}^{(1)} \cdot \psi_{N-1,j} = 0 \tag{6.29a}$$

for the boundary of $x = 0$, and

$$c_{N,2}^{(1)} \cdot \psi_{2,j} + \sum_{k=1,k\neq 2,N-1}^{N} c_{N,k}^{(1)} \cdot \psi_{k,j} + c_{N,N-1}^{(1)} \cdot \psi_{N-1,j} = 0 \tag{6.29b}$$

for the boundary of $x = 1$, and

$$\bar{c}_{1,2}^{(1)} \cdot \psi_{i,2} + \sum_{k=1,k\neq 2,M-1}^{M} \bar{c}_{1,k}^{(1)} \cdot \psi_{i,k} + \bar{c}_{1,M-1}^{(1)} \cdot \psi_{i,M-1} = 0 \tag{6.29c}$$

for the boundary of $y = 0$, and

$$\bar{c}_{M,2}^{(1)} \cdot \psi_{i,2} + \sum_{k=1,k\neq 2,M-1}^{M} \bar{c}_{M,k}^{(1)} \cdot \psi_{i,k} + \bar{c}_{M,M-1}^{(1)} \cdot \psi_{i,M-1} = 0 \tag{6.29d}$$

for the boundary of $y = 1$.

Substituting Equation 6.28 into Equations 6.29*a* and 6.29*b*, we can obtain

$$\psi_{2,j} = \frac{1}{AXN} \cdot \left[\sum_{k=3}^{N-2} \left(c_{1,k}^{(1)} \cdot c_{N,N-1}^{(1)} - c_{N,k}^{(1)} \cdot c_{1,N-1}^{(1)} \right) \cdot \psi_{k,j} \right] \tag{6.30a}$$

$$\psi_{N-1,j} = \frac{1}{AXN} \cdot \left[\sum_{k=3}^{N-2} \left(c_{N,k}^{(1)} \cdot c_{1,2}^{(1)} - c_{1,k}^{(1)} \cdot c_{N,2}^{(1)} \right) \cdot \psi_{k,j} \right] \tag{6.30b}$$

for $j = 2, 3, \ldots, M-1$

where

$$AXN = c_{N,2}^{(1)} \cdot c_{1,N-1}^{(1)} - c_{1,2}^{(1)} \cdot c_{N,N-1}^{(1)}$$

Similarly, substituting Equation 6.28 into Equations 6.29*c* and 6.29*d* gives

$$\psi_{i,2} = \frac{1}{AYM} \cdot \left[\sum_{k=3}^{M-2} \left(\bar{c}_{1,k}^{(1)} \cdot \bar{c}_{M,M-1}^{(1)} - \bar{c}_{M,k}^{(1)} \cdot \bar{c}_{1,M-1}^{(1)} \right) \cdot \psi_{i,k} \right] \tag{6.31a}$$

$$\psi_{i,M-1} = \frac{1}{AYM} \cdot \left[\sum_{k=3}^{M-2} \left(\bar{c}_{M,k}^{(1)} \cdot \bar{c}_{1,2}^{(1)} - \bar{c}_{1,k}^{(1)} \cdot \bar{c}_{M,2}^{(1)} \right) \cdot \psi_{i,k} \right] \tag{6.31b}$$

for $i = 2, 3, \ldots, N-1$

where $AYM = \bar{c}_{M,2}^{(1)} \cdot \bar{c}_{1,M-1}^{(1)} - \bar{c}_{1,2}^{(1)} \cdot \bar{c}_{M,M-1}^{(1)}$. Equations 6.30 and 6.31 provide the solutions at a layer adjacent to the boundary. It is noted that when this approach is used, Equation 6.21 can only be applied for the interior points $3 \le i \le N-2$, $3 \le j \le M-2$. This approach has been successfully applied by Shu and Richards (1992*a, b*) to generate accurate DQ solutions.

One-layer approach: For this approach, only the Dirichlet condition (Equation 6.28) is implemented at the boundary points. The Neumann condition is built into the DQ weighting coefficient matrices, which will then be applied to discretize the stream function equation.

Consider the DQ analog of the derivative $\partial\psi/\partial x$ at the grid points on a line y_j parallel to the x-axis

$$\begin{Bmatrix} (\partial\psi/\partial x)_{1,j} \\ (\partial\psi/\partial x)_{2,j} \\ \cdot \\ \cdot \\ \cdot \\ (\partial\psi/\partial x)_{N-1,j} \\ (\partial\psi/\partial x)_{N,j} \end{Bmatrix} = \begin{bmatrix} c_{1,1}^{(1)} & c_{1,2}^{(1)} & \cdots & c_{1,N-1}^{(1)} & c_{1,N}^{(1)} \\ c_{2,1}^{(1)} & c_{2,2}^{(1)} & \cdots & c_{2,N-1}^{(1)} & c_{2,N}^{(1)} \\ \cdot & \cdot & \cdots & \cdot & \cdot \\ \cdot & \cdot & \cdots & \cdot & \cdot \\ \cdot & \cdot & \cdots & \cdot & \cdot \\ c_{N-1,1}^{(1)} & c_{N-1,2}^{(1)} & \cdots & c_{N-1,N-1}^{(1)} & c_{N-1,N}^{(1)} \\ c_{N,1}^{(1)} & c_{N,2}^{(1)} & \cdots & c_{N,N-1}^{(1)} & c_{N,N}^{(1)} \end{bmatrix} \begin{Bmatrix} \psi_{1,j} \\ \psi_{2,j} \\ \cdot \\ \cdot \\ \cdot \\ \psi_{N-1,j} \\ \psi_{N,j} \end{Bmatrix} \tag{6.32}$$

To satisfy the derivative condition of ψ (Equation 6.25*a*), Equation 6.32 can be modified by zeroing the first and last rows of the matrix

$$\begin{Bmatrix} (\partial\psi/\partial x)_{1,j} \\ (\partial\psi/\partial x)_{2,j} \\ \cdot \\ \cdot \\ \cdot \\ (\partial\psi/\partial x)_{N-1,j} \\ (\partial\psi/\partial x)_{N,j} \end{Bmatrix} = \begin{bmatrix} 0 & 0 & \cdot\cdot\cdot & 0 & 0 \\ c_{2,1}^{(1)} & c_{2,2}^{(1)} & \cdot\cdot\cdot & c_{2,N-1}^{(1)} & c_{2,N}^{(1)} \\ \cdot & \cdot & \cdot\cdot\cdot & \cdot & \cdot \\ \cdot & \cdot & \cdot\cdot\cdot & \cdot & \cdot \\ \cdot & \cdot & \cdot\cdot\cdot & \cdot & \cdot \\ c_{N-1,1}^{(1)} & c_{N-1,2}^{(1)} & \cdot\cdot\cdot & c_{N-1,N-1}^{(1)} & c_{N-1,N}^{(1)} \\ 0 & 0 & \cdot\cdot\cdot & 0 & 0 \end{bmatrix} \begin{Bmatrix} \psi_{1,j} \\ \psi_{2,j} \\ \cdot \\ \cdot \\ \cdot \\ \psi_{N-1,j} \\ \psi_{N,j} \end{Bmatrix} \tag{6.33}$$

The modified matrix in Equation 6.33 is noted as $[\tilde{A}^1]$ while the original matrix in Equation 6.32 is represented by $[A^1]$. The matrix form of Equation 6.33 can be written as

$$\{\partial\psi/\partial x\}_j = [\tilde{A}^1]\{\psi\}_j \tag{6.34}$$

Similarly, the matrix form of the DQ analog of the derivative $\partial^2\psi/\partial x^2$ at the grid points on a line y_j parallel to the *x*-axis can be written as

$$\{\partial^2\psi/\partial x^2\}_j = [\tilde{A}^2]\{\psi\}_j \tag{6.35}$$

where $[\tilde{A}^2]$ is the modified DQ weighting coefficient matrix for the second order derivative $\partial^2\psi/\partial x^2$. On the other hand, we note that the second order derivative $\partial^2\psi/\partial x^2$ can be obtained by differentiating the first order derivative $\partial\psi/\partial x$. Thus, the matrix $[\tilde{A}^2]$ can be computed by

$$[\tilde{A}^2] = [A^1]\cdot[\tilde{A}^1] \tag{6.36}$$

The above process can also be applied to modify the DQ weighting coefficient matrices in the *y* direction. Let $[B^1]$ and $[B^2]$ be the original DQ weighting coefficient matrices related to $\partial\psi/\partial y$ and $\partial^2\psi/\partial y^2$, $[\tilde{B}^1]$ and $[\tilde{B}^2]$ be the modified weighting coefficient matrices. Similar to Equation 6.36, we have

$$[\tilde{B}^2] = [B^1]\cdot[\tilde{B}^1] \tag{6.37}$$

where

$$[B^1] = \begin{bmatrix} \bar{c}^{(1)}_{1,1} & \bar{c}^{(1)}_{1,2} & \cdots & \bar{c}^{(1)}_{1,M-1} & \bar{c}^{(1)}_{1,M} \\ \bar{c}^{(1)}_{2,1} & \bar{c}^{(1)}_{2,2} & \cdots & \bar{c}^{(1)}_{2,M-1} & \bar{c}^{(1)}_{2,M} \\ \cdot & \cdot & \cdots & \cdot & \cdot \\ \cdot & \cdot & \cdots & \cdot & \cdot \\ \cdot & \cdot & \cdots & \cdot & \cdot \\ \bar{c}^{(1)}_{M-1,1} & \bar{c}^{(1)}_{M-1,2} & \cdots & \bar{c}^{(1)}_{M-1,M-1} & \bar{c}^{(1)}_{M-1,M} \\ \bar{c}^{(1)}_{M,1} & \bar{c}^{(1)}_{M,2} & \cdots & \bar{c}^{(1)}_{M,M-1} & \bar{c}^{(1)}_{M,M} \end{bmatrix}$$

$$[\tilde{B}^1] = \begin{bmatrix} 0 & 0 & \cdots & 0 & 0 \\ \bar{c}^{(1)}_{2,1} & \bar{c}^{(1)}_{2,2} & \cdots & \bar{c}^{(1)}_{2,M-1} & \bar{c}^{(1)}_{2,M} \\ \cdot & \cdot & \cdots & \cdot & \cdot \\ \cdot & \cdot & \cdots & \cdot & \cdot \\ \cdot & \cdot & \cdots & \cdot & \cdot \\ \bar{c}^{(1)}_{M-1,1} & \bar{c}^{(1)}_{M-1,2} & \cdots & \bar{c}^{(1)}_{M-1,M-1} & \bar{c}^{(1)}_{M-1,M} \\ 0 & 0 & \cdots & 0 & 0 \end{bmatrix}$$

It is noted that the modified DQ weighting coefficient matrices $[\tilde{A}^2]$ and $[\tilde{B}^2]$ are only applied to Equation 6.21. For other equations, the original DQ weighting coefficient matrices should be employed. Similar to the finite element method, when this approach is applied, the derivative (Neumann) condition is automatically and exactly satisfied. And since only the Dirichlet condition is implemented at the boundary points, Equation 6.21 has to be applied to the interior points, $2 \le i \le N-1$, $2 \le j \le M-1$. Shu and Xue (1998) compared this approach with the two-layer approach and found that both approaches can achieve almost the same level of accuracy but the two-layer approach requires less computational effort.

The one-layer approach can be implemented in a simple and efficient way. This approach directly substitutes the derivative conditions into the DQ discretization of the second order derivatives. To show this, we consider the discretization of $\partial^2\psi/\partial x^2$ at a mesh point (x_i, y_j). We know that

$$\frac{\partial^2\psi}{\partial x^2} = \frac{\partial}{\partial x}\left(\frac{\partial\psi}{\partial x}\right) \tag{6.38}$$

The DQ discretization of $\partial^2\psi/\partial x^2$ can hence be done in two steps. In the first step, we have

$$\left(\frac{\partial^2\psi}{\partial x^2}\right)_{i,j} = \sum_{k1=1}^{N} c_{i,k1}^{(1)} \cdot \left(\frac{\partial\psi}{\partial x}\right)_{k1,j} \tag{6.39}$$

Substituting the derivative condition 6.25*a* into the above equation gives

$$\left(\frac{\partial^2\psi}{\partial x^2}\right)_{i,j} = \sum_{k1=2}^{N-1} c_{i,k1}^{(1)} \cdot \left(\frac{\partial\psi}{\partial x}\right)_{k1,j} \tag{6.40}$$

In the second step, the first order derivative in Equation 6.40 is further discretized by the DQ method, which gives

$$\left(\frac{\partial^2\psi}{\partial x^2}\right)_{i,j} = \sum_{k=1}^{N}\sum_{k1=2}^{N-1} c_{i,k1}^{(1)} \cdot c_{k1,k}^{(1)} \cdot \psi_{k,j} = \sum_{k=1}^{N} \tilde{c}_{i,k}^{(2)} \cdot \psi_{k,j} \tag{6.41}$$

where $\tilde{c}_{i,k}^{(2)} = \sum_{k1=2}^{N-1} c_{i,k1}^{(1)} \cdot c_{k1,k}^{(1)}$. Clearly, $\tilde{c}_{i,k}^{(2)}$ is an element of the matrix $[\tilde{A}^2]$. It is indicated that the above method can be easily applied to cases where the derivative condition is not zero.

6.3.2.3 Implementation of Boundary Condition for Temperature

The boundary condition for T at $x = 0$ and $x = 1$ can be easily implemented by

$$T_{1,j} = 1, \quad T_{N,j} = 0, \quad \text{for } j = 1, 2, \ldots, M \tag{6.42a}$$

The derivative conditions for T at $y = 0$ and $y = 1$ are discretized by the DQ method and then combined to give

$$T_{i,1} = \frac{1}{AYT} \cdot \left[\sum_{k=2}^{M-1} \left(\bar{c}_{1,k}^{(1)} \cdot \bar{c}_{M,M}^{(1)} - \bar{c}_{M,k}^{(1)} \cdot \bar{c}_{1,M}^{(1)}\right) \cdot T_{i,k}\right] \tag{6.42b}$$

$$T_{i,M} = \frac{1}{AYT} \cdot \left[\sum_{k=2}^{M-1} \left(\bar{c}_{M,k}^{(1)} \cdot \bar{c}_{1,1}^{(1)} - \bar{c}_{1,k}^{(1)} \cdot \bar{c}_{M,1}^{(1)}\right) \cdot T_{i,k}\right] \tag{6.42c}$$

for $i = 2, 3, \ldots, N-1$

where

$$AYT = \bar{c}_{M,1}^{(1)} \cdot \bar{c}_{1,M}^{(1)} - \bar{c}_{1,1}^{(1)} \cdot \bar{c}_{M,M}^{(1)}$$

It is noted that at each boundary point, there is only one condition for T. Therefore, Equation 6.23 has to be applied at all the interior points, $2 \le i \le N-1$, $2 \le j \le M-1$.

6.3.3 Solution Procedures

In this section, we discuss some important aspects in the DQ computation of the vorticity-stream function formulation.

Mesh point distribution: The effect of mesh point distribution on the eigenvalues of the DQ discretization matrices has been discussed in Chapter 4. That discussion can give us a guidance on practical applications. For the PDQ method, although there is no restriction on the choice of the mesh point distribution, it is recommended that a non-uniform mesh be used. Shu and Richards (1990) found that the Chebyshev-Gauss-Lobatto grid can generally provide very accurate DQ solutions. For the sample problem, the mesh point distribution is chosen as

$$x_i = \frac{1}{2}\left(1 - \cos\frac{i-1}{N-1}\pi\right), \quad i = 1, 2, \ldots, N \tag{6.43a}$$

$$y_j = \frac{1}{2}\left(1 - \cos\frac{j-1}{M-1}\pi\right), \quad j = 1, 2, \ldots, M \tag{6.43b}$$

Solution techniques for discrete governing equations: The solution techniques for the DQ resultant equations have been described in Chapter 5. These techniques can be applied to solve Equations 6.21-6.23.

Equation 6.21 is an algebraic equation system, which can be solved by a direct method or an iterative method. Since the DQ method can lead to accurate numerical results by merely using a few mesh points, the use of direct method is very attractive for low and moderate Reynolds number flows. Among all the direct methods, the *LU* decomposition is the best choice. In the earlier application of the DQ method, the *LU* decomposition was often used to solve Equation 6.21 (Shu and Richards 1992*a*, *b*, Shu, Khoo and Yeo 1994, Shu, Chew and Liu 1997, etc.). For High Reynolds number flows, the boundary layer is very thin, and more mesh points are needed. For this case, direct methods may not be efficient. Instead, we can use iterative methods. The SOR method is an efficient iterative method for solving algebraic equations. Its performance when applied to the DQ algebraic equations has been discussed in detail by Shu, Yeo and Yao (1998). The implementation of SOR method has been described in Chapter 5.

The vorticity and temperature equations, Equations 6.22 and 6.23, are ordinary differential equations, which can be solved by explicit methods introduced in Chapter 5. In earlier works, the 4-stage Runge-Kutta method was often used. The explicit method is very simple and easy to implement. But it has a serious limitation on the time step size, especially when a large number of mesh points is used. To overcome such a limitation on the time step size, we can further discretize the time derivatives in Equations 6.22 and 6.23, and apply an iterative method to solve the resultant algebraic equations. Again, the SOR method is a good choice for this process. Note that Equations 6.22 and 6.23 are time-dependent equations, which can be applied to unsteady and steady problems. For steady problems, we just march the solution in time until the steady state solution is obtained.

Convergence criteria: Basically, two kinds of convergence criterion can be used. One is the relative error defined by

$$\text{err}(f,i,j) = \left| \frac{f^{n+1}(i,j) - f^{n}(i,j)}{f^{n}(i,j)} \right| \tag{6.44}$$

where n represents the time level, f is a flow variable which can be the vorticity, stream function, temperature. With Equation 6.44, the convergence criterion can be set as

$$\max_{i,j} \text{err}(f,i,j) \le \varepsilon \tag{6.45}$$

where ε is the threshold and is normally in the range of 10^{-3}-10^{-6}, depending on the problem considered. The other alternative is to use the residuals of Equations 6.21-6.23 to establish a convergence criterion. The convergence criterion for this method is

$$\max_{i,j} \text{Res}(f,i,j) \le \varepsilon \tag{6.46}$$

where $\text{Res}(f,i,j)$ represents the residual of equation for f at mesh point (x_i, y_j). For example, for the stream function equation, $\text{Res}(\psi,i,j)$ is

$$\text{Res}(\psi,i,j) = \omega_{i,j} - \sum_{k=1}^{N} c_{i,k}^{(2)} \cdot \psi_{k,j} - \sum_{k=1}^{M} \bar{c}_{j,k}^{(2)} \cdot \psi_{i,k} \tag{6.47}$$

Computational sequence: The computational sequence can be summarized as follows

(1) Input data such as the number of grid points, convergence threshold, Reynolds number, geometric parameters, and so on.

(2) Mesh generation: calculate the coordinates of the mesh points.

(3) Compute the DQ (PDQ or FDQ) weighting coefficients.

(4) If Equation 6.21 is solved by LU decomposition, form the matrix and then perform the LU decomposition. The resultant matrices $[L]$ and $[U]$ are stored for subsequent applications.

(5) Give an initial guess of the flow field since the governing equations are nonlinear.

(6) Solve the vorticity Equation 6.22.

(7) Solve the stream function Equation 6.21 using the values of the vorticity obtained in step (6).

(8) Compute the velocity components u and v from Equation 6.17.

(9) Solve for the temperature from Equation 6.23.

(10) Update the boundary conditions for ω, ψ and T.

(11) Check the convergence criteria. If they are satisfied, stop the computation and output the results. Otherwise, use the newly obtained results as an initial guess and repeat the computation from step (6).

6.3.4 Some Numerical Examples

The DQ method has been successfully applied to simulate many incompressible viscous flows. The driven cavity flow was first considered by Shu (1991) to validate the PDQ method. After that, Shu and his colleagues successfully simulated the natural convection in a square cavity (Shu, Khoo, Yeo and Chew 1994), the flow past a circular cylinder (Shu 1991, Shu and Richards 1992*a*), the flow in Czochralski crystal growth (Shu, Chew and Liu 1997), the natural convection in a concentric annulus (Shu 1999) and others. Other applications of the DQ method in this area include the work of Striz and Chen (1994) for the simulation of a lid-driven cavity flow, Kukreti and Siddiqi (1997) for the analysis of fluid storage tanks, and Zhang, Khoo and Yeo (1997) for the numerical study of the effect of free surface and water depth on the stability of wakes. In the following, we will only show two numerical examples involving the flow past a circular cylinder and the natural convection in a concentric annulus.

6.3.4.1 The Flow Past A Circular Cylinder

For the flow past a circular cylinder, the cylindrical coordinate system can be used. In particular, the following transformation

$$x = e^{\eta} \cos\xi, \; y = e^{\eta} \sin\xi \tag{6.48}$$

is used to map the physical domain into a rectangular domain. In addition, to avoid dealing with large values of ψ occurring in the far field and also to facilitate the numerical implementation of the far field boundary conditions, the stream function ψ

is decomposed into two parts such as $\psi = \psi_{in} + \overline{\psi}$, where ψ_{in} is chosen as the value for the inviscid flow, i.e.,

$$\psi_{in} = \left(e^{\eta} - e^{-\eta}\right)\sin\xi \tag{6.49}$$

As a consequence, the governing equations for the problem become

$$\overline{\psi}_{\xi\xi} + \overline{\psi}_{\eta\eta} = e^{2\eta}\omega \tag{6.50}$$

$$e^{2\eta}\omega_t + (\overline{\psi}_{\xi} + v_1)\omega_{\eta} - (\overline{\psi}_{\eta} + u_1)\omega_{\xi} = \frac{2}{\mathrm{Re}}(\omega_{\xi\xi} + \omega_{\eta\eta}) \tag{6.51}$$

where the subscripts represent the derivatives, and

$$u_1 = (e^{\eta} + e^{-\eta})\sin\xi, \qquad v_1 = (e^{\eta} - e^{-\eta})\cos\xi$$

On the surface of the body, the no-slip boundary conditions are

$$\overline{\psi} = 0,\ \overline{\psi}_{\eta} = -2\sin\xi \qquad \text{on } \eta = 0 \tag{6.52a}$$

$$\omega = \overline{\psi}_{\eta\eta} \qquad \text{on } \eta = 0 \tag{6.52b}$$

and the boundary conditions in the far field are

$$\overline{\psi} = 0,\ \overline{\psi}_{\eta} = 0 \qquad \text{on } \eta = \infty \tag{6.53a}$$

$$\omega = e^{-2\eta}\overline{\psi}_{\eta\eta} \qquad \text{on } \eta = \infty \tag{6.53b}$$

For numerical simulation, the infinite boundary in the η direction is truncated to $\eta_{\max} = 3.0$. For the steady state solution of the problem, the most sensitive parameters for checking the accuracy of the numerical simulation is the calculation of the parameters defining the structure of the wake behind the cylinder. The cylinder and the geometrical parameters of the closed wake are shown in Figure 6.3.

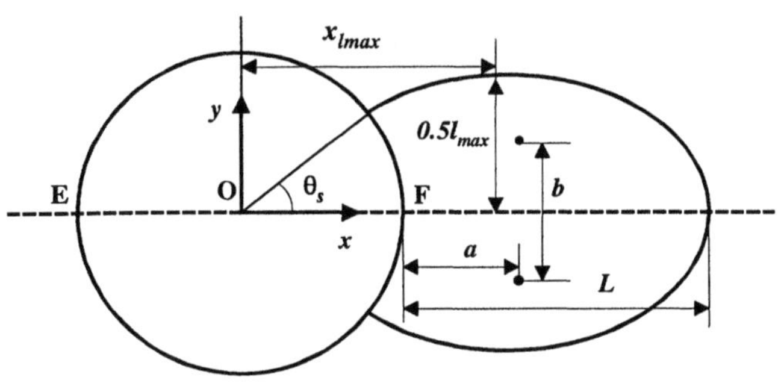

Figure 6.3 Geometrical parameters of the closed wake behind a cylinder

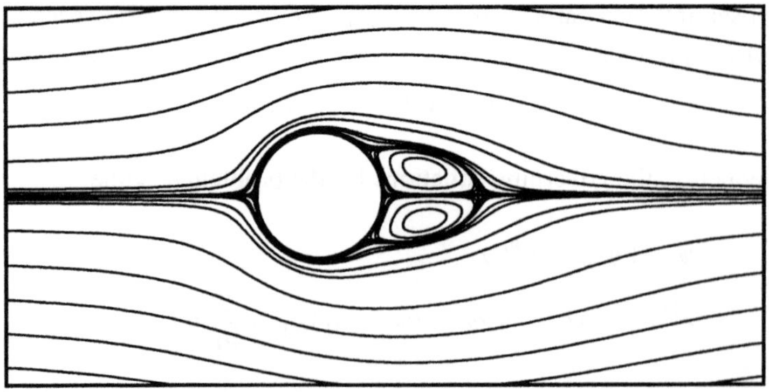

Figure 6.4 Computed streamlines past a circular cylinder, Re = 20 (Shu 1991)

The PDQ method is applied in both the ξ and η directions. Table 6.1 lists the details of the parameters of the wake eddy pair. Also included in the table are the experimental data (Tritton 1959, Coutanceau and Bouard 1977) and other numerical results (Gresho et al. 1984, Dennis and Chang 1970). The DQ results in Table 6.1 were obtained by using a mesh size of 25×21. It is noted that for the experimental data, C_D (drag coefficient) is obtained from the work of Tritton (1959), while other parameters are obtained from the work of Coutanceau and Bouard (1977) with $\lambda = 0$, where λ is the ratio between the cylinder and tank diameter. From Table 6.1, it can be seen that the DQ results are closer to the experimental data than those of Gresho et al. although these authors put the outer boundary further away from the cylinder surface and used a larger number of mesh points. This example demonstrates that the DQ method is a robust and efficient numerical technique. The streamlines for Re=20 computed using the PDQ method is displayed in Figure 6.4. The symmetric eddy pair is clearly shown in the figure.

Table 6.1 Geometric parameters of the closed wake behind a cylinder (Shu 1991)

Re	*References*	L	a	b	x_{lmax}	l_{max}	θ_s	C_D
20	*Experiment*	0.93	0.33	0.47	0.66	0.80	44.8^0	2.1243
	DQ	0.92	0.352	0.41	0.68	0.74	43.7^0	2.1220
	Gresho et al.	0.94	-	-	-	-	43.7^0	2.0450
25	*Experiment*	1.21	0.44	0.51	0.75	0.85	48.0^0	1.8176
	DQ	1.21	0.424	0.475	0.73	0.82	46.6^0	1.8336
	Gresho et al.	1.15	0.38	0.47	0.67	0.81	43.7^0	2.2600

6.3.4.2 The Natural Convection in A Concentric Annulus

For this problem, it is also preferable to use the cylindrical coordinate system. In the cylindrical coordinate system, the governing equations can be rewritten as

$$\nabla^2 \psi = \omega \tag{6.54}$$

$$\frac{\partial \omega}{\partial t} + \left[u \frac{\partial \omega}{\partial r} + \frac{v}{r}\frac{\partial \omega}{\partial \theta} \right] = \Pr \nabla^2 \omega - \Pr \cdot Ra \left[\sin\theta \frac{\partial T}{\partial r} + \frac{1}{r}\cos\theta \frac{\partial T}{\partial \theta} \right] \tag{6.55}$$

$$\frac{\partial T}{\partial t} + u\frac{\partial T}{\partial r} + \frac{v}{r}\frac{\partial T}{\partial \theta} = \nabla^2 T \tag{6.56}$$

where ∇^2 is the Laplacian operator. The boundary conditions on the two impermeable isothermal walls are given by

$$\psi = \frac{\partial \psi}{\partial r} = 0, \quad \omega = \frac{\partial^2 \psi}{\partial r^2}, \quad T = 1 \tag{6.57a}$$

on the inner cylinder and

$$\psi = \frac{\partial \psi}{\partial r} = 0, \quad \omega = \frac{\partial^2 \psi}{\partial r^2}, \quad T = 0 \tag{6.57b}$$

on the outer cylinder. In the θ direction, the periodic condition is specified. For the DQ computation, the derivatives in the r direction are discretized using the PDQ method, while the derivatives in the θ direction are approximated using the FDQ method.

To validate the DQ results, the values for the average equivalent heat conductivity are computed. The average equivalent conductivity is defined as

$$\bar{k}_{eq} = -\frac{\ln(RR)}{2\pi} \oint \frac{\partial T}{\partial n} \cdot ds \tag{6.58}$$

where $RR = R_o / R_i$ is the radius ratio, R_i and R_o are respectively the inner and outer cylinder radii. Equation 6.58 can be further reduced to

$$\bar{k}_{eqi} = -\frac{\ln(RR)}{2\pi(RR-1)} \int_0^{2\pi} \frac{\partial T}{\partial r} \cdot d\theta \tag{6.59a}$$

for the inner cylinder, and

$$\bar{k}_{eqo} = -\frac{RR \cdot \ln(RR)}{2\pi(RR-1)} \int_0^{2\pi} \frac{\partial T}{\partial r} \cdot d\theta \tag{6.59b}$$

for the outer cylinder. For the computation, Pr and L/D_i are set as 0.7 and 0.8 respectively, where L is the gap between the cylinders, i.e., $L = R_0 - R_i$. L is also the reference length used for non-dimensionalization.

Table 6.2 Comparison of average equivalent conductivities (Shu 1999)

$\bar{k}_{eq}$	*Reference*	*Ra*					
		10^2	10^3	3×10^3	6×10^3	10^4	5×10^4
$\bar{k}_{eqi}$	DQ	1.001	1.082	1.397	1.715	1.980	2.965
	Kuehn et al.	1.000	1.081	1.404	1.736	2.010	3.024
$\bar{k}_{eqo}$	DQ	1.001	1.082	1.397	1.715	1.980	2.965
	Kuehn et al.	1.002	1.084	1.402	1.735	2.005	2.973

Table 6.2 compares the average equivalent conductivities computed by the DQ method with those given by Kuehn and Goldstein (1976) for Ra = 10^2, 10^3, 3×10^3, 6×10^3, 10^4 and 5×10^4. The mesh sizes used for the DQ results are 3×8, 5×9, 13×13, 17×15, 19×15 and 29×21 respectively for Ra = 10^2, 10^3, 3×10^3, 6×10^3, 10^4 and 5×10^4. From the table, it can be observed that the computed average equivalent conductivities by the DQ method for the inner and outer cylinders are the same. This confirms the theoretical analysis. Since there is no energy loss for the whole system, the average equivalent conductivities for the inner and outer cylinders should be the same in theory. Figure 6.5 shows the streamlines and the isotherms of the DQ results for Ra = 5×10^4. The separation of the inner- and outer-cylinder thermal boundary layer and the symmetry of the flow pattern can be seen clearly.

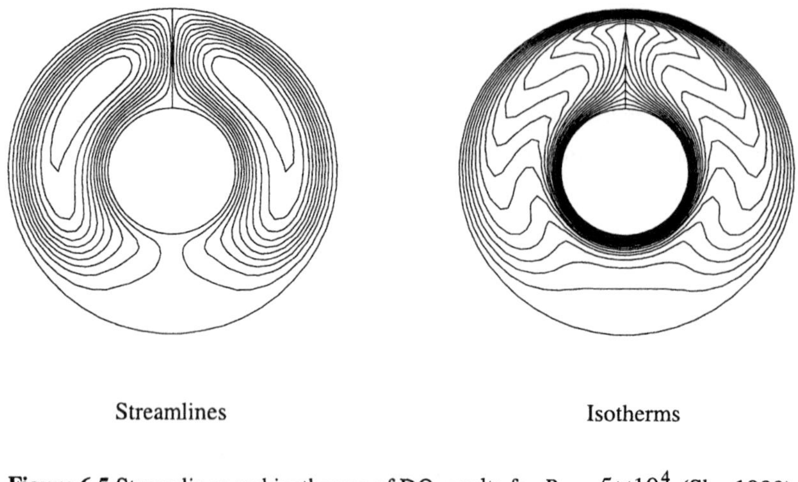

Streamlines Isotherms

Figure 6.5 Streamlines and isotherms of DQ results for Ra = 5×10^4 (Shu 1999)

6.4 Solution of Incompressible Navier-Stokes Equations in Primitive Variable Form

As mentioned in the introduction, the vorticity-stream function formulation is very useful for two-dimensional problems since it can greatly save the amount of computational effort required. However, this formulation cannot be extended to the three-dimensional case. To solve a three-dimensional problem, one has to use the primitive variable form of the incompressible Navier-Stokes equations. In this section, we will show how the DQ method can be applied to solve the incompressible Navier-Stokes equations in primitive variable form.

6.4.1 Introduction

To illustrate the solution procedure, the two-dimensional lid-driven cavity flow is considered. For the two-dimensional problem, Equation 6.14 is simplified to

$$\frac{\partial u}{\partial x}+\frac{\partial v}{\partial y}=0 \tag{6.60}$$

$$\frac{\partial u}{\partial t}+u\frac{\partial u}{\partial x}+v\frac{\partial u}{\partial y}=-\frac{\partial p}{\partial x}+\frac{1}{\mathrm{Re}}\nabla^2 u \tag{6.61}$$

$$\frac{\partial v}{\partial t}+u\frac{\partial v}{\partial x}+v\frac{\partial v}{\partial y}=-\frac{\partial p}{\partial y}+\frac{1}{\mathrm{Re}}\nabla^2 v \tag{6.62}$$

The boundary conditions for the problem are

$$u=0,\ v=0, \quad \text{at } x=0,1,\ 0<y<1 \tag{6.63a}$$

$$u=0,\ v=0, \quad \text{at } y=0,\ 0\le x\le 1 \tag{6.63b}$$

$$u=1,\ v=0, \quad \text{at } y=1,\ 0\le x\le 1 \tag{6.63c}$$

On the other hand, we note from Equation 6.60 that when the continuity equation is satisfied on the four boundaries, the following Neumann conditions for velocity can be derived

$$\frac{\partial u}{\partial x}=0\text{, on the left and right boundaries, } x=0,1,\ \ 0<y<1 \tag{6.64a}$$

$$\frac{\partial v}{\partial y}=0\text{, on the bottom and top boundaries, } y=0,1,\ \ 0\le x\le 1 \tag{6.64b}$$

6.4.2 Pressure Correction Method

The pressure correction method is an iterative process for solving Equations 6.60-6.62. It starts with an initial guess of the pressure and velocity fields. Unlike the conventional application of the pressure correction method, in this work, all the spatial derivatives are discretized by the DQ method which is based on a non-staggered grid.

With an initial guess of the pressure field and the velocity field at the time level n, the application of the DQ method to the spatial derivatives and the first order forward difference scheme to the temporal derivatives of Equations 6.61 and 6.62 gives

$$u_{i,j}^{*} = u_{i,j}^{n} + \Delta t \left[A_{i,j}^{n} - \sum_{k=1}^{N} c_{i,k}^{(1)} p_{k,j}^{*} \right] \tag{6.65}$$

$$v_{i,j}^{*} = v_{i,j}^{n} + \Delta t \left[B_{i,j}^{n} - \sum_{k=1}^{M} \bar{c}_{j,k}^{(1)} p_{i,k}^{*} \right] \tag{6.66}$$

where

$$A_{i,j}^{n} = \frac{1}{\mathrm{Re}} \left(\sum_{k=1}^{N} c_{i,k}^{(2)} u_{k,j}^{n} + \sum_{k=1}^{M} \bar{c}_{j,k}^{(2)} u_{i,k}^{n} \right) - u_{i,j}^{n} \sum_{k=1}^{N} c_{i,k}^{(1)} u_{k,j}^{n} - v_{i,j}^{n} \sum_{k=1}^{M} \bar{c}_{j,k}^{(1)} u_{i,k}^{n}$$

$$B_{i,j}^{n} = \frac{1}{\mathrm{Re}} \left(\sum_{k=1}^{N} c_{i,k}^{(2)} v_{k,j}^{n} + \sum_{k=1}^{M} \bar{c}_{j,k}^{(2)} v_{i,k}^{n} \right) - u_{i,j}^{n} \sum_{k=1}^{N} c_{i,k}^{(1)} v_{k,j}^{n} - v_{i,j}^{n} \sum_{k=1}^{M} \bar{c}_{j,k}^{(1)} v_{i,k}^{n}$$

$c_{i,j}^{(1)}$ and $c_{i,j}^{(2)}$ are respectively the weighting coefficients of the first and second order derivatives in the x-direction, while $\bar{c}_{i,j}^{(1)}$ and $\bar{c}_{i,j}^{(2)}$ are respectively the weighting coefficients of the first and second order derivatives in the y-direction.

The intermediate velocity fields $u*$ and $v*$ at time level $n+1$, given from Equations 6.65, 6.66, may not satisfy the continuity equation, Equation 6.60. To satisfy Equation 6.60, a correction is needed for the velocity and pressure fields, that is

$$u^{n+1} = u^{*} + u' \tag{6.67a}$$

$$v^{n+1} = v^{*} + v' \tag{6.67b}$$

$$p^{n+1} = p^{*} + p' \tag{6.67c}$$

With the corrected values of u^{n+1}, v^{n+1} and p^{n+1}, it is assumed that Equations 6.60-6.62 are satisfied. Thus, in the same manner, we can obtain

$$u_{i,j}^{n+1} = u_{i,j}^{n} + \Delta t\left[A_{i,j}^{n} - \sum_{k=1}^{N} c_{i,k}^{(1)} p_{k,j}^{n+1}\right] \tag{6.68}$$

$$v_{i,j}^{n+1} = v_{i,j}^{n} + \Delta t\left[B_{i,j}^{n} - \sum_{k=1}^{M} \bar{c}_{j,k}^{(1)} p_{i,k}^{n+1}\right] \tag{6.69}$$

$$\sum_{k=1}^{N} c_{i,k}^{(1)} u'_{k,j} + \sum_{k=1}^{M} \bar{c}_{j,k}^{(1)} v'_{i,k} = -S_{i,j}^{*} \tag{6.70}$$

where

$$S_{i,j}^{*} = \sum_{k=1}^{N} c_{i,k}^{(1)} u_{k,j}^{*} + \sum_{k=1}^{M} \bar{c}_{j,k}^{(1)} v_{i,k}^{*}$$

Obviously, subtracting Equation 6.65 from Equation 6.68 gives

$$u'_{i,j} = -\Delta t \sum_{k=1}^{N} c_{i,k}^{(1)} p'_{k,j} \tag{6.71}$$

and subtracting Equation 6.66 from Equation 6.69 leads to

$$v'_{i,j} = -\Delta t \sum_{k=1}^{M} \bar{c}_{j,k}^{(1)} p'_{i,k} \tag{6.72}$$

Note that the velocity corrections u' and v' on the boundary are always zero during the iteration process. Thus, when Equations 6.71, 6.72 are substituted into Equation 6.70, we can get

$$\sum_{k=1}^{N} W_{i,k}^{(2)} p'_{k,j} + \sum_{k=1}^{M} \overline{W}_{j,k}^{(2)} p'_{i,k} = S_{i,j}^{*} / \Delta t \tag{6.73}$$

where

$$W_{i,k}^{(2)} = \sum_{k1=2}^{N-1} c_{i,k1}^{(1)} c_{k1,k}^{(1)}\,,\quad \overline{W}_{j,k}^{(2)} = \sum_{k1=2}^{M-1} \bar{c}_{j,k1}^{(1)} \bar{c}_{k1,k}^{(1)}$$

The use of Equations 6.65-6.67, 6.71-6.73 constitutes an iterative process. At the beginning, an initial guess of the velocity and pressure fields is made. Then from Equations 6.65 and 6.66, the intermediate velocities u^* and v^* are computed. The u^* and v^* obtained are therewith substituted into Equation 6.73 to solve for the pressure correction p'. From the pressure correction p', the velocity correction u' and v' can be calculated from Equations 6.71 and 6.72. Next, by using Equation 6.67, the improved velocity and pressure fields can be obtained. After that, we need to check whether the convergence criteria are satisfied. If so, the computation is stopped. Otherwise, the improved velocity and pressure fields are taken as the new initial guess and the above process is repeated until the convergence criteria are satisfied.

During the above iterative process, there are two major concerns. One is the numerical implementation of the boundary condition for the pressure correction equation since the physical condition for the pressure correction does not exist. The other concern pertains to the enforcement of the continuity condition on the boundary. As mentioned previously, the DQ discretization is based on a non-staggered grid. Thus, unlike the low order finite volume method that is based on a staggered grid, the continuity equation on the boundary is not automatically satisfied in the DQ computation. As will be shown subsequently, the satisfying of the continuity equation on the boundary is critical for the numerical simulation. If this condition is not satisfied, it implies that there are many holes on the boundary, through which mass flux can flow in or flow out. Hence, the physical problem has been altered. To avoid this, we must ensure that the continuity equation, i.e., Equation 6.64, is accurately satisfied on the boundary. In the following, we will present two approaches to specify the boundary condition for the pressure correction equation, and to ensure that the continuity equation (Equation 6.64) is satisfied on the boundary.

6.4.3 Two Approaches to Specify Boundary Condition for p' and to Enforce Continuity Condition on the Boundary

6.4.3.1 Approach 1

It is observed from Equation 6.67 that p^* is the initial guess for the pressure while p' is the pressure correction. If for a problem, the pressure is given on the boundary, p^* should always be chosen as the given value. Then for this case, p' should be taken as zero on the boundary. For the general case, the pressure is usually not given on the boundary, but it can be computed from the momentum equations. For the sample problem shown above, the pressure on the boundary can be computed from

$$\frac{\partial p}{\partial x}=\frac{1}{\mathrm{Re}}\frac{\partial^2 u}{\partial x^2}, \text{ at } x=0,\ 1,\ 0<y<1 \tag{6.74a}$$

$$\frac{\partial p}{\partial y}=\frac{1}{\mathrm{Re}}\frac{\partial^2 v}{\partial y^2}, \text{ at } y=0,\ 1,\ 0\le x\le 1 \tag{6.74b}$$

Using the DQ method to discretize the spatial derivatives in Equation 6.74, the pressure on the boundary can be calculated from

$$p_{1,j} = \frac{1}{AXN}\left[(c_{N,N}^{(1)}C_{1,j} - c_{1,N}^{(1)}C_{N,j}) + \sum_{k=2}^{N-1}(c_{N,k}^{(1)}c_{1,N}^{(1)} - c_{1,k}^{(1)}c_{N,N}^{(1)})p_{k,j}\right] \tag{6.75a}$$

$$p_{N,j} = -\frac{1}{AXN}\left[(c_{N,1}^{(1)}C_{1,j} - c_{1,1}^{(1)}C_{N,j}) + \sum_{k=2}^{N-1}(c_{N,k}^{(1)}c_{1,1}^{(1)} - c_{1,k}^{(1)}c_{N,1}^{(1)})p_{k,j}\right] \tag{6.75b}$$

$$p_{i,1} = \frac{1}{AYM}\left[(\bar{c}_{M,M}^{(1)}D_{i,1} - \bar{c}_{1,M}^{(1)}D_{i,M}) + \sum_{k=2}^{M-1}(\bar{c}_{M,k}^{(1)}\bar{c}_{1,M}^{(1)} - \bar{c}_{1,k}^{(1)}\bar{c}_{M,M}^{(1)})p_{i,k}\right] \tag{6.75c}$$

$$p_{i,M} = -\frac{1}{AYM}\left[(\bar{c}_{M,1}^{(1)}D_{i,1} - \bar{c}_{1,1}^{(1)}D_{i,M}) + \sum_{k=2}^{M-1}(\bar{c}_{M,k}^{(1)}\bar{c}_{1,1}^{(1)} - \bar{c}_{1,k}^{(1)}\bar{c}_{M,1}^{(1)})p_{i,k}\right] \tag{6.75d}$$

for $i = 1, 2, \ldots\ldots, N;\ j = 1, 2, \ldots\ldots, M$

where

$$AXN = c_{1,1}^{(1)}c_{N,N}^{(1)} - c_{N,1}^{(1)}c_{1,N}^{(1)},\quad AYM = \bar{c}_{1,1}^{(1)}\bar{c}_{M,M}^{(1)} - \bar{c}_{M,1}^{(1)}\bar{c}_{1,M}^{(1)}$$

$$C_{1,j} = \frac{1}{\mathrm{Re}}\sum_{k=1}^{N}c_{1,k}^{(2)}u_{k,j},\quad C_{N,j} = \frac{1}{\mathrm{Re}}\sum_{k=1}^{N}c_{N,k}^{(2)}u_{k,j}$$

$$D_{i,1} = \frac{1}{\mathrm{Re}}\sum_{k=1}^{M}\bar{c}_{1,k}^{(2)}v_{i,k},\quad D_{i,M} = \frac{1}{\mathrm{Re}}\sum_{k=1}^{M}\bar{c}_{M,k}^{(2)}v_{i,k}$$

When the pressure on the boundary is determined from Equation 6.75, the pressure correction on the boundary can be specified as $p' = 0$. Then Equation 6.73 can be solved by the SOR method. As shown in the following results and discussion, having the proper boundary condition for the pressure correction is not sufficient, as this may still lead to inaccurate numerical results. To get an accurate solution, one has to enforce the continuity equation on the boundary. In this approach, Equation 6.64 is enforced after the improved velocity field is obtained from Equation 6.67.

Equation 6.63 provides one boundary condition for the velocity at each boundary point. So, when Equation 6.64 is also applied, there will be two boundary conditions for the velocity. One condition can be considered as the equation for the boundary point itself, and the other condition (derivative condition) is converted to an equation for its neighboring point. After DQ discretization of the derivatives in Equation 6.64, the velocity on the boundary and its neighboring points can be computed from

$$u_{1,j} = 0,\ u_{N,j} = 0,\ u_{i,1} = 0,\ u_{i,M} = 1 \tag{6.76a}$$

$$v_{1,j} = 0,\ v_{N,j} = 0,\ v_{i,1} = 0,\ v_{i,M} = 0 \tag{6.76b}$$

for $j = 1, 2, \ldots, M;\ i = 2, 3, \ldots, N\text{-}1$

$$u_{2,j} = \frac{1}{c_{1,2}^{(1)}c_{N,N-1}^{(1)} - c_{N,2}^{(1)}c_{1,N-1}^{(1)}}\sum_{k=3}^{N-2}(c_{N,k}^{(1)}c_{1,N-1}^{(1)} - c_{1,k}^{(1)}c_{N,N-1}^{(1)})u_{k,j} \tag{6.77a}$$

$$u_{N-1,j} = \frac{1}{c^{(1)}_{N,2}c^{(1)}_{1,N-1} - c^{(1)}_{1,2}c^{(1)}_{N,N-1}} \sum_{k=3}^{N-2} (c^{(1)}_{N,k}c^{(1)}_{1,2} - c^{(1)}_{1,k}c^{(1)}_{N,2})u_{k,j} \tag{6.77b}$$

for $j = 2, 3, \ldots, M-1$

$$v_{i,2} = \frac{1}{\bar{c}^{(1)}_{1,2}\bar{c}^{(1)}_{M,M-1} - \bar{c}^{(1)}_{M,2}\bar{c}^{(1)}_{1,M-1}} \sum_{k=3}^{M-2} (\bar{c}^{(1)}_{M,k}\bar{c}^{(1)}_{1,M-1} - \bar{c}^{(1)}_{1,k}\bar{c}^{(1)}_{M,M-1})v_{i,k} \tag{6.78a}$$

$$v_{i,M-1} = \frac{1}{\bar{c}^{(1)}_{M,2}\bar{c}^{(1)}_{1,M-1} - \bar{c}^{(1)}_{1,2}\bar{c}^{(1)}_{M,M-1}} \sum_{k=3}^{M-2} (\bar{c}^{(1)}_{M,k}\bar{c}^{(1)}_{1,2} - \bar{c}^{(1)}_{1,k}\bar{c}^{(1)}_{M,2})v_{i,k} \tag{6.78b}$$

for $i = 2, 3, \ldots\ldots, N-1$

It is noted that in this approach, the momentum equations are satisfied at all points including the boundary points, but the continuity equation may not be satisfied at grid points along $i = 2$, $i = N-1$, $j = 2$ and $j = M-1$ lines. Equation 6.67*c* is only applied to the interior mesh points. Since the pressure on the boundary is directly computed from the momentum equations, the pressure field obtained by this approach is a true field.

6.4.3.2 Approach 2

In this approach, the satisfying of the continuity equation on the boundary and the implementation of the boundary condition for the pressure correction are combined. In other words, the boundary condition for the pressure correction is derived from the continuity equation on the boundary.

Substituting Equations 6.67*a*, 6.67*b* into Equation 6.64, we obtain

$$\frac{\partial u'}{\partial x} = -\frac{\partial u^*}{\partial x}, \text{ at } x = 0, 1, \ 0 < y < 1 \tag{6.79a}$$

$$\frac{\partial v'}{\partial y} = -\frac{\partial v^*}{\partial y}, \text{ at } y = 0, 1, \ 0 \le x \le 1 \tag{6.79b}$$

Using Equation 6.71 and DQ discretization for the derivatives, Equation 6.79*a* gives the following condition for p' at the boundary of $i = 1$ and $i = N$,

$$W^{(2)}_{1,1}p'_{1,j} + W^{(2)}_{1,N}p'_{N,j} = \frac{1}{\Delta t}\sum_{k=1}^{N} c^{(1)}_{1,k}u^*_{k,j} - \sum_{k=2}^{N-1} W^{(2)}_{1,k}p'_{k,j} \tag{6.80a}$$

$$W^{(2)}_{N,1}p'_{1,j} + W^{(2)}_{N,N}p'_{N,j} = \frac{1}{\Delta t}\sum_{k=1}^{N} c^{(1)}_{N,k}u^*_{k,j} - \sum_{k=2}^{N-1} W^{(2)}_{N,k}p'_{k,j} \tag{6.80b}$$

Similarly, using Equation 6.72 and DQ discretization, Equation 6.79*b* provides

$$\overline{W}_{1,1}^{(2)} p_{i,1}^{'} + \overline{W}_{1,M}^{(2)} p_{i,M}^{'} = \frac{1}{\Delta t} \sum_{k=1}^{M} \overline{c}_{1,k}^{(1)} v_{i,k}^{*} - \sum_{k=2}^{M-1} \overline{W}_{1,k}^{(2)} p_{i,k}^{'} \tag{6.81a}$$

$$\overline{W}_{M,1}^{(2)} p_{i,1}^{'} + \overline{W}_{M,M}^{(2)} p_{i,M}^{'} = \frac{1}{\Delta t} \sum_{k=1}^{M} \overline{c}_{M,k}^{(1)} v_{i,k}^{*} - \sum_{k=2}^{M-1} \overline{W}_{M,k}^{(2)} p_{i,k}^{'} \tag{6.81b}$$

When the pressure correction at all interior points are computed from Equation 6.73, Equation 6.80 can be used to update the pressure correction on the boundaries of $i = 1$ and $i = N$ while Equation 6.81 can be utilized to determine the pressure correction on the boundaries of $j = 1$ and $j = M$. It should be indicated that, in this approach, the continuity equation and Equation 6.67*c* are applied at all points including the boundary points, but the momentum equations are only applied at all interior mesh points. The pressure field obtained by this approach is a relative field.

6.4.4 Computational Sequence

Like the computation in the vorticity-stream function formulation, Equation 6.43 can be used to generate a mesh point distribution for the solution of the primitive variable form of the incompressible Navier-Stokes equations. In addition, Equation 6.45 or 6.46 can also be utilized to set up a convergence criterion for the present case. The computational sequence is listed as follows.

(1) Input data such as the number of grid points, convergence threshold, Reynolds number, geometric parameters, and assume an initial flow field.

(2) Solve the momentum equations for u^* using Equation 6.65 and v^* using Equation 6.66 at all interior grid points.

(3) Solve the pressure correction equation for p' using Equation 6.73 at all interior grid points.

(4) Update the pressure correction p' on the boundary using Equations 6.80 and 6.81. This step is only applicable to Approach 2 since for Approach 1, p' is always zero on the boundary.

(5) Correct the pressure p^{n+1} using Equation 6.67*c* at all interior grid points.

(6) Update the pressure p^{n+1} at the boundary points. For Approach 1, use Equation 6.75, while for Approach 2, use Equation 6.67*c*.

(7) Calculate the velocity correction u' using Equation 6.71 and v' using Equation 6.72 for all the interior grid points.

(8) Correct the velocity u^{n+1} using Equation 6.67*a* and v^{n+1} using Equation 6.67*b* for all the interior mesh points.

(9) Update the velocity u^{n+1} using Equation 6.77 and v^{n+1} using Equation 6.78 for grid points adjacent to the boundary. This step is only applied to Approach 1.

(10) Check whether the convergence criterion is satisfied. If so, the computation is stopped. Otherwise, set $u^n = u^{n+1}$, $v^n = v^{n+1}$, $p^* = p^{n+1}$, return to step (2) and repeat the procedure until the convergence criterion is satisfied.

6.4.5 Sample Application and Comments on the Two Approaches

We will use the lid-driven cavity flow as an example to show the importance of enforcing the continuity condition on the boundary, and to illustrate the performance of the two approaches. The PDQ method is applied to discretize all the spatial derivatives.

6.4.5.1 Importance of Enforcing Continuity Condition on the Boundary

As mentioned previously, when the pressure correction method is applied using DQ discretization, the continuity equation on the boundary may not be satisfied. In this section, we will show that if the continuity condition on the boundary is not enforced, the residual for the continuity equation on the boundary defined by

$$R_w = \left(\frac{\partial u}{\partial x} + \frac{\partial v}{\partial y} \right)_{wall} \tag{6.82}$$

may result in large values. We will demonstrate this by applying Approach 1 without using Equations 6.77, 6.78 to simulate the sample problem. It is indicated that Equations 6.77, 6.78 are derived from the continuity equation on the boundary. So, when these equations are not applied, it implies that the continuity equation on the boundary is not enforced. From the numerical experiment, it was found that for this case, the values of R_w at some boundary points are very large. At different boundary points, the values of R_w are different. It was also found that the value of R_w depends on the mesh size. Figure 6.6 shows the residual distribution for the continuity equation along the left ($x = 0$) and bottom ($y = 0$) boundaries for $\text{Re} = 400$. It can be observed from the figure that along the left boundary, the values of R_w are very small when $y \le 0.5$. However, as y is increased above 0.5 along the left boundary, the value of R_w is increased greatly. When the mesh size is increased, the value of R_w along the left boundary is slightly decreased, which seems to approach a grid-independent value. On the other hand, as the mesh size is increased, the value of R_w along the bottom boundary is increased considerably. For the other two boundaries ($x = 1$ and $y = 1$), the residual distributions have similar behavior, that is, at most boundary points, the value of R_w is very large and depends on the mesh size.

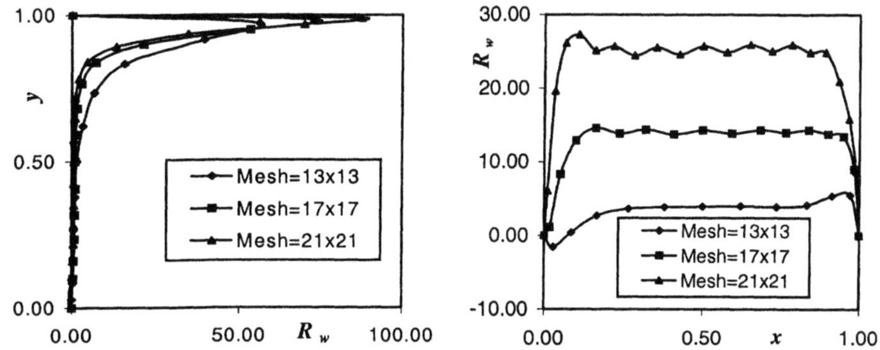

(*a*) Along left boundary ($x = 0$) (*b*) Along bottom boundary ($y = 0$)

Figure 6.6 Residual distribution of continuity equation along the boundary (Re=400)

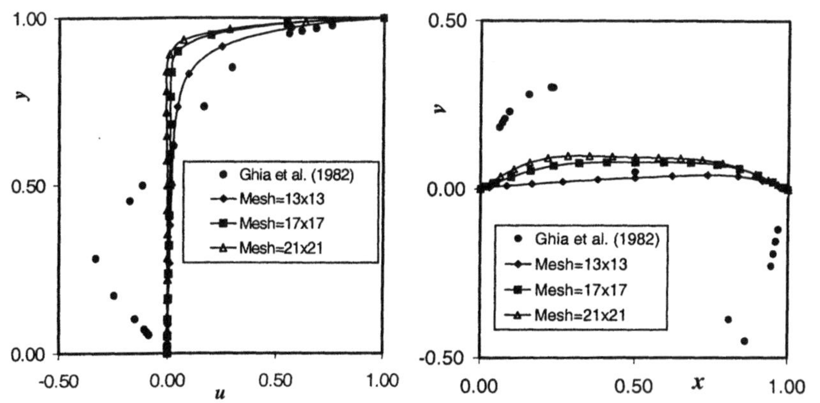

(*a*) *u*-profile along vertical centerline (*b*) *v*-profile along horizontal centerline

Figure 6.7 Velocity profiles along geometric centerlines when continuity equation is not enforced on the boundary (Re = 400)

When the value of R_w is not zero at a boundary point, it implies that this boundary point acts like a hole, through which there is a mass transfer. The flux of mass transfer depends on the value of R_w and the direction of mass transfer depends on the sign of R_w. So, when the values of R_w at many boundary points are very large, it means that there are many holes on the boundary which allow mass transfer to occur. Therefore, the physical problem is altered, and the numerical results obtained cannot be used to approximate the true physical solution. Furthermore, since the values of R_w depend on the mesh size, the numerical results obtained from different mesh sizes could be different, reflecting the fact that different problems are being solved. This can be confirmed by Figure 6.7, in which the velocity profiles along the horizontal and vertical centerlines are presented. It can be seen clearly from the figure that as compared to the results of Ghia et al. (1982), the present results are clearly incorrect. When different mesh sizes are used, the converged results are different, indicating that a grid-independent solution cannot be achieved.

The above study strongly suggests that in order to obtain a true solution, the continuity condition on the boundary should be enforced.

6.4.5.2 Comments on Performance of Two Approaches

When the continuity condition is enforced on the boundary, in other words, when Approach 1 is applied to solve the sample problem, it was found that the numerical results are very accurate and agree well with the solutions of Ghia et al. (1982). This can be observed from Figure 6.8, which compares the velocity profiles of Approach 1 with those of Ghia et al. (1982) for Re = 400. The DQ results generated using Approach 1 were obtained by employing a mesh size of 21×21, and the results of Ghia et al. (1982) were given from a multi-grid finite difference approach using a mesh size of 129×129.

We have also compared the performance of Approach 1 and Approach 2 when they were applied to the sample problem. From numerical experiments, it was found that both approaches have almost the same level of efficiency, but Approach 1 requires slightly less computational effort.

It was also found that Approach 2 is very sensitive to the treatment of singularity. Although the geometry of the sample problem is very simple, there is a singularity for the u-velocity at the two upper corner points. When the boundary condition for u is directly implemented, it was found that Approach 1 is able of generating very accurate results for both the velocity profiles and the vorticity distribution, but Approach 2 can only generate accurate results for the velocity profiles. The vorticity contours obtained from Approach 2 show many wiggles. The irregular distribution of vorticity generated by Approach 2 is actually due to the effect of singularity points. This phenomenon has also been highlighted by Ku et al. (1987) in their work which uses the pseudospectral method to simulate the same problem. To remove the effect of singularity points on the numerical solution, Ku et al. (1987) suggested a smooth distribution for the u-velocity near the two upper corner points. At the two corner points, the u-velocity is set to 0, but at their immediate neighboring points on the top boundary, u is adjusted to 0.3 instead of 1. For other points on the top boundary, u is kept to 1. Ku et al. (1987) showed that this artificial treatment can greatly improve the accuracy of the numerical results. When we adopted the same treatment as used by Ku et al. (1987) to remove the effect of singularity, the results of Approach 2 are greatly improved, but the results of Approach 1 show no obvious improvement. In addition, it was found that the pressure field obtained by Approach 1 is more accurate than that obtained from Approach 2. This is reasonable because Approach 1 provides the true pressure field while Approach 2 gives a relative pressure field.

It should be indicated that although Approach 1 is generally capable of providing more accurate numerical solution, it cannot ensure that the residuals of the continuity equation at all the interior points are sufficiently small. As shown in the above section, for Approach 1, the velocity along the line just adjacent to the boundary is computed from the continuity equation on the boundary. So, for the interior points along this line, the continuity equation is actually not satisfied.

Although the two approaches are demonstrated by a two-dimensional example, they can be easily applied to three-dimensional flows.

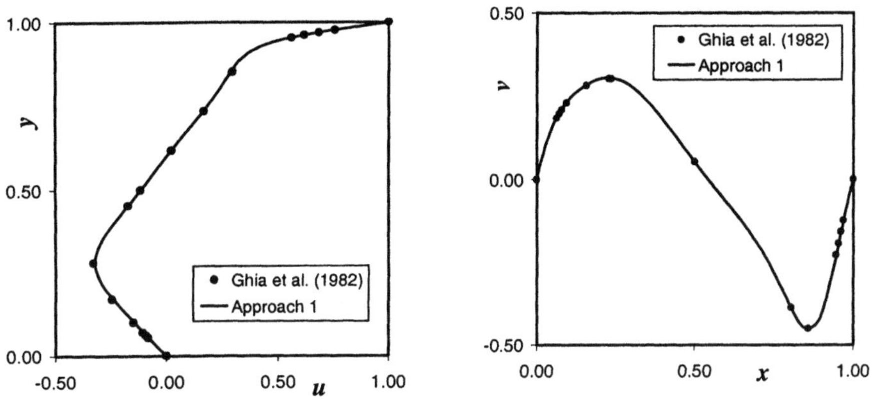

(*a*) *u*-profile along vertical centerline (*b*) *v*-profile along horizontal centerline

Figure 6.8 Velocity profiles along geometric centerlines obtained by Approach 1 (Re = 400)

Chapter 7

Application of Differential Quadrature Method to Structural and Vibration Analysis

7.1 Introduction

As shown in the previous chapters, the differential quadrature method has a feature in that it can obtain very accurate numerical results by using just a few grid points. This feature has a particular merit in its application to structural and vibration analysis. For example, the vibration of a thin plate is governed by a fourth order partial differential equation. When a numerical method is applied to discretize the spatial derivatives, the partial differential equation can be reduced to a set of algebraic equations. The eigenvalues of the resultant algebraic equation system provide the vibrational frequencies of the problem. Usually, the number of interior grid points is equal to the dimension of the resultant algebraic equation system, thus providing the same number of eigenfrequencies. Among all the computed eigenfrequencies, only low frequencies are of practical interest. As we know, low order methods such as finite differences and finite elements are only capable of obtaining accurate numerical results by using a large number of grid points. So, when low order methods are applied, they need to use a large number of grid points to obtain highly accurate values for low frequencies. As a result, a lot of virtual storage and computational effort are required. On the other hand, when the DQ method is applied, the low frequencies can be obtained very accurately by using a considerably smaller number of grid points due to the feature of the method. As a consequence, very little computational effort and virtual storage are needed when the DQ method is used.

The pioneering work in the application of the DQ method to the structural mechanics problems was given by Striz, Jang and Bert (1988) and Jang, Bert and Striz (1989). Since then, the application of the DQ method covers almost all the areas of structural and vibration analysis of beams, arches, shafts, plates and shells.

The group from University of Oklahoma (Bert, Wang and Striz 1993, 1994*a*, *b*, Bert and Malik 1996*a*, *b*, *c*, 1997*a*, *b*, Bert and Kang 1996, Chen, Striz and Bert 1997, Farsa, Kukreti and Bert 1993*a*, *b*, Kang and Bert 1997, Kang, Bert and Striz 1995, 1996*a*, *b*, Kukreti, Farsa and Bert 1992, 1993, 1996, Malik and Bert 1995, 1996, Striz, Wang and Bert 1995, Wang and Bert 1993, Wang, Bert and Striz 1993, 1994, Wang, Striz and Bert 1993, 1994) extensively studied the deflection, buckling and vibration problems of beams, shafts, arches and various plates using the DQ method. Other researchers also reported excellent results in the application of the DQ method to beam and plate problems. Du, Lim and Lin (1994, 1995) and Du, Liew and Lim (1996) conducted the buckling, bending and vibration analysis of beams and plates. Laura and his group (Laura and Gutierrez 1993, 1994, 1995, Gutierrez and Laura 1995*a*, *b*, 1999, Gutierrez et al. 1995, 1996, 1998) studied the vibration of beams and plates under various conditions. Lin, Lim and Du (1994*a*, *b*) considered the static analysis of plates with nonlinear supports. Wang (1995), Wang, Yang, Xiao (1995), and Gu and Wang (1997) performed vibration analysis of circular and laminated plates. Chen, Zhong and He (1997) showed the DQ analysis of anisotropic plates. Shu and Du (1997*a*, *b*), Shu and Wang (1999) illustrated in detail the implementation of boundary conditions to the DQ vibration analysis of beams and plates. The buckling analysis of beams and composite plates was reported by Moradi and Taheri (1998, 1999), Pandey and Sherbourne (1991), Sherbourne and Pandey (1991). Recently, Liew and his group (Han and Liew 1997*a*, *b*, *c*, 1998, 1999, Liew and Han 1997*a*, *b*, 1998, Liew, Han and Xiao 1996, 1997, Liew, Han, Xiao and Du 1996) reported some excellent results in structural and vibration analysis of reissner/mindlin plates. Liew and Teo (1998, 1999), Teo and Liew (1999) also extended the DQ application to three-dimensional bending and vibration analysis of rectangular plates.

Apart from the application of the DQ method to beams and plates, some effort was also made to analyze the free vibration of shells. Shu (1996*a*, *b*) reported excellent results for the free vibration analysis of isotropic and composite laminated conical shells. The vibration analysis of conical shells was also shown by Li and Lam (1998), Ng, Li, Lam and Loy (1999), Wu and Hung (1999) using an asymptotic theory. For free vibration analysis of isotropic and composite laminated cylindrical shells and panels, some accurate results were given by Bert and Malik (1996*d*), Loy, Lam and Shu (1997), Shu and Du (1997*c*), Loy, Lam, Li and Liu (1999). The buckling analysis of circular cylindrical shells was shown by Mirfakhraei and Redekop (1998). In addition, a computer program was provided by Zhang, Mirfakhraei and Redekop (1998) for the analysis of a toroidal shell.

For structural and vibration analysis of beams and plates with complex geometries, domain decomposition and coordinate transformation techniques were developed. The multi-domain DQ approach is also known as the DQ element method by some researchers (Chen 1997, 1998*a*, *b*, 1999*a*, *b*, Liu and Liew 1998*a*, 1999, Striz, Chen and Bert 1994, 1997, Wang and Gu 1997, Wang, Wang and Chen 1998). The application of the DQ method in a curvilinear coordinate system has been reported by Bert and Malik (1996*a*), Han and Liew (1997*c*), Zhong (1998). We will describe the multi-domain DQ method and coordinate transformation technique in Chapter 9.

It should be indicated that the DQ application in the structural and vibration analysis is still under active research, so, the above survey is very incomplete.

In this chapter, due to the limitation of space, we do not intend to cover all the aspects of DQ application to structural and vibration analysis. Instead, we concentrate on the basic solution procedures for structural and vibration analysis of sample beam, plate and shell problems. The focus will be on the implementation of the boundary conditions since this is a challenging problem in the application of the DQ method to structural mechanics problems. For example, the governing equations for the thin beam and plate are of fourth order. Then at each boundary point, there are two boundary conditions. To properly implement these two boundary conditions, four approaches, namely, the δ-technique (Jang, Bert and Striz 1989), the modified weighting coefficient matrix approach (Wang and Bert 1993, Malik and Bert 1996), the approach of directly substituting the boundary conditions into the discrete governing equations (Shu and Du 1997*a*), and the general approach (Shu and Du 1997*b*), will be introduced. We wish that these basic descriptions would serve as a guidance and motivation for readers to conduct further research in the area.

7.2 Differential Quadrature Analysis of Beams

In this section, the DQ method is applied to the structural and vibration analysis of beams. For simplicity, we will only consider the Bernoulli-Euler beam.

7.2.1 Governing Equations and Boundary Conditions

For a beam, three problems are often encountered, i.e., the bending, vibration and column buckling analysis. For a Bernoulli-Euler beam of varying cross-section with length *L*, the non-dimensional governing equations are

$$s(X)\frac{d^4W}{dX^4}+2\frac{ds(X)}{dX}\frac{d^3W}{dX^3}+\frac{d^2s(X)}{dX^2}\frac{d^2W}{dX^2}+\frac{L^4q(x)}{EI_0}=0 \tag{7.1}$$

for the bending analysis, and

$$s(X)\frac{d^4W}{dX^4}+2\frac{ds(X)}{dX}\frac{d^3W}{dX^3}+\frac{d^2s(X)}{dX^2}\frac{d^2W}{dX^2}-\Omega^2W=0 \tag{7.2}$$

for the vibration analysis, and

$$s(X)\frac{d^4W}{dX^4}+2\frac{ds(X)}{dX}\frac{d^3W}{dX^3}+\frac{d^2s(X)}{dX^2}\frac{d^2W}{dX^2}+\frac{PL^2}{EI_0}\frac{d^2W}{dX^2}=0 \tag{7.3}$$

for the column buckling analysis, where $s(X)=EI/EI_0$, $\Omega^2=\rho AL^4\omega^2/EI_0$, $X=x/L$, *EI* is the beam's flexural rigidity, ρA is the mass per unit length, $q(x)$ is the external distributed load, ω is the dimensional frequency, *P* is the axial compressive load. For a beam of varying cross-section, *EI* and *A* are functions of the coordinate *x*. The governing equation for a beam is a 4th order ordinary differential equation. For a well-posed problem, it requires four boundary conditions. These can be obtained by specifying two boundary conditions at the end $X = 0$, and another two boundary conditions at the end $X = 1$. Basically, there are three types of boundary conditions. For vibration analysis, these boundary conditions are given as

Simply supported end (SS)

$$W = 0 \quad \text{and} \quad \frac{d^2W}{dX^2} = 0 \tag{7.4}$$

Clamped end (C)

$$W = 0 \quad \text{and} \quad \frac{dW}{dX} = 0 \tag{7.5}$$

Free end (F)

$$\frac{d^2W}{dX^2} = 0 \quad \text{and} \quad \frac{d^3W}{dX^3} = 0 \tag{7.6}$$

7.2.2 Numerical Discretization

For the numerical computation, the continuous solution is approximated by the functional values at discrete points. Now, we assume that the computational domain $0 \le X \le 1$ is divided into $(N-1)$ intervals with coordinates of the grid points given as $X_1, X_2, \ldots, X_N$. Although a uniform mesh can be used for the analysis, it is recommended that a non-uniform mesh be used. Here, we adopt the mesh point distribution used by Shu (1991),

$$X_i = \frac{1}{2}\left[1 - \cos\left(\frac{i-1}{N-1} \cdot \pi\right)\right], \quad i = 1, 2, \ldots, N \tag{7.7}$$

With the coordinates of mesh points given by Equation 7.7, the PDQ or FDQ weighting coefficients can be easily computed. These weighting coefficients can then be used to discretize Equations 7.1-7.3. Using the DQ method, Equation 7.1 can be discretized as

$$s^{(2)}(X_i)\sum_{k=1}^{N} c_{ik}^{(2)} W_k + 2s^{(1)}(X_i)\sum_{k=1}^{N} c_{ik}^{(3)} \cdot W_k + s(X_i)\sum_{k=1}^{N} c_{ik}^{(4)} W_k = -\frac{L^4}{EI_0} q(X_i) \tag{7.8}$$

Similarly, Equation 7.2 is discretized as

$$s^{(2)}(X_i)\sum_{k=1}^{N} c_{ik}^{(2)} W_k + 2s^{(1)}(X_i)\sum_{k=1}^{N} c_{ik}^{(3)} \cdot W_k + s(X_i)\sum_{k=1}^{N} c_{ik}^{(4)} W_k = \Omega^2 W_i \tag{7.9}$$

and Equation 7.3 is approximated by

$$s^{(2)}(X_i)\sum_{k=1}^{N} c_{ik}^{(2)} W_k + 2s^{(1)}(X_i)\sum_{k=1}^{N} c_{ik}^{(3)} \cdot W_k + s(X_i)\sum_{k=1}^{N} c_{ik}^{(4)} W_k = -\frac{PL^2}{EI_0}\sum_{k=1}^{N} c_{ik}^{(2)} W_k \tag{7.10}$$

where $W_i, i = 1, 2, ..., N$, is the functional value at the grid point X_i, $s^{(2)}(X_i)$ and $s^{(1)}(X_i)$ are respectively the second and first order derivatives of $s(X)$ at X_i, $c_{ik}^{(n)}, n = 2, 3, 4$ are the DQ weighting coefficients of the nth order derivative. With proper implementation of the boundary conditions, which will be described in the next section, Equation 7.8 can be put in matrix form as

$$[A_d]\{W\} = \{b\} \tag{7.11}$$

where $[A_d]$ is a matrix, $\{W\}$ is a vector of unknowns and $\{b\}$ is a known vector. The algebraic Equation 7.11 can be solved using the techniques shown in Chapter 5. It can be solved using direct methods such as the *LU* decomposition, or iterative methods such as the SOR approach. In a similar manner, Equation 7.9 can be written as

$$[A_v]\{W\} = \Omega^2\{W\} \tag{7.12}$$

and Equation 7.10 can be put to

$$[A_b]\{W\} = -\frac{PL^2}{EI_0}[B]\{W\} \tag{7.13}$$

where $[A_v], [A_b]$ and $[B]$ are matrices. Equations 7.12 and 7.13 are eigenvalue systems. The frequency of a free vibration problem can be obtained from the eigenvalues of Equation 7.12, while the buckling load of a column buckling problem can be computed from the eigenvalues of Equation 7.13. The eigenvalues of Equations 7.12 and 7.13 can be calculated using standard solvers such as the *QR* algorithm.

7.2.3 Implementation of Boundary Conditions

We will use the vibration problem to illustrate the implementation of boundary conditions.

7.2.3.1 The δ-technique

The δ-technique was proposed by Jang, Bert and Striz (1989) to eliminate the difficulties in implementing two conditions at a single boundary point. It applies the

Dirichlet condition ($W = 0$) at the boundary point itself, and the derivative condition at its adjacent point which is at a distance δ from the boundary point.

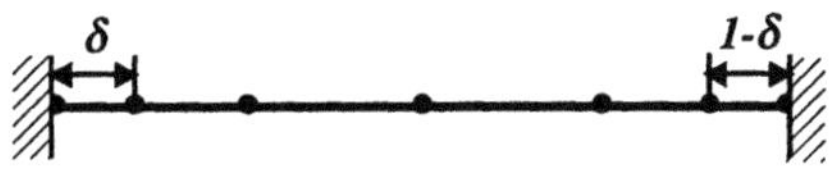

Figure 7.1 Illustration of δ-points for a beam problem

As shown in Figure 7.1, the δ-points are actually the grid points $X_2 = \delta$ and $X_{N-1} = 1-\delta$. As an example, when the simply supported condition is specified at the two ends, the derivative condition $\partial^2 W / \partial X^2 = 0$ is discretized at the mesh points X_2 and X_{N-1}, which gives

$$\sum_{k=1}^{N} c_{2,k}^{(2)} \cdot W_k = 0 \tag{7.14a}$$

$$\sum_{k=1}^{N} c_{N-1,k}^{(2)} \cdot W_k = 0 \tag{7.1ba}$$

Clearly, the derivative condition is not accurately approximated by the δ-technique. There are two major drawbacks to the δ-technique. One drawback arises from the implementation of the derivative condition at the δ-point. Since it is an approximation to the true boundary condition that should be implemented at the boundary, one can expect that the numerical results may depend on the choice of the δ-value. To obtain an accurate numerical solution, the values of δ should be chosen to be very small (possibly not greater than 0.0001). However, this small value of δ would result in the second drawback of this technique. When one mesh spacing (δ) is much smaller than the others, the DQ weighting coefficient matrices may become highly ill-conditioned, which then causes the solution to oscillate. As a result, the numerical solution is less accurate.

7.2.3.2 Modification of Weighting Coefficient Matrices

This technique was presented by Wang and Bert (1993) to overcome the drawbacks of the δ-technique. For this approach, only one boundary condition is numerically implemented. The other boundary condition (derivative condition) is built into the DQ weighting coefficient matrices. To demonstrate this approach, we consider the following boundary conditions (C-SS)

$$W = 0,\ dW / dX = 0,\ \text{at } X = 0 \tag{7.15a}$$

$$W = 0,\ d^2W / dX^2 = 0,\ \text{at } X = 1 \tag{7.15b}$$

Let $[A^{(n)}]$ and $[\tilde{A}^{(n)}]$ be respectively the original and modified weighting coefficient matrices of the nth order derivative. Since

$$\frac{d^n W}{dX^n} = \frac{d}{dX}\left(\frac{d^{n-1}W}{dX^{n-1}}\right) \tag{7.16}$$

we then have

$$[A^{(n)}] == [A^{(1)}][A^{(n-1)}] = [A^{(2)}][A^{(n-2)}] = \cdots = [A^{(n-1)}][A^{(1)}] \tag{7.17}$$

Now, consider the DQ analog of the derivative dW/dX at all mesh points

$$\begin{Bmatrix} (dW/dX)_1 \\ (dW/dX)_2 \\ \cdot \\ \cdot \\ \cdot \\ (dW/dX)_{N-1} \\ (dW/dX)_N \end{Bmatrix} = \begin{bmatrix} c_{1,1}^{(1)} & c_{1,2}^{(1)} & \cdots & c_{1,N-1}^{(1)} & c_{1,N}^{(1)} \\ c_{2,1}^{(1)} & c_{2,2}^{(1)} & \cdots & c_{2,N-1}^{(1)} & c_{2,N}^{(1)} \\ \cdot & \cdot & \cdots & \cdot & \cdot \\ \cdot & \cdot & \cdots & \cdot & \cdot \\ \cdot & \cdot & \cdots & \cdot & \cdot \\ c_{N-1,1}^{(1)} & c_{N-1,2}^{(1)} & \cdots & c_{N-1,N-1}^{(1)} & c_{N-1,N}^{(1)} \\ c_{N,1}^{(1)} & c_{N,2}^{(1)} & \cdots & c_{N,N-1}^{(1)} & c_{N,N}^{(1)} \end{bmatrix} \begin{Bmatrix} W_1 \\ W_2 \\ \cdot \\ \cdot \\ \cdot \\ W_{N-1} \\ W_N \end{Bmatrix} \tag{7.18}$$

To satisfy the derivative condition of Equation 7.15a, Equation 7.18 should be modified by zeroing the first row of the matrix

$$\begin{Bmatrix} (dW/dX)_1 \\ (dW/dX)_2 \\ \cdot \\ \cdot \\ \cdot \\ (dW/dX)_{N-1} \\ (dW/dX)_N \end{Bmatrix} = \begin{bmatrix} 0 & 0 & \cdots & 0 & 0 \\ c_{2,1}^{(1)} & c_{2,2}^{(1)} & \cdots & c_{2,N-1}^{(1)} & c_{2,N}^{(1)} \\ \cdot & \cdot & \cdots & \cdot & \cdot \\ \cdot & \cdot & \cdots & \cdot & \cdot \\ \cdot & \cdot & \cdots & \cdot & \cdot \\ c_{N-1,1}^{(1)} & c_{N-1,2}^{(1)} & \cdots & c_{N-1,N-1}^{(1)} & c_{N-1,N}^{(1)} \\ c_{N,1}^{(1)} & c_{N,2}^{(1)} & \cdots & c_{N,N-1}^{(1)} & c_{N,N}^{(1)} \end{bmatrix} \begin{Bmatrix} W_1 \\ W_2 \\ \cdot \\ \cdot \\ \cdot \\ W_{N-1} \\ W_N \end{Bmatrix} \tag{7.19}$$

Obviously, the matrix in Equation 7.18 is $[A^{(1)}]$, and the matrix in Equation 7.19 is $[\tilde{A}^{(1)}]$. To satisfy the boundary condition (7.15a), the second order weighting coefficient matrix $[\bar{A}^{(2)}]$ can be derived from Equation 7.17

$$[\bar{A}^{(2)}] = [A^{(1)}][\tilde{A}^{(1)}] \tag{7.20}$$

In order to satisfy the boundary condition (7.15*b*) at $X = 1$, $[\overline{A}^{(2)}]$ should be modified by zeroing its last row, that is,

$$[\widetilde{A}^{(2)}] = \begin{bmatrix} \overline{c}_{1,1}^{(2)} & \overline{c}_{1,2}^{(2)} & \cdots & \overline{c}_{1,N-1}^{(2)} & \overline{c}_{1,N}^{(2)} \\ \overline{c}_{2,1}^{(2)} & \overline{c}_{2,2}^{(2)} & \cdots & \overline{c}_{2,N-1}^{(2)} & \overline{c}_{2,N}^{(2)} \\ \cdot & \cdot & \cdots & \cdot & \cdot \\ \cdot & \cdot & \cdots & \cdot & \cdot \\ \cdot & \cdot & \cdots & \cdot & \cdot \\ \overline{c}_{N-1,1}^{(2)} & \overline{c}_{N-1,2}^{(2)} & \cdots & \overline{c}_{N-1,N-1}^{(2)} & \overline{c}_{N-1,N}^{(2)} \\ 0 & 0 & \cdots & 0 & 0 \end{bmatrix} \quad (7.21)$$

It is noted that the derivative conditions given in Equation 7.15 have been completely built into the matrix $[\widetilde{A}^{(2)}]$. With $[\widetilde{A}^{(2)}]$, the modified weighting coefficient matrices $[\widetilde{A}^{(3)}]$ and $[\widetilde{A}^{(4)}]$ can be computed by

$$[\widetilde{A}^{(3)}] = [A^{(1)}] \cdot [\widetilde{A}^{(2)}] \quad (7.22)$$

$$[\widetilde{A}^{(4)}] = [A^{(2)}] \cdot [\widetilde{A}^{(2)}] \quad (7.23)$$

In the above process, the weighting coefficient matrices are modified through the matrix form. This process is very simple since it just zeros some elements of the matrix. However, for the non-homogeneous derivative condition, this process cannot be simply implemented. One has to determine the elements of the matrix by solving an ill-conditioned equation system (Malik and Bert 1996). To eliminate the difficulties encountered in the above process, we propose the use of the summation form to determine the elements of the modified weighting coefficient matrices. For the general case, we suppose that the non-homogeneous boundary conditions are

$$W = a,\ dW/dX = b,\ \text{at } X = 0 \quad (7.24a)$$

$$W = c,\ d^2W/dX^2 = d\ ,\ \text{at } X = 1 \quad (7.24b)$$

where a, b, c and d are constants. At first, we consider the DQ discretization of the second order derivative at a mesh point X_i

$$W_{XX}(X_i) = \sum_{k1=1}^{N} c_{i,k1}^{(1)} \cdot W_X(X_{k1}) \quad (7.25)$$

Substituting the derivative condition (Equation 7.24*a*) into Equation 7.25 and further discretizing the first order derivative, we obtain

$$W_{XX}(X_i) = c_{i,1}^{(1)} \cdot b + \sum_{k=1}^{N} \sum_{k1=2}^{N} c_{i,k1}^{(1)} \cdot c_{k1,k}^{(1)} \cdot W_k = c_{i,1}^{(1)} \cdot b + \sum_{k=1}^{N} \tilde{c}_{i,k}^{(2)} \cdot W_k \tag{7.26}$$

Obviously, the elements of the weighting coefficient matrix for the second order derivative are calculated by

$$\tilde{c}_{i,k}^{(2)} = \sum_{k1=2}^{N} c_{i,k1}^{(1)} c_{k1,k}^{(1)} \tag{7.27}$$

In a similar manner, the discretization of the third and fourth order derivatives can be written as

$$W_{XXX}(X_i) = c_{i,N}^{(1)} d + b \sum_{k1=1}^{N-1} c_{i,k1}^{(1)} c_{k1,1}^{(1)} + \sum_{k=1}^{N} \tilde{c}_{i,k}^{(3)} W_k \tag{7.28}$$

$$W_{XXXX}(X_i) = c_{i,N}^{(2)} d + b \sum_{k1=1}^{N-1} c_{i,k1}^{(2)} c_{k1,1}^{(1)} + \sum_{k=1}^{N} \tilde{c}_{i,k}^{(4)} W_k \tag{7.29}$$

where $\tilde{c}_{i,k}^{(3)} = \sum_{k1=1}^{N-1} c_{i,k1}^{(1)} \tilde{c}_{k1,k}^{(2)}$, $\tilde{c}_{i,k}^{(4)} = \sum_{k1=1}^{N-1} c_{i,k1}^{(2)} \tilde{c}_{k1,k}^{(2)}$, $\tilde{c}_{i,k}^{(3)}$ and $\tilde{c}_{i,k}^{(4)}$ are the elements of the modified weighting coefficient matrices of the third and fourth order derivatives. Clearly, the summation form can easily implement the non-homogeneous condition and compute the elements of the modified weighting coefficient matrices. This process is particularly useful in the bending analysis.

It should be indicated that similar to the finite element method, when this approach is applied, the derivative condition is automatically and exactly satisfied. And since only one condition is implemented at the boundary point, the discrete governing equation has to be applied to the interior points, $2 \le i \le N-1$. So, the dimension of the equation system using this approach is $(N-2)\times(N-2)$. This approach has been successfully applied to solve some beam problems with very good accuracy. However, as indicated by Shu and Du (1997*a*), there are some limitations to the application of this approach. One limitation is in the implementation of the clamped-clamped (C-C) boundary conditions. For this case, the approach fails to compute some of the frequencies.

7.2.3.3 Direct Substitution of Boundary Conditions into Discrete Governing Equations

This approach was proposed by Shu and Du (1997*a*) to implement the simply supported, clamped conditions and their combinations. The essence of the approach is that the Dirichlet condition is implemented at the boundary point, while the derivative condition is discretized by the DQ method. The discretized derivative conditions at the two ends are then combined to give the solutions W_2 and W_{N-1}.

The expressions for W_2 and W_{N-1} are then substituted into the discrete governing equation which is applied to the interior points $3 \le i \le N-2$. The dimension of the equation system using this approach is $(N-4)\times(N-4)$.

For any combination of the clamped and simply supported conditions at the two ends, the discrete boundary conditions using the DQ method can be written as

$$W_1 = 0 \tag{7.30a}$$

$$\sum_{k=1}^{N} c_{1,k}^{(n0)} \cdot W_k = 0 \tag{7.30b}$$

$$W_N = 0 \tag{7.30c}$$

$$\sum_{k=1}^{N} c_{N,k}^{(n1)} \cdot W_k = 0 \tag{7.30d}$$

where $n0$ and $n1$ may be taken as either 1 or 2. By choosing the values of $n0$ and $n1$, Equation 7.30 can give the following four sets of boundary conditions,

$n0 = 1,\ n1 = 1$	———	clamped-clamped
$n0 = 1,\ n1 = 2$	———	clamped-simply supported
$n0 = 2,\ n1 = 1$	———	simply supported-clamped
$n0 = 2,\ n1 = 2$	———	simply supported-simply supported

Equations 7.30*a* and 7.30*c* can be easily substituted into Equation 7.9. This is not the case for Equations 7.30*b* and 7.30*d*. However, one can couple these two equations together to give two solutions, W_2 and W_{N-1}, as

$$W_2 = \frac{1}{AXN} \cdot \sum_{k=3}^{N-2} AXK1 \cdot W_k \tag{7.31a}$$

$$W_{N-1} = \frac{1}{AXN} \cdot \sum_{k=3}^{N-2} AXKN \cdot W_k \tag{7.31b}$$

where

$$AXK1 = c_{1,k}^{(n0)} \cdot c_{N,N-1}^{(n1)} - c_{1,N-1}^{(n0)} \cdot c_{N,k}^{(n1)}$$

$$AXKN = c_{1,2}^{(n0)} \cdot c_{N,k}^{(n1)} - c_{1,k}^{(n0)} \cdot c_{N,2}^{(n1)}$$

$$AXN = c_{N,2}^{(n1)} \cdot c_{1,N-1}^{(n0)} - c_{1,2}^{(n0)} \cdot c_{N,N-1}^{(n1)}$$

According to Equation 7.31, W_2 and W_{N-1} are expressed in terms of W_3, W_4, ..., W_{N-2}, and can be easily substituted into Equation 7.9. It should be noted that Equation 7.30 provides four boundary equations. In total, we have N unknowns W_1, ..., W_N. In order to close the system, the discretized governing Equation 7.9 has to be applied at (N–4) mesh points. This can be done by applying Equation 7.9 at grid points X_3, X_4, ..., X_{N-2}. Substituting Equations 7.30*a*, 7.30*c* and 7.31 into Equation 7.9 gives

$$s^{(2)}(X_i)\sum_{k=3}^{N-2} C_1 W_k + 2s^{(1)}(X_i)\sum_{k=3}^{N-2} C_2 W_k + s(X_i)\sum_{k=3}^{N-2} C_3 W_k = \Omega^2 W_i \quad (7.32)$$

for $i = 3, 4, \ldots, N-2$

where

$$C_1 = c_{i,k}^{(2)} - \frac{c_{i,2}^{(2)} \cdot AXK1 + c_{i,N-1}^{(2)} \cdot AXKN}{AXN}$$

$$C_2 = c_{i,k}^{(3)} - \frac{c_{i,2}^{(3)} \cdot AXK1 + c_{i,N-1}^{(3)} \cdot AXKN}{AXN}$$

$$C_3 = c_{i,k}^{(4)} - \frac{c_{i,2}^{(4)} \cdot AXK1 + c_{i,N-1}^{(4)} \cdot AXKN}{AXN}$$

It is noted that Equation 7.32 has (N–4) equations with (N–4) unknowns, which can be written in matrix form given by Equation 7.12.

It should be indicated that this approach can be extended to implement the free conditions or the combination of the free and clamped/simply supported conditions. For example, when the F-F boundary conditions are implemented, we need to discretize the four derivative conditions at first. These four discrete conditions are then coupled to give four solutions W_1, W_2, W_{N-1} and W_N in terms of $W_3, \ldots, W_{N-2}$. The expressions of W_1, W_2, W_{N-1} and W_N are substituted into Equation 7.9 to form the final eigenvalue equation system.

7.2.4 Numerical Example: Free Vibration Analysis of A Uniform Beam

The free vibration analysis of a uniform beam ($s(X) = 1$) is taken as an example to show the high efficiency and accuracy of the DQ method. For numerical computations, the PDQ method is used to discretize the derivatives, and the method of directly substituting the boundary conditions into the discrete governing equations is applied to implement the boundary conditions. The mesh point distribution is given by Equation 7.7. Table 7.1 lists the natural frequencies of the first 5 modes for three sets of boundary conditions, that is, simply supported-simply supported (SS-SS), clamped-clamped (C-C), and clamped-simply supported (C-SS) conditions. The DQ results were provided by Shu and Du (1997*a*) using 15 mesh points. The exact results given by Blevins (1984) are also included in Table 7.1 for comparison. It can be observed from Table 7.1 that, for all the three cases, the DQ results are very accurate although only 15 mesh points are used.

Table 7.1 Comparison of natural frequencies of a uniform beam (Shu & Du 1997*a*)

	Ω_1	Ω_2	Ω_3	Ω_4	Ω_5
		simply supported-simply supported			
Blevins (1984)	9.8696	39.4784	88.8264	157.9137	246.7401
PDQ ($N = 15$)	9.8696	39.4784	88.8249	158.0619	248.4716
		clamped-clamped			
Blevins (1984)	22.3733	61.6728	120.9034	199.8594	298.5555
PDQ ($N = 15$)	22.3733	61.6728	120.9021	199.9365	299.3886
		clamped-simply supported			
Blevins (1984)	15.4182	49.9648	104.2477	178.2697	272.0310
PDQ ($N = 15$)	15.4182	49.9648	104.2471	178.4642	273.1126

7.3 Differential Quadrature Analysis of Thin Plates

In this section, we will illustrate the application of the DQ method to the structural and vibration analysis of thin plates. For simplicity, only the rectangular plate is considered.

7.3.1 Governing Equations and Boundary Conditions

Deflection, free vibration and buckling are three typical problems for a plate. The non-dimensional governing equations for these three cases can be written as

Plate deflection

$$\frac{\partial^4 W}{\partial X^4} + 2\lambda^2 \frac{\partial^4 W}{\partial X^2 \partial Y^2} + \lambda^4 \frac{\partial^4 W}{\partial Y^4} = \frac{a^4 q(X,Y)}{D} \tag{7.33}$$

Transverse vibration of thin, isotropic plates

$$\frac{\partial^4 W}{\partial X^4} + 2\lambda^2 \frac{\partial^4 W}{\partial X^2 \partial Y^2} + \lambda^4 \frac{\partial^4 W}{\partial Y^4} = \Omega^2 W \tag{7.34}$$

Buckling of a plate under uniaxial compression

$$\frac{\partial^4 W}{\partial X^4} + 2\lambda^2 \frac{\partial^4 W}{\partial X^2 \partial Y^2} + \lambda^4 \frac{\partial^4 W}{\partial Y^4} = \frac{N_x a^2}{D} \frac{\partial^2 W}{\partial X^2} \tag{7.35}$$

where W is the dimensionless mode shape function, $q(X,Y)$ is the external distributed load, Ω is the dimensionless frequency, $X = x/a$ and $Y = y/b$ are dimensionless coordinates, a and b are the lengths of the plate edges, $\lambda = a/b$ is the aspect ratio, and N_x is the uniaxial load. Further, $\Omega = \omega a^2 \sqrt{\rho / D}$, where ω is the dimensional circular frequency, $D = Eh^3/[12(1-\nu^2)]$ is the flexural rigidity, E, ν, ρ and h are Young's modulus, Poisson's ratio, the density of the plate material, and the plate thickness, respectively. It should be mentioned that the above equations do not cover all the cases. For example, for free vibration of the anisotropic plates, Equation 7.34 has to be changed to include more terms. Equation 7.35 can be modified to consider buckling under different compressions.

The governing equation for a thin plate is a 4th order partial differential equation with respect to X and Y. It requires two boundary conditions at each edge. There are three basic boundary conditions. For free vibration analysis, these boundary conditions are

<u>Simply supported edge (SS)</u>

$$W = 0, \qquad \frac{\partial^2 W}{\partial X^2} = 0 \tag{7.36a}$$

at $X = 0$ and 1, and

$$W = 0, \qquad \frac{\partial^2 W}{\partial Y^2} = 0 \tag{7.36b}$$

at $Y = 0$ and 1.

<u>Clamped edge (C)</u>

$$W = 0, \qquad \frac{\partial W}{\partial X} = 0 \tag{7.37a}$$

at $X = 0$ and 1, and

$$W = 0, \qquad \frac{\partial W}{\partial Y} = 0 \tag{7.37b}$$

at $Y = 0$ and 1.

Free edge (F)

$$\frac{\partial^2 W}{\partial X^2} + \nu \cdot \lambda^2 \cdot \frac{\partial^2 W}{\partial Y^2} = 0, \qquad \frac{\partial^3 W}{\partial X^3} + (2-\nu) \cdot \lambda^2 \cdot \frac{\partial^3 W}{\partial X \partial Y^2} = 0 \tag{7.38a}$$

at $X = 0$ and 1, and

$$\lambda^2 \cdot \frac{\partial^2 W}{\partial Y^2} + \nu \cdot \frac{\partial^2 W}{\partial X^2} = 0, \qquad \lambda^2 \cdot \frac{\partial^3 W}{\partial Y^3} + (2-\nu) \cdot \frac{\partial^3 W}{\partial X^2 \partial Y} = 0 \tag{7.38b}$$

at $Y = 0$ and 1, and

$$\frac{\partial^2 W}{\partial X \partial Y} = 0 \tag{7.38c}$$

at the corner of two adjacent free edges.

7.3.2 Numerical Discretization

The computational domain of a rectangular plate is $0 \le X \le 1$, $0 \le Y \le 1$. For numerical computation, we need to perform a mesh generation first. As for the case of the plate, the mesh generation in the X and Y directions can be given by

$$X_i = \frac{1}{2}\left[1 - \cos\left(\frac{i-1}{N-1} \cdot \pi\right)\right], \quad i = 1, 2, \ldots, N \tag{7.39a}$$

$$Y_j = \frac{1}{2}\left[1 - \cos\left(\frac{j-1}{M-1} \cdot \pi\right)\right], \quad j = 1, 2, \ldots, M \tag{7.39b}$$

where N and M are the number of the grid points in the X and Y direction respectively. With the coordinates of the mesh points given by Equation 7.39, the PDQ or FDQ weighting coefficients can be easily computed. These weighting coefficients can be used to discretize Equations 7.33-7.35. Let $c_{i,j}^{(n)}$ be the DQ weighting coefficients of the nth order derivative in the X direction, and $\bar{c}_{i,j}^{(m)}$ be the DQ weighting coefficients of the mth order derivative in the Y direction. Using the DQ method, Equation 7.33 can be discretized as

$$\sum_{k=1}^{N} c_{i,k}^{(4)} W_{k,j} + 2\lambda^2 \sum_{k1=1}^{N} \sum_{k2=1}^{M} c_{i,k1}^{(2)} \bar{c}_{j,k2}^{(2)} W_{k1,k2} + \lambda^4 \sum_{k=1}^{M} \bar{c}_{j,k}^{(4)} W_{i,k} = \frac{a^4 q_{i,j}}{D} \tag{7.40}$$

and Equation 7.34 is discretized as

$$\sum_{k=1}^{N} c_{i,k}^{(4)} W_{k,j} + 2\lambda^2 \sum_{k1=1}^{N} \sum_{k2=1}^{M} c_{i,k1}^{(2)} \bar{c}_{j,k2}^{(2)} W_{k1,k2} + \lambda^4 \sum_{k=1}^{M} \bar{c}_{j,k}^{(4)} W_{i,k} = \Omega^2 W_{i,j} \tag{7.41}$$

In a similar manner, Equation 7.35 is discretized as

$$\sum_{k=1}^{N} c_{i,k}^{(4)} W_{k,j} + 2\lambda^2 \sum_{k1=1}^{N} \sum_{k2=1}^{M} c_{i,k1}^{(2)} \bar{c}_{j,k2}^{(2)} W_{k1,k2} + \lambda^4 \sum_{k=1}^{M} \bar{c}_{j,k}^{(4)} W_{i,k} = \frac{a^2 N_x}{D} \sum_{k=1}^{N} c_{i,k}^{(2)} W_{k,j} \tag{7.42}$$

where $W_{i,j}$ is the functional value at the grid point (X_i, Y_j). After implementation of the boundary conditions, Equation 7.40 can be written in matrix form as Equation 7.11. Similarly, Equation 7.41 can be put in the same form as Equation 7.12, and Equation 7.42 can also be put into a similar form as Equation 7.13. The solution techniques for the three matrix forms have been described in the section on beams.

7.3.3 Implementation of Boundary Conditions

The free vibration problem is used to demonstrate the implementation of the boundary conditions.

7.3.3.1 The δ-technique

The δ-technique for beams can be directly extended to plates. Physically, there are two conditions at a boundary point. The δ-technique only implements one condition (Dirichlet type) at the boundary point. The derivative condition is discretized at the δ-point which is very close to the boundary. For the beam, there are only two δ-points. However, for a plate, there are many δ-points. The δ-points for a plate are distributed along the four mesh lines given by $i = 2,\ i = N-1,\ j = 2,\ j = M-1$. Therefore, the number of δ–points for a plate depends on the number of mesh points, N and M. For plates, the δ–technique has the same drawbacks as previously discussed for beams.

7.3.3.2 Modification of Weighting Coefficient Matrices

For plates, modification of the weighting coefficient matrices in each direction can be done in the same manner as shown in the section for beams. In the following, we will take the SS-C-C-SS condition as an example to show how the weighting coefficient matrices can be modified by the summation form. Here, SS-C-C-SS represents the simply supported (SS) condition at $X = 0$, clamped (C) condition at $Y = 0$, clamped (C) condition at $X = 1$, and simply supported (SS) condition at $Y = 1$. Note that this sequence of boundary conditions for a plate is applied in the whole book. The SS-C-C-SS condition for a rectangular plate can be written as

$$W = 0, \quad \frac{\partial^2 W}{\partial X^2} = 0, \text{ at } X = 0 \tag{7.43a}$$

$$W = 0, \quad \frac{\partial W}{\partial X} = 0, \text{ at } X = 1 \tag{7.43b}$$

$$W = 0, \quad \frac{\partial W}{\partial Y} = 0, \text{ at } Y = 0 \tag{7.43c}$$

$$W = 0, \quad \frac{\partial^2 W}{\partial Y^2} = 0, \text{ at } Y = 1 \tag{7.43d}$$

Let $\tilde{c}^{(2)}_{i,j}$ and $\tilde{c}^{(4)}_{i,j}$ be the modified weighting coefficients of the second and fourth order derivatives in the X direction, $\tilde{\tilde{c}}^{(2)}_{i,j}$ and $\tilde{\tilde{c}}^{(4)}_{i,j}$ be the modified weighting coefficients of the second and fourth order derivatives in the Y direction. When the derivative condition in Equation 7.43 is built into the weighting coefficient matrix, $\tilde{c}^{(2)}_{i,j}$, $\tilde{c}^{(4)}_{i,j}$, $\tilde{\tilde{c}}^{(2)}_{i,j}$ and $\tilde{\tilde{c}}^{(4)}_{i,j}$ can be determined from

$$\tilde{c}^{(2)}_{i,j} = \sum_{k=1}^{N-1} c^{(1)}_{i,k} c^{(1)}_{k,j} \tag{7.44a}$$

$$\tilde{c}^{(4)}_{i,j} = \sum_{k=2}^{N} c^{(2)}_{i,k} \tilde{c}^{(2)}_{k,j} \tag{7.44b}$$

$$\tilde{\tilde{c}}^{(2)}_{i,j} = \sum_{k=2}^{M} \bar{c}^{(1)}_{i,k} \bar{c}^{(1)}_{k,j} \tag{7.44c}$$

$$\tilde{\tilde{c}}^{(4)}_{i,j} = \sum_{k=1}^{M-1} \bar{c}^{(2)}_{i,k} \tilde{\tilde{c}}^{(2)}_{k,j} \tag{7.44d}$$

With the modified weighting coefficients given by Equation 7.44, Equation 7.41 can be changed to

$$\sum_{k=1}^{N} \tilde{c}^{(4)}_{i,k} W_{k,j} + 2\lambda^2 \sum_{k1=1}^{N} \sum_{k2=1}^{M} \tilde{c}^{(2)}_{i,k1} \tilde{\tilde{c}}^{(2)}_{j,k2} W_{k1,k2} + \lambda^4 \sum_{k=1}^{M} \tilde{\tilde{c}}^{(4)}_{j,k} W_{i,k} = \Omega^2 W_{i,j} \tag{7.45}$$

Equation 7.45 is applied at the interior points, $2 \le i \le N-1, 2 \le j \le M-1$, and only the Dirichlet condition ($W = 0$) is implemented. Like the beam, when two opposite edges of a plate are clamped, this approach may result in the loss of some important frequencies.

7.3.3.3 Direct Substitution of Boundary Conditions into Discrete Governing Equations

The idea of this approach is the same as that for the beam. It was presented by Shu and Du (1997*a*) to implement the simply supported and clamped conditions. The derivatives in the boundary condition are also discretized by the DQ method. The discrete form of any combination of the clamped and simply supported conditions can be written as

$$W_{1,j} = 0,\ W_{N,j} = 0,\ W_{i,1} = 0\ ,\ W_{i,M} = 0 \text{ at all boundary points} \tag{7.46a}$$

$$\sum_{k=1}^{N} c_{1,k}^{(n0)} \cdot W_{k,j} = 0 \quad \text{at } X\text{=}0 \tag{7.46b}$$

$$\sum_{k=1}^{N} c_{N,k}^{(n1)} \cdot W_{k,j} = 0 \quad \text{at } X\text{=}1 \tag{7.46c}$$

$$\sum_{k=1}^{M} \bar{c}_{1,k}^{(m0)} \cdot W_{i,k} = 0 \quad \text{at } Y\text{=}0 \tag{7.46d}$$

$$\sum_{k=1}^{M} \bar{c}_{M,k}^{(m1)} \cdot W_{i,k} = 0 \quad \text{at } Y\text{=}1 \tag{7.46e}$$

where $n0$, $n1$, $m0$ and $m1$ are taken as either 1 or 2: 1 is used for the clamped edge condition and 2 is used for the simply supported edge condition. $n0$, $n1$, $m0$ and $m1$ correspond to the edges of $X = 0$, $X = 1$, $Y = 0$, $Y = 1$, respectively. It is noted that Equation 7.46*a* corresponds to the Dirichlet boundary condition at the four edges of the plate, and Equations 7.46*b*, 7.46*c*, 7.46*d* and 7.46*e* result from the derivative boundary conditions. Obviously, Equation 7.46*a* can be easily substituted into Equation 7.41. However, Equations 7.46*b*, 7.46*c*, 7.46*d* and 7.46*e* cannot be directly substituted into Equation 7.41. This difficulty can be easily overcome. Using the same procedure as for the beam, Equations 7.46*b*, 7.46*c* can be coupled to give two solutions $W_{2,j}$ and $W_{N-1,j}$, which are located at the grid points shown by the symbol **O** in Figure 7.2,

$$W_{2,j} = \frac{1}{AXN} \cdot \sum_{k=3}^{N-2} AXK1 \cdot W_{k,j} \tag{7.47a}$$

$$W_{N-1,j} = \frac{1}{AXN} \cdot \sum_{k=3}^{N-2} AXKN \cdot W_{k,j} \tag{7.47b}$$

$$\text{for } j = 3, 4, \ldots, M-2$$

where

$$AXN = c_{N,2}^{(n1)} \cdot c_{1,N-1}^{(n0)} - c_{1,2}^{(n0)} \cdot c_{N,N-1}^{(n1)}$$

$$AXK1 = c_{1,k}^{(n0)} \cdot c_{N,N-1}^{(n1)} - c_{1,N-1}^{(n0)} \cdot c_{N,k}^{(n1)}$$

$$AXKN = c_{1,2}^{(n0)} \cdot c_{N,k}^{(n1)} - c_{1,k}^{(n0)} \cdot c_{N,2}^{(n1)}$$

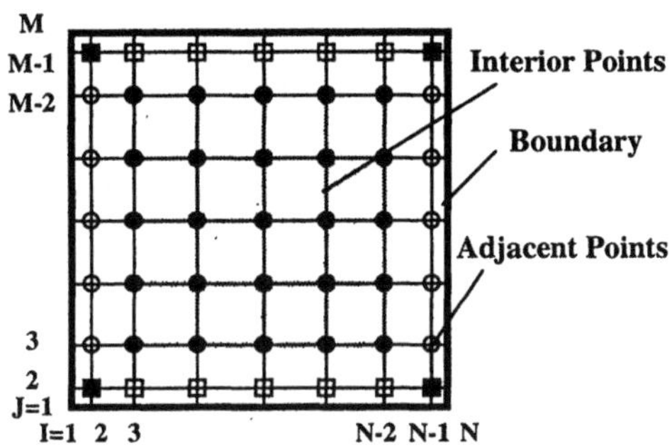

Figure 7.2 Illustration of interior and adjacent points for a rectangular plate

Similarly, Equations 7.46*d* and 7.46*e* can be coupled to give two solutions $W_{i,2}$ and $W_{i,M-1}$, which are located at the grid points shown by the symbol □ in Figure 7.2,

$$W_{i,2} = \frac{1}{AYM} \cdot \sum_{k=3}^{M-2} AYK1 \cdot W_{i,k} \tag{7.48a}$$

$$W_{i,M-1} = \frac{1}{AYM} \cdot \sum_{k=3}^{M-2} AYKM \cdot W_{i,k} \tag{7.48b}$$

$$\text{for } i = 3, 4, \ldots, N-2$$

where

$$AYM = \bar{c}_{M,2}^{(m1)} \cdot \bar{c}_{1,M-1}^{(m0)} - \bar{c}_{1,2}^{(m0)} \cdot \bar{c}_{M,M-1}^{(m1)}$$

$$AYK1 = \bar{c}_{1,k}^{(m0)} \cdot \bar{c}_{M,M-1}^{(m1)} - \bar{c}_{1,M-1}^{(m0)} \cdot \bar{c}_{M,k}^{(m1)}$$

$$AYKM = \bar{c}_{1,2}^{(m0)} \cdot \bar{c}_{M,k}^{(m1)} - \bar{c}_{1,k}^{(m0)} \cdot \bar{c}_{M,2}^{(m1)}$$

For the points near the four corners shown by the symbol ■ in Figure 7.2, the four Equations 7.46*b*, 7.46*c*, 7.46*d* and 7.46*e* have to be coupled to provide the following four solutions

$$W_{2,2} = \frac{1}{AXN} \cdot \frac{1}{AYM} \cdot \sum_{k1=3}^{N-2} \sum_{k2=3}^{M-2} AXK1 \cdot AYK1 \cdot W_{k1,k2} \tag{7.49a}$$

$$W_{N-1,2} = \frac{1}{AXN} \cdot \frac{1}{AYM} \cdot \sum_{k1=3}^{N-2} \sum_{k2=3}^{M-2} AXKN \cdot AYK1 \cdot W_{k1,k2} \quad (7.49b)$$

$$W_{2,M-1} = \frac{1}{AXN} \cdot \frac{1}{AYM} \cdot \sum_{k1=3}^{N-2} \sum_{k2=3}^{M-2} AXK1 \cdot AYKM \cdot W_{k1,k2} \quad (7.49c)$$

$$W_{N-1,M-1} = \frac{1}{AXN} \cdot \frac{1}{AYM} \cdot \sum_{k1=3}^{N-2} \sum_{k2=3}^{M-2} AXKN \cdot AYKM \cdot W_{k1,k2} \quad (7.49d)$$

For Equation 7.49, the index k in the *AXK1* and *AXKN* expressions is replaced by $k1$, and the index k in the *AYK1* and *AYKM* expressions is replaced by $k2$. With Equations 7.46*a*, 7.47, 7.48 and 7.49, all the boundary conditions can be directly substituted into Equation 7.41. As a result, the final eigenvalue equation system becomes

$$\sum_{k=3}^{N-2} C_1 W_{k,j} + 2\lambda^2 \sum_{k1=3}^{N-2} \sum_{k2=3}^{M-2} C_2 W_{k1,k2} + \lambda^4 \sum_{k=3}^{M-2} C_3 W_{i,k} = \Omega^2 W_{i,j} \quad (7.50)$$

where

$$C_1 = c_{i,k}^{(4)} - \frac{c_{i,2}^{(4)} \cdot AXK1 + c_{i,N-1}^{(4)} \cdot AXKN}{AXN},$$

$$C_2 = c_{i,k1}^{(2)} \cdot \bar{c}_{j,k2}^{(2)} - \frac{\left(AXK1 \cdot c_{i,2}^{(2)} + AXKN \cdot c_{i,N-1}^{(2)}\right)}{AXN} \cdot \bar{c}_{j,k2}^{(2)} -$$
$$\frac{\left(AYK1 \cdot \bar{c}_{j,2}^{(2)} + AYKM \cdot \bar{c}_{j,M-1}^{(2)}\right)}{AYM} \cdot c_{i,k1}^{(2)} +$$
$$\frac{\left(AXK1 \cdot AYK1 \cdot c_{i,2}^{(2)} \cdot \bar{c}_{j,2}^{(2)} + AXKN \cdot AYK1 \cdot c_{i,N-1}^{(2)} \cdot \bar{c}_{j,2}^{(2)}\right)}{AXN \cdot AYM} +$$
$$\frac{\left(AXK1 \cdot AYKM \cdot c_{i,2}^{(2)} \cdot \bar{c}_{j,M-1}^{(2)} + AXKN \cdot AYKM \cdot c_{i,N-1}^{(2)} \cdot \bar{c}_{j,M-1}^{(2)}\right)}{AXN \cdot AYM}$$

$$C_3 = \bar{c}_{j,k}^{(4)} - \frac{\bar{c}_{j,2}^{(4)} \cdot AYK1 + \bar{c}_{j,M-1}^{(4)} \cdot AYKM}{AYM}$$

Since Equation 7.50 has $(N-4)\times(M-4)$ unknowns, it should be applied at $(N-4)\times(M-4)$ interior points to close the system. This can be done by applying it at the interior points, $3 \le i \le N-2$, $3 \le j \le M-2$, as shown in Figure 7.2.

A Fortran program using this approach for the free vibration analysis of rectangular plates is attached in Appendix *B*. It should be indicated that Equations 7.46*b* and 7.46*c* or Equations 7.46*d* and 7.46*e* can be coupled to give two solutions at different locations. Shu and Chen (1999) studied this problem in detail, and found that the location of the two solutions used in the present approach is the best choice.

7.3.3.4 General Approach

This approach was proposed by Shu and Du (1997*b*) for the implementation of general boundary conditions. For this approach, the two physical conditions at a boundary point are discretized, if any, by the DQ method. Then, one discrete condition is considered as the equation for the boundary point itself, and the other discrete condition is taken as the equation for its adjacent interior point. To illustrate this approach, we consider a plate configuration of C-SS-F-F. After performing the DQ discretization, the discretized boundary conditions along the four edges of a rectangular plate can be written as

$$W_{1,j} = 0 \tag{7.51a}$$

$$\sum_{k=1}^{N} c_{1,k}^{(1)} \cdot W_{k,j} = 0 \tag{7.51b}$$

for the edge of $X = 0$, and

$$\sum_{k=1}^{N} c_{N,k}^{(2)} \cdot W_{k,j} + \nu \cdot \lambda^2 \cdot \sum_{k=1}^{M} \bar{c}_{j,k}^{(2)} \cdot W_{N,k} = 0 \tag{7.52a}$$

$$\sum_{k=1}^{N} c_{N,k}^{(3)} \cdot W_{k,j} + (2-\nu) \cdot \lambda^2 \cdot \sum_{k1=1}^{N} \sum_{k2=1}^{M} c_{N,k1}^{(1)} \cdot \bar{c}_{j,k2}^{(2)} \cdot W_{k1,k2} = 0 \tag{7.52b}$$

for the edge of $X = 1$, and

$$W_{i,1} = 0 \tag{7.53a}$$

$$\sum_{k=1}^{M} \bar{c}_{1,k}^{(2)} \cdot W_{i,k} = 0 \tag{7.53b}$$

for the edge of $Y = 0$, and

$$\lambda^2 \cdot \sum_{k=1}^{M} \bar{c}_{M,k}^{(2)} \cdot W_{i,k} + \nu \cdot \sum_{k=1}^{N} c_{i,k}^{(2)} \cdot W_{k,M} = 0 \tag{7.54a}$$

$$\lambda^3 \cdot \sum_{k=1}^{M} \bar{c}_{M,k}^{(3)} \cdot W_{i,k} + (2-\nu) \cdot \lambda^2 \cdot \sum_{k1=1}^{N} \sum_{k2=1}^{M} c_{i,k1}^{(2)} \cdot \bar{c}_{M,k2}^{(1)} \cdot W_{k1,k2} = 0 \tag{7.54b}$$

for the edge of $Y = 1$.

It is noted that Equations 7.52 and 7.54 involve all the functional values in the whole computational domain. Thus, it is very difficult to couple Equations 7.51 and 7.52 or Equations 7.53 and 7.54 together to obtain analytical expressions for the solution of $W_{2,j}$ and $W_{N-1,j}$ or $W_{i,2}$ and $W_{i,M-1}$, as previously done in the approach of directly substituting the boundary conditions into the discrete governing equations. For the two boundary equations at each boundary point, Equations 7.51*a*, 7.52*a*, 7.53*a* and 7.54*a* are considered as the equations for the boundary points themselves, while Equations 7.51*b*, 7.52*b*, 7.53*b* and 7.54*b* are considered as the equations for their adjacent interior points. For example, Equation 7.51*b* can be taken as the equation for the adjacent interior points $i = 2, j = 2, 3, \ldots, M-1$. At the four corners, the discretized boundary conditions can be written as

$$W_{1,1} = 0, \; W_{N,1} = 0, \; W_{1,M} = 0 \tag{7.55a}$$

$$\sum_{k1=1}^{N} \sum_{k2=1}^{M} c_{N,k1}^{(1)} \cdot \bar{c}_{M,k2}^{(1)} \cdot W_{k1,k2} = 0 \tag{7.55b}$$

Clearly, Equation 7.55*b* involves all the functional values in the whole computational domain. For the present approach, the functional values in the whole domain are decomposed into two portions. One portion is based on the interior points where the governing equations are applied. The functional values of this portion can be represented by a vector $\{W_I\}$. The other portion consists of the remaining functional values located at the boundary points and their adjacent interior points, which can be denoted by a vector $\{W_B\}$. The *ij* component of $\{W_I\}$ can be defined by

$$\{W_I\}_{ij} = W_{i,j} \tag{7.56}$$

where $ij = (M-4)\cdot(i-3) + j - 2$, $3 \le i \le N-2, 3 \le j \le M-2$. The *ij* components of $\{W_B\}$ are defined by

$$\{W_B\}_{ij} = W_{i,j} \tag{7.57}$$

where

$$ij = \begin{cases} (i-1)\cdot M + j & \text{when } 1 \le i \le 2, 1 \le j \le M \\ 2M + 4\cdot(i-3) + j & \text{when } 3 \le i \le N-2, 1 \le j \le 2 \\ 2M + 4\cdot(i-3) + 2 + j - M + 2 & \text{when } 3 \le i \le N-2, M-1 \le j \le M \\ 2M + 4\cdot(N-4) + (i-N+1)\cdot M + j & \text{when } N-1 \le i \le N, 1 \le j \le M \end{cases}$$

With the above definitions, Equation 7.41 can be written in the following matrix form

$$[A_{IB}]\cdot\{W_B\} + [A_{II}]\cdot\{W_I\} = \Omega^2 \cdot \{W_I\} \tag{7.58}$$

Similarly, Equations 7.51-7.55 can be put in the following matrix form

$$[A_{BB}]\cdot\{W_B\}+[A_{BI}]\cdot\{W_I\}=0 \tag{7.59}$$

Substituting Equation 7.59 into Equation 7.58 gives the final eigenvalue equation system as

$$([A_{II}]-[A_{IB}]\cdot[A_{BB}]^{-1}\cdot[A_{BI}])\cdot\{W_I\}=\Omega^2\cdot\{W_I\} \tag{7.60}$$

It is noted that the dimensions of matrices $[A_{II}]$, $[A_{BB}]$, $[A_{IB}]$ and $[A_{BI}]$ are $(N-4)\times(M-4)$ by $(N-4)\times(M-4)$, $(4N+4M-16)$ by $(4N+4M-16)$, $(N-4)\times(M-4)$ by $(4N+4M-16)$, and $(4N+4M-16)$ by $(N-4)\times(M-4)$, respectively. It was demonstrated by Shu and Du (1997*b*) that this approach is a general method for implementing the boundary conditions. It can be applied to any type of boundary conditions including the combination of the clamped and simply supported edge conditions.

It should be indicated that although the above approaches are demonstrated by uniform boundary conditions, they can be easily applied to non-uniform and mixed boundary conditions. Some works on implementing mixed and non-uniform boundary conditions have been done by Shu and Wang (1999), Siddiqi and Kukreti (1998).

7.3.4 Numerical Example: Free Vibration Analysis of Square Plates

We will use the free vibration analysis of a square plate as an example to demonstrate the high accuracy of the DQ results, and the effect of mesh point distribution on the free corner condition. The derivatives are discretized by the PDQ method, and the mesh point distribution is given by Equation 7.39. We will only consider three plate configurations, that is, SS-SS-SS-SS, C-C-C-C, and F-F-F-F. For the SS-SS-SS-SS and C-C-C-C plate configurations, both the method of direct substitution of boundary conditions into the discrete governing equations and the general approach can be applied to implement the boundary conditions. It was found that both methods provide exactly the same numerical results. However, the general approach requires more virtual storage and computational effort. For the F-F-F-F plate configuration, the boundary condition can only be implemented by the general approach. Table 7.2 shows the natural frequencies of the first five modes for SS-SS-SS-SS, C-C-C-C, and F-F-F-F plate configurations. The DQ results were given by Shu and Du (1997*b*) using a mesh size of 12×12. For the purpose of comparison, the results of Leissa (1973), Leissa and Narita (1984) are also included in Table 7.2. Note that Leissa's results are given from the Rayleigh-Ritz method, which are exact for the SS-SS-SS-SS plate configuration. The results of Leissa and Narita (1984) are also obtained by using Rayleigh-Ritz method with more terms for each symmetry case. Thus, it is believed that the results of Leissa and Narita (1984) are more accurate than those of Leissa (1973). It can be observed from Table 7.2 that the DQ results for the SS-SS-SS-SS and C-C-C-C configurations agree very well with the data of Leissa (1973). However, for the F-F-F-F configuration, the DQ results exhibit some errors.

Table 7.2 Comparison of natural frequencies of a square plate, $N = M = 12$ (Shu and Du 1997*b*)

	Ω_1	Ω_2	Ω_3	Ω_4	Ω_5
			SS-SS-SS-SS		
Leissa (1973)	19.7392	49.3480	49.3480	78.9568	98.6960
PDQ ($\alpha = \beta = 1$)	19.7392	49.3495	49.3495	78.9586	98.4154
			C-C-C-C		
Leissa (1973)	35.992	73.413	73.413	108.270	131.640
PDQ ($\alpha = \beta = 1$)	35.986	73.399	73.399	108.230	131.418
			F-F-F-F		
Leissa (1973)	13.489	19.789	24.432	35.024	35.024
Leissa and Narita (1984)	13.468	19.596	24.271	34.801	34.801
PDQ ($\alpha = \beta = 1$)	7.947	19.600	23.293	30.590	31.409
PDQ ($\alpha = \beta = 0$)	13.454	19.597	24.271	34.815	34.817

Shu and Du (1997*b*) found that for the plate configuration with at least one free corner, the DQ results are very sensitive to the mesh point distribution. When the mesh point distribution is obtained from Equation 7.39, the DQ results are quite erroneous. However, when the mesh points are further clustered towards the boundary, the accuracy of the DQ results is greatly improved. To stretch the grid, Shu (1991) proposed the following mesh point distribution

$$X_i = (1-\alpha)(3\xi_i^2 - 2\xi_i^3) + \alpha\xi_i, \qquad i=1, 2, \ldots, N \qquad (7.61a)$$

$$Y_j = (1-\beta)(3\eta_j^2 - 2\eta_j^3) + \beta\eta_j, \qquad j=1, 2, \ldots, M \qquad (7.61b)$$

where α and β are the stretching parameters, ξ_i and η_j are the coordinates of the basic mesh points given by Equation 7.39. Obviously, when $\alpha = \beta = 1$, the mesh point distribution given by Equation 7.61 is exactly the same as that given by Equation 7.39. When α and β approach zero, the mesh points are strongly clustered towards the boundaries. Shu and Du (1997*b*) found that when α and β are taken as zero, the DQ results using a mesh size of 12×12 are very accurate for the F-F-F-F configuration. This can be clearly seen from Table 7.2. Table 7.2 shows that the DQ results of $\alpha = \beta = 0$ are closer to the data of Leissa and Narita (1984) than the results of Leissa (1973). This may indicate that the DQ results are very accurate. For the other plate configurations with at least one free corner, the mesh point distribution given by Equation 7.61 with $\alpha = \beta = 0$ can also generate accurate DQ results.

7.4 Differential Quadrature Analysis of Shells

In this section, we will describe the application of the DQ method to the vibration analysis of shells. Since the cylindrical shell is a special case of conical shells, the description will mainly focus on the conical shell.

7.4.1 Governing Equations and Boundary Conditions

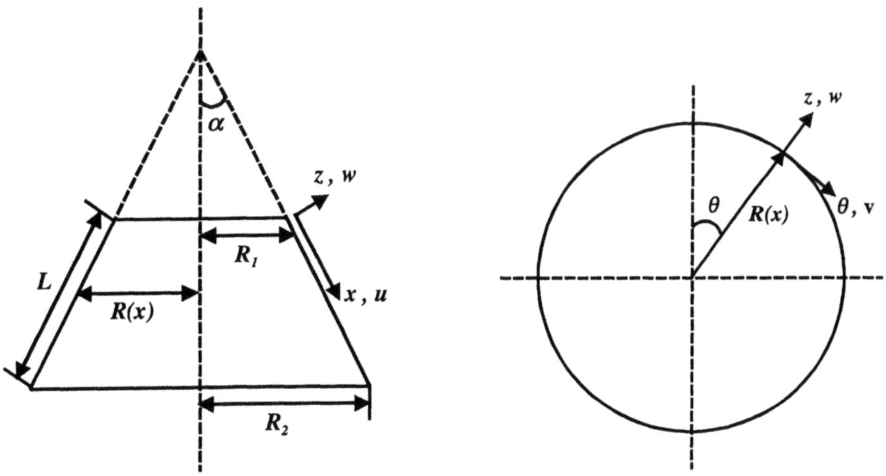

Figure 7.3 Geometry of a conical shell

Consider a conical shell as shown in Figure 7.3. R_1 and R_2 are respectively the radii of the cone at its small and large edges, α is the semivertex angle of the cone and $L = (R_2 - R_1)/\sin\alpha$ is the cone length along its generator. The conical shell is referred to a coordinate system (x, θ, z) as shown in Figure 7.3. The components of the displacement of the shell with reference to this coordinate system are denoted by u, v and w in the x, θ and z directions, respectively. The radius of the cone at any point along its length is given by

$$R(x) = R_1 + x\sin\alpha \tag{7.62}$$

Note that when $\alpha = 0$, the conical shell becomes a cylindrical shell. The equilibrium equations of motion in terms of the force and moment resultants can be written as

$$L_x(u,v,w) - \rho_t \frac{\partial^2 u}{\partial t^2} = 0 \tag{7.63a}$$

$$L_\theta(u,v,w) - \rho_t \frac{\partial^2 v}{\partial t^2} = 0 \tag{7.63b}$$

$$L_z(u,v,w) - \rho_t \frac{\partial^2 w}{\partial t^2} = 0 \tag{7.63c}$$

where

$$L_x = \frac{\partial N_x}{\partial x} + \frac{\sin\alpha}{R}(N_x - N_\theta) + \frac{1}{R}\frac{\partial N_{x\theta}}{\partial \theta} \tag{7.64a}$$

$$L_\theta = \frac{\partial N_{x\theta}}{\partial x} + \frac{1}{R}\frac{\partial N_\theta}{\partial \theta} + \frac{2\sin\alpha}{R} N_{x\theta} + \frac{\cos\alpha}{R} Q_\theta \tag{7.64b}$$

$$L_z = \frac{\partial Q_x}{\partial x} + \frac{1}{R}\frac{\partial Q_\theta}{\partial \theta} + \frac{\sin\alpha}{R} Q_x - \frac{\cos\alpha}{R} N_\theta \tag{7.64c}$$

$$\rho_t = \int_{-h/2}^{h/2} \rho \, dz \tag{7.64d}$$

Here ρ and ρ_t are the density and density per unit length respectively, N_j and M_j are the force and moment resultants defined respectively as

$$N = (N_x, N_\theta, N_{x\theta})^T = \int_{-h/2}^{h/2} (\sigma_x, \sigma_\theta, \sigma_{x\theta})^T \, dz \tag{7.65a}$$

$$M = (M_x, M_\theta, M_{x\theta})^T = \int_{-h/2}^{h/2} (\sigma_x, \sigma_\theta, \sigma_{x\theta})^T z \, dz \tag{7.65b}$$

The transverse shear force resultants can be given from M_x, M_θ and $M_{x\theta}$ by

$$Q_x = \frac{1}{R}\frac{\partial}{\partial x}[RM_x] - \frac{M_\theta \sin\alpha}{R} + \frac{1}{R}\frac{\partial M_{x\theta}}{\partial \theta} \tag{7.66a}$$

$$Q_\theta = \frac{1}{R}\frac{\partial}{\partial x}[RM_{x\theta}] + \frac{M_{x\theta} \sin\alpha}{R} + \frac{1}{R}\frac{\partial M_\theta}{\partial \theta} \tag{7.66b}$$

For a thin and generally orthotropic layer, the stresses defined in Equation 7.65 are governed by

$$\begin{pmatrix} \sigma_x \\ \sigma_\theta \\ \sigma_{x\theta} \end{pmatrix} = \begin{bmatrix} \overline{Q}_{11} & \overline{Q}_{12} & \overline{Q}_{16} \\ \overline{Q}_{12} & \overline{Q}_{22} & \overline{Q}_{26} \\ \overline{Q}_{16} & \overline{Q}_{26} & \overline{Q}_{66} \end{bmatrix} \begin{pmatrix} e_x \\ e_\theta \\ e_{x\theta} \end{pmatrix} \tag{7.67}$$

where the elasticity matrix $[\overline{Q}]$ is defined as

$$[\overline{Q}] = [T][Q][T]^T \tag{7.68}$$

in which $[T]$ is the transformation matrix between the principal material coordinates and the shell's coordinates and is defined as

$$[T] = \begin{bmatrix} \cos^2\phi & \sin^2\phi & -2\sin\phi\cdot\cos\phi \\ \sin^2\phi & \cos^2\phi & 2\sin\phi\cdot\cos\alpha\ \phi \\ \sin\phi\cdot\cos\phi & -\sin\phi\cdot\cos\phi & \cos^2\phi-\sin^2\phi \end{bmatrix} \tag{7.69}$$

where ϕ is the angular orientation of the fibers. The material constants in $[Q]$ are defined as

$$Q_{11} = \frac{E_x}{1-\nu_{x\theta}\cdot\nu_{\theta x}} \tag{7.70a}$$

$$Q_{12} = \frac{\nu_{x\theta}E_\theta}{1-\nu_{x\theta}\cdot\nu_{\theta x}} \tag{7.70b}$$

$$Q_{22} = \frac{E_\theta}{1-\nu_{x\theta}\cdot\nu_{\theta x}} \tag{7.70c}$$

$$Q_{66} = G_{x\theta} \tag{7.70d}$$

where E_x and E_θ are Young's moduli, $G_{x\theta}$ is the shear modulus, and $\nu_{x\theta}$ and $\nu_{\theta x}$ are Poisson's ratios. In Equation 7.67, e_x, e_θ and $e_{x\theta}$ are the strains in the axial direction, circumferential direction and the shear strain on the $x\theta$ plane respectively. Following Love's first approximation theory, these strain components are defined as linear functions of the thickness coordinate z

$$e_x = e_1 + zk_1 \tag{7.71a}$$

$$e_\theta = e_2 + zk_2 \tag{7.71b}$$

$$e_{x\theta} = \gamma + 2z\tau \tag{7.71c}$$

where e_1, e_2 and γ are the reference surface strains, k_1, k_2 and τ are the reference surface curvatures. The reference surface strains and curvatures are defined as

$$e = \begin{pmatrix} e_1 \\ e_2 \\ \gamma \end{pmatrix} = \begin{bmatrix} \dfrac{\partial u}{\partial x} \\ \dfrac{1}{R}\dfrac{\partial v}{\partial \theta} + \dfrac{u \sin\alpha}{R} + \dfrac{w \cos\alpha}{R} \\ \dfrac{1}{R}\dfrac{\partial u}{\partial \theta} + \dfrac{\partial v}{\partial x} - \dfrac{v \sin\alpha}{R} \end{bmatrix} \tag{7.72a}$$

$$k = \begin{pmatrix} k_1 \\ k_2 \\ 2\tau \end{pmatrix} = \begin{bmatrix} -\dfrac{\partial^2 w}{\partial x^2} \\ -\dfrac{1}{R^2}\dfrac{\partial^2 w}{\partial \theta^2} + \dfrac{\cos\alpha}{R^2}\dfrac{\partial v}{\partial \theta} - \dfrac{\sin\alpha}{R}\dfrac{\partial w}{\partial x} \\ 2\left(-\dfrac{1}{R}\dfrac{\partial^2 w}{\partial x \partial \theta} + \dfrac{\sin\alpha}{R^2}\dfrac{\partial w}{\partial \theta} + \dfrac{\cos\alpha}{R}\dfrac{\partial v}{\partial x} - \dfrac{v \sin\alpha \cdot \cos\alpha}{R^2} \right) \end{bmatrix} \tag{7.72b}$$

By substituting Equations 7.67-7.72 into Equation 7.65, the force and moment resultants can be obtained by

$$\begin{bmatrix} N \\ M \end{bmatrix} = \begin{bmatrix} A & B \\ B & D \end{bmatrix} \cdot \begin{bmatrix} e \\ k \end{bmatrix} \tag{7.73}$$

where A_{ij}, B_{ij} and D_{ij} are the extensional, coupling and bending stiffnesses, which are defined respectively as

$$(A_{ij}, B_{ij}, D_{ij}) = \int_{-h/2}^{h/2} \overline{Q}_{ij} (1, z, z^2)\, dz \tag{7.74}$$

For a shell that is composed of different layers of orthotropic materials, these stiffnesses can be written as

$$A_{ij} = \sum_{k=1}^{N_l} \overline{Q}_{ij}^k (h_k - h_{k-1}) \tag{7.75a}$$

$$B_{ij} = \frac{1}{2} \sum_{k=1}^{N_l} \overline{Q}_{ij}^k (h_k^2 - h_{k-1}^2) \tag{7.75b}$$

$$D_{ij} = \frac{1}{3} \sum_{k=1}^{N_l} \overline{Q}_{ij}^k (h_k^3 - h_{k-1}^3) \tag{7.75c}$$

where N_l denotes the number of layers in the laminated shell, h_k and h_{k-1} denote the distances from the shell reference surface to the outer and inner surfaces of the kth layer. For an isotropic shell, $B_{ij} = 0$, A_{ij} and D_{ij} can be simply computed by

$$A_{ij} = Q_{ij} h, \quad i, j = 1, 2, 6 \tag{7.76a}$$

$$D_{ij} = \frac{1}{12} Q_{ij} h^3, \quad i, j = 1, 2, 6 \tag{7.76b}$$

where h is the shell thickness. Substituting Equation 7.73 into Equation 7.63 leads to

$$L_{11}u + L_{12}v + L_{13}w = \rho h \frac{\partial^2 u}{\partial t^2} \tag{7.77a}$$

$$L_{21}u + L_{22}v + L_{23}w = \rho h \frac{\partial^2 v}{\partial t^2} \tag{7.77b}$$

$$L_{31}u + L_{32}v + L_{33}w = \rho h \frac{\partial^2 w}{\partial t^2} \tag{7.77c}$$

where

$$L_{11} = A_{11}\frac{\partial^2}{\partial x^2} + \frac{A_{11}\sin\alpha}{R}\frac{\partial}{\partial x} - \frac{A_{22}\sin^2\alpha}{R^2} + \frac{A_{66}}{R^2}\frac{\partial^2}{\partial \theta^2}$$

$$L_{12} = \frac{(A_{12} + A_{66})}{R}\frac{\partial^2}{\partial x \partial \theta} - \frac{(A_{22} + A_{66})\sin\alpha}{R^2}\frac{\partial}{\partial \theta} + \frac{(B_{12} + 2B_{66})\cos\alpha}{R^2}\frac{\partial^2}{\partial x \partial \theta}$$
$$- \frac{(B_{12} + B_{22} + 2B_{66})\sin\alpha\cos\alpha}{R^3}\frac{\partial}{\partial \theta}$$

$$L_{13} = \frac{A_{12}\cos\alpha}{R}\frac{\partial}{\partial x} - \frac{A_{22}\sin\alpha\cos\alpha}{R^2} - B_{11}\frac{\partial^3}{\partial x^3} - \frac{(B_{12} + 2B_{66})}{R^2}\frac{\partial^3}{\partial x \partial \theta^2}$$
$$- \frac{B_{11}\sin\alpha}{R}\frac{\partial^2}{\partial x^2} + \frac{(B_{12} + B_{22} + 2B_{66})\sin\alpha}{R^3}\frac{\partial^2}{\partial \theta^2} + \frac{B_{22}\sin^2\alpha}{R^2}\frac{\partial}{\partial x}$$

$$L_{21} = \frac{(A_{12} + A_{66})}{R}\frac{\partial^2}{\partial x \partial \theta} + \frac{(A_{22} + A_{66})\sin\alpha}{R^2}\frac{\partial}{\partial \theta}$$
$$+ \frac{(B_{12} + B_{66})\cos\alpha}{R^2}\frac{\partial^2}{\partial x \partial \theta} + \frac{(B_{22} - B_{66})\sin\alpha\cos\alpha}{R^3}\frac{\partial}{\partial \theta}$$

$$L_{22} = A_{66}\left[\frac{\partial^2}{\partial x^2}+\frac{\sin\alpha}{R}\frac{\partial}{\partial x}-\frac{\sin^2\alpha}{R^2}\right]+\left(\frac{A_{22}}{R^2}+\frac{2B_{22}\cos\alpha}{R^3}+\frac{D_{22}\cos^2\alpha}{R^4}\right)\frac{\partial^2}{\partial\theta^2}$$

$$+\frac{B_{66}\cos\alpha}{R}\left[\frac{3\partial^2}{\partial x^2}-\frac{\sin\alpha}{R}\frac{\partial}{\partial x}+\frac{\sin^2\alpha}{R^2}\right]+\frac{2D_{66}\cos^2\alpha}{R^2}\left[\frac{\partial^2}{\partial x^2}-\frac{2\sin\alpha}{R}\frac{\partial}{\partial x}+\frac{2\sin^2\alpha}{R^2}\right]$$

$$L_{23} = \left(\frac{A_{22}\cos\alpha}{R^2}+\frac{B_{22}\cos^2\alpha}{R^3}-\frac{4D_{66}\cos\alpha\sin^2\alpha}{R^4}\right)\frac{\partial}{\partial\theta}-\left(\frac{B_{22}}{R^3}+\frac{D_{22}\cos\alpha}{R^4}\right)\frac{\partial^3}{\partial\theta^3}$$

$$-\left(\frac{B_{22}\sin\alpha}{R^2}+\frac{(D_{22}-4D_{66})\sin\alpha\cdot\cos\alpha}{R^3}\right)\frac{\partial^2}{\partial x\partial\theta}$$

$$-\left(\frac{(D_{12}+2D_{66})\cos\alpha}{R^2}+\frac{(B_{12}+2B_{66})}{R}\right)\frac{\partial^3}{\partial x^2\partial\theta}$$

$$L_{31} = -\frac{A_{12}\cos\alpha}{R}\frac{\partial}{\partial x}-\frac{A_{22}\sin\alpha\cdot\cos\alpha}{R^2}+B_{11}\frac{\partial^3}{\partial x^3}+\frac{(B_{12}+2B_{66})}{R^2}\frac{\partial^3}{\partial x\partial\theta^2}$$

$$+\frac{2B_{11}\sin\alpha}{R}\frac{\partial^2}{\partial x^2}-\frac{B_{22}\sin^2\alpha}{R^2}\frac{\partial}{\partial x}+\frac{(B_{22}-2B_{66})\sin\alpha}{R^3}\frac{\partial^2}{\partial\theta^2}+\frac{B_{22}\sin^3\alpha}{R^3}$$

$$L_{32} = -\left(\frac{A_{22}\cos\alpha}{R^2}+\frac{\left(B_{22}\cos2\alpha-2B_{66}\sin^2\alpha\right)}{R^3}-\frac{\left(2D_{12}+2D_{22}+8D_{66}\right)\cos\alpha\sin^2\alpha}{R^4}\right)\frac{\partial}{\partial\theta}$$

$$+\left(\frac{B_{22}}{R^3}+\frac{D_{22}\cos\alpha}{R^4}\right)\frac{\partial^3}{\partial\theta^3}+\left(\frac{(B_{12}+2B_{66})}{R}+\frac{(D_{12}+4D_{66})\cos\alpha}{R^2}\right)\frac{\partial^3}{\partial x^2\partial\theta}$$

$$-\left(\frac{(B_{22}+2B_{66})\sin\alpha}{R^2}+\frac{(D_{22}+2D_{12}+8D_{66})\sin\alpha\cdot\cos\alpha}{R^3}\right)\frac{\partial^2}{\partial x\partial\theta}$$

$$L_{33} = -\frac{A_{22}\cos^2\alpha}{R^2}+\frac{2B_{12}\cos\alpha}{R}\frac{\partial^2}{\partial x^2}+\frac{2B_{22}\cos\alpha}{R^3}\frac{\partial^2}{\partial\theta^2}+\frac{B_{22}\cos\alpha\sin^2\alpha}{R^3}$$

$$-D_{11}\frac{\partial^4}{\partial x^4}-\frac{2(D_{12}+2D_{66})}{R^2}\frac{\partial^4}{\partial x^2\partial\theta^2}-\frac{D_{22}}{R^4}\frac{\partial^4}{\partial\theta^4}$$

$$-\frac{2D_{11}\sin\alpha}{R}\frac{\partial^3}{\partial x^3}+\frac{2(D_{12}+4D_{66})\sin\alpha}{R^3}\frac{\partial^3}{\partial x\partial\theta^2}+\frac{D_{22}\sin^2\alpha}{R^2}\frac{\partial^2}{\partial x^2}$$

$$-\frac{2(D_{12}+D_{22}+4D_{66})\sin^2\alpha}{R^4}\frac{\partial^2}{\partial\theta^2}-\frac{D_{22}\sin^3\alpha}{R^3}\frac{\partial}{\partial x}$$

A general relation for the displacement fields of a thin conical shell with general boundary conditions can be written as

$$u = U(x)\cdot\cos(n\theta)\cdot\cos(\omega t) \tag{7.78a}$$

$$v = V(x)\cdot\sin(n\theta)\cdot\cos(\omega t) \tag{7.78b}$$

$$w = W(x)\cdot\cos(n\theta)\cdot\cos(\omega t) \tag{7.78c}$$

Substituting Equation 7.78 into Equation 7.77, the governing equations can be rewritten as

$$S_{110}U + S_{111}U^{(1)} + S_{112}U^{(2)} + S_{120}V + S_{121}V^{(1)} + S_{130}W + S_{131}W^{(1)} + S_{132}W^{(2)} + S_{133}W^{(3)} = -\rho h\omega^2 U \tag{7.79a}$$

$$S_{210}U + S_{211}U^{(1)} + S_{220}V + S_{221}V^{(1)} + S_{222}V^{(2)} + S_{230}W + S_{231}W^{(1)} + S_{232}W^{(2)} = -\rho h\omega^2 V \tag{7.79b}$$

$$S_{310}U + S_{311}U^{(1)} + S_{312}U^{(2)} + S_{313}U^{(3)} + S_{320}V + S_{321}V^{(1)} + S_{322}V^{(2)} + S_{330}W + S_{331}W^{(1)} + S_{332}W^{(2)} + S_{333}W^{(3)} + S_{334}W^{(4)} = -\rho h\omega^2 W \tag{7.79c}$$

where $U^{(i)}, V^{(i)}$ and $W^{(i)}$ denote respectively the ith order derivative of U, V and W. S_{ijk} are the constant coefficients defined as follows

$$S_{110} = -\frac{A_{66}n^2}{R^2} - \frac{A_{22}\sin^2\alpha}{R^2}$$

$$S_{111} = \frac{A_{11}\sin\alpha}{R}$$

$$S_{112} = A_{11}$$

$$S_{120} = -\frac{(B_{12} + B_{22} + 2B_{66})n\cdot\sin\alpha\cdot\cos\alpha}{R^3} - \frac{(A_{22} + A_{66})n\cdot\sin\alpha}{R^2}$$

$$S_{121} = \frac{(B_{12} + 2B_{66})n\cdot\cos\alpha}{R^2} + \frac{(A_{12} + A_{66})n}{R}$$

$$S_{130} = -\frac{(B_{12} + B_{22} + 2B_{66})n^2\cdot\sin\alpha}{R^3} - \frac{A_{22}\sin\alpha\cdot\cos\alpha}{R^2}$$

$$S_{131} = \frac{(B_{12} + 2B_{66})n^2}{R^2} + \frac{B_{22}\sin^2\alpha}{R^2} + \frac{A_{12}\cos\alpha}{R}$$

$$S_{132} = -\frac{B_{11}\sin\alpha}{R}$$

$$S_{133} = -B_{11}$$

$$S_{210} = -\frac{(B_{22} - B_{66})n \cdot \sin\alpha \cdot \cos\alpha}{R^3} - \frac{(A_{22} + A_{66})n \cdot \sin\alpha}{R^2}$$

$$S_{211} = -\left(\frac{B_{12}n \cdot \cos\alpha}{R^2} + \frac{B_{66}n \cdot \cos\alpha}{R^2} + \frac{A_{12}n}{R} + \frac{A_{66}n}{R}\right)$$

$$S_{220} = -\frac{(D_{22}n^2 - 4D_{66}\sin^2\alpha)\cos^2\alpha}{R^4} - \frac{(2B_{22}n^2 - B_{66}\sin^2\alpha)\cos\alpha}{R^3} - \frac{(A_{22}n^2 + A_{66}\sin^2\alpha)}{R^2}$$

$$S_{221} = -\frac{4D_{66}\sin\alpha \cdot \cos^2\alpha}{R^3} - \frac{B_{66}\sin\alpha \cdot \cos\alpha}{R^2} + \frac{A_{66}\sin\alpha}{R}$$

$$S_{222} = A_{66} + \frac{2D_{66}\cos^2\alpha}{R^2} + \frac{3B_{66}\cos\alpha}{R}$$

$$S_{230} = -\left(\frac{(D_{22}n^2 - 4D_{66}\sin^2\alpha)n\cos\alpha}{R^4} + \frac{B_{22}n(\cos^2\alpha + n^2)}{R^3} + \frac{A_{22}n\cos\alpha}{R^2}\right)$$

$$S_{231} = \frac{(D_{22} - 4D_{66})n\cos\alpha \cdot \sin\alpha}{R^3} + \frac{B_{22}n\sin\alpha}{R^2}$$

$$S_{232} = \frac{(D_{12} + 2D_{66})n\cos\alpha}{R^2} + \frac{(B_{12} + 2B_{66})n}{R}$$

$$S_{310} = \frac{(-B_{22}n^2 + 2B_{66}n^2 + B_{22}\sin^2\alpha)\sin\alpha}{R^3} - \frac{A_{22}\sin\alpha \cdot \cos\alpha}{R^2}$$

$$S_{311} = -\left(\frac{B_{12}n^2 + 2B_{66}n^2 + B_{22}\sin^2\alpha}{R^2} + \frac{A_{12}\cos\alpha}{R}\right)$$

$$S_{312} = \frac{2B_{11}\sin\alpha}{R}$$

$$S_{313} = B_{11}$$

$$S_{320} = \frac{(-D_{22}n^2 + 2D_{12}\sin^2\alpha + 2D_{22}\sin^2\alpha + 8D_{66}\sin^2\alpha)n\cos\alpha}{R^4} - \frac{A_{22}n\cos\alpha}{R^2} + \frac{(-B_{22}n^2 - B_{22}\cos^2\alpha + B_{22}\sin^2\alpha + 2B_{66}\sin^2\alpha)n}{R^3}$$

$$S_{321} = -\frac{(D_{22} + 2D_{12} + 8D_{66})n\cos\alpha \cdot \sin\alpha}{R^3} - \frac{(B_{22} + 2B_{66})n\sin\alpha}{R^2}$$

$$S_{322} = \frac{(D_{12} + 4D_{66})n\cos\alpha}{R^2} + \frac{(B_{12} + 2B_{66})n}{R}$$

$$S_{330} = \frac{(-D_{22}n^2 + 2D_{12}\sin^2\alpha + 2D_{22}\sin^2\alpha + 8D_{66}\sin^2\alpha)n^2}{R^4}$$

$$+ \frac{(-2n^2 + \sin^2\alpha)B_{22}\cos\alpha}{R^3} - \frac{A_{22}\cos^2\alpha}{R^2}$$

$$S_{331} = -\frac{(D_{22}\sin^2\alpha + 2D_{12}n^2 + 8D_{66}n^2)\sin\alpha}{R^3}$$

$$S_{332} = \frac{2D_{12}n^2}{R^2} + \frac{4D_{66}n^2}{R^2} + \frac{D_{22}\sin^2\alpha}{R^2} + \frac{2B_{12}\cos\alpha}{R}$$

$$S_{333} = -\frac{2D_{11}\sin\alpha}{R}$$

$$S_{334} = -D_{11}$$

Equations 7.79 are the governing equations to be solved by the DQ method. The following two types of boundary conditions are usually encountered in shell analysis.

<u>Simply supported edge (SS)</u>

$$V = 0,\ W = 0,\ N_x = 0,\ M_x = 0 \tag{7.80}$$

Using Equations 7.72 and 7.73, Equation 7.80 can be simplified as

$$V = 0,\ W = 0 \tag{7.81a}$$

$$U^{(1)} + a_1U + a_2V + a_3W + a_4W^{(1)} + a_5W^{(2)} = 0 \tag{7.81b}$$

$$W^{(2)} + b_1U + b_2U^{(1)} + b_3V + b_4W + b_5W^{(1)} = 0 \tag{7.81c}$$

where a_i and b_i are the constant coefficients defined as

$$a_1 = \frac{A_{12}\sin\alpha}{A_{11}R}$$

$$a_2 = \frac{B_{12}n\cos\alpha}{A_{11}R^2} + \frac{A_{12}n}{A_{11}R}$$

$$a_3 = \frac{A_{12}\cos\alpha}{A_{11}R} + \frac{B_{12}n^2}{A_{11}R^2}$$

$$a_4 = -\frac{B_{12}\sin\alpha}{A_{11}R}$$

$$a_5 = -\frac{B_{11}}{A_{11}}$$

$$b_1 = -\frac{B_{12}\sin\alpha}{D_{11}R}$$

$$b_2 = -\frac{B_{11}}{D_{11}}$$

$$b_3 = -\frac{D_{12}n\cos\alpha}{D_{11}R^2} - \frac{B_{12}n}{D_{11}R}$$

$$b_4 = -\frac{B_{12}\cos\alpha}{D_{11}R} - \frac{D_{12}n^2}{D_{11}R^2}$$

$$b_5 = \frac{D_{12}\sin\alpha}{D_{11}R}$$

Clamped edge (C)

$$V = 0,\ W = 0,\ U = 0,\ W^{(1)} = 0 \tag{7.82}$$

It can be seen from Equations 7.81 and 7.82 that there are two boundary conditions for W and one boundary condition for U and V at each edge.

7.4.2 Numerical Discretization

For the numerical computation, the mesh point distribution can be taken as

$$x_i = \frac{L}{2}\left(1 - \cos\frac{i-1}{N-1}\pi\right), \qquad i = 1, 2, \ldots, N \tag{7.83}$$

With Equation 7.83, the DQ weighting coefficients can be easily calculated. Let $c_{i,j}^{(n)}$ be the DQ weighting coefficients of the nth order derivative. After DQ discretization, Equation 7.79 becomes

$$S_{110}U_i + \sum_{k=1}^{N}(S_{111}c_{i,k}^{(1)} + S_{112}c_{i,k}^{(2)})U_k + S_{120}V_i + \sum_{k=1}^{N}S_{121}c_{i,k}^{(1)}V_k$$

$$+ S_{130}W_i + \sum_{k=1}^{N}(S_{131}c_{i,k}^{(1)} + S_{132}c_{i,k}^{(2)} + S_{133}c_{i,k}^{(3)})W_k = -\rho h\omega^2 U_i \qquad (7.84a)$$

$$S_{210}U_i + \sum_{k=1}^{N}S_{211}c_{i,k}^{(1)}U_k + S_{220}V_i + \sum_{k=1}^{N}(S_{221}c_{i,k}^{(1)} + S_{222}c_{i,k}^{(2)})V_k$$

$$+ S_{230}W_i + \sum_{k=1}^{N}(S_{231}c_{i,k}^{(1)} + S_{232}c_{i,k}^{(2)})W_k = -\rho h\omega^2 V_i \qquad (7.84b)$$

$$S_{310}U_i + \sum_{k=1}^{N}(S_{311}c_{i,k}^{(1)} + S_{312}c_{i,k}^{(2)} + S_{313}c_{i,k}^{(3)})U_k + S_{320}V_i + \sum_{k=1}^{N}(S_{321}c_{i,k}^{(1)} + S_{322}c_{i,k}^{(2)})V_k$$

$$+ S_{330}W_i + \sum_{k=1}^{N}(S_{331}c_{i,k}^{(1)} + S_{332}c_{i,k}^{(2)} + S_{333}c_{i,k}^{(3)} + S_{334}c_{i,k}^{(4)})W_k = -\rho h\omega^2 W_i \qquad (7.84c)$$

It is noted that Equations 7.84*a*, 7.84*b*, 7.84*c* are the governing equations for variables *U, V* and *W* respectively. Since there is one boundary condition for *U* and *V* respectively, and two boundary conditions for *W* at each edge, Equations 7.84*a* and 7.84*b* have to be applied at *N*–2 interior points, $2 \le i \le N-1$, and Equation 7.84*c* should be applied at *N*–4 interior points, $3 \le i \le N-2$.

7.4.3 Implementation of Boundary Conditions

For simplicity, the conical shells with simply supported small and large edges (SS-SS) are chosen to demonstrate the implementation of the boundary conditions. After applying the DQ discretization, the boundary conditions at the small edge (R_1) can be written as

$$V_1 = 0, \qquad W_1 = 0 \qquad (7.85a)$$

$$(c_{1,1}^{(1)} + a_1^1)U_1 + \sum_{k=2}^{N}c_{1,k}^{(1)}U_k + a_2^1 V_1 + a_3^1 W_1 + \sum_{k=1}^{N}(a_4^1 c_{1,k}^{(1)} + a_5^1 c_{1,k}^{(2)})W_k = 0 \qquad (7.85b)$$

$$b_1^1 U_1 + b_2^1 \sum_{k=1}^{N}c_{1,k}^{(1)}U_k + b_3^1 V_1 + b_4^1 W_1 + \sum_{k=1}^{N}(c_{1,k}^{(2)} + b_5^1 c_{1,k}^{(1)})W_k = 0 \qquad (7.85c)$$

Similarly, the boundary conditions at the large edge (R_2) can be discretized as

$$V_N = 0, \; W_N = 0 \qquad (7.86a)$$

$$(c_{N,N}^{(1)} + a_1^N)U_N + \sum_{k=1}^{N-1} c_{N,k}^{(1)} U_k + a_2^N V_N + a_3^N W_N$$

$$+ \sum_{k=1}^{N} (a_4^N c_{N,k}^{(1)} + a_5^N c_{N,k}^{(2)}) W_k = 0 \tag{7.86b}$$

$$b_1^N U_N + b_2^N \sum_{k=1}^{N} c_{N,k}^{(1)} U_k + b_3^N V_N + b_4^N W_N + \sum_{k=1}^{N} (c_{N,k}^{(2)} + b_5^N c_{N,k}^{(1)}) W_k = 0 \tag{7.86c}$$

where a_i^1 and b_i^1 represent the values of a_i and b_i at the small edge, while a_i^N and b_i^N denote the values of a_i and b_i at the large edge. It is noted that Equations 7.85, 7.86 provide 8 boundary equations. Among them, V_1, V_N and W_1, W_N are considered to be the equations for V and W at two boundary points x_1 and x_N. Equations 7.85*b* and 7.86*b* are taken to be the equations for U at the boundary points x_1 and x_N. Since the equations for W at the boundary points are already given by W_1 and W_N, Equations 7.85*c* and 7.86*c* can be considered as the equations for W at two adjacent interior points, x_2 and x_{N-1}.

To derive the eigenvalue equation system, the functional values in the whole computational domain are decomposed into two portions. One portion is based on the interior points where the governing equations are applied. The functional values of this portion can be represented by a vector $\{Z_I\}$ given by

$$\{Z_I\}^T = \{U_2, U_3, \cdots, U_{N-1}, V_2, V_3, \cdots, V_{N-1}, W_3, W_4, \cdots, W_{N-2}\}^T \tag{7.87}$$

The other portion consists of the remaining functional values denoted by a vector $\{Z_B\}$, which is given by

$$\{Z_B\}^T = \{U_1, U_N, V_1, V_N, W_1, W_2, W_{N-1}, W_N\}^T \tag{7.88}$$

With the above definitions, Equation 7.84 can be written in the following matrix form

$$[A_{IB}] \cdot \{Z_B\} + [A_{II}] \cdot \{Z_I\} = -\rho h \omega^2 \cdot \{Z_I\} \tag{7.89}$$

Similarly, Equations 7.85, 7.86 can be put in the following matrix form

$$[A_{BB}] \cdot \{Z_B\} + [A_{BI}] \cdot \{Z_I\} = 0 \tag{7.90}$$

Substituting Equation 7.90 into Equation 7.89 gives the final eigenvalue equation system as

$$([A_{II}]-[A_{IB}]\cdot[A_{BB}]^{-1}\cdot[A_{BI}])\cdot\{Z_I\}=-\rho h\omega^2\cdot\{Z_I\} \qquad (7.91)$$

It is noted that the dimensions of matrices $[A_{II}],[A_{BB}],[A_{IB}]$ and $[A_{BI}]$ are (3N–8)× (3N–8), 8 ×8, (3N–8)×8, and 8×(3N–8) respectively.

When an isotropic conical shell is considered, the implementation of the boundary conditions can be simplified. For isotropic conical shells, $B_{ij}=0$, $i,j=1,2,6$, thus, $a_4=a_5=b_1=b_2=0$. Then using Equation 7.85*a*, Equations 7.85*b* and 7.85*c* can be reduced to

$$(c_{1,1}^{(1)}+a_1^1)U_1+c_{1,N}^{(1)}U_N=-\sum_{k=2}^{N-1}c_{1,k}^{(1)}U_k \qquad (7.92a)$$

$$(c_{1,2}^{(2)}+b_5^1c_{1,2}^{(1)})W_2+(c_{1,N-1}^{(2)}+b_5^1c_{1,N-1}^{(1)})W_{N-1}=-\sum_{k=3}^{N-2}(c_{1,k}^{(2)}+b_5^1c_{1,k}^{(1)})W_k \qquad (7.92b)$$

Similarly, Equations 7.86*b* and 7.86*c* can be simplified as

$$c_{N,1}^{(1)}U_1+(c_{N,N}^{(1)}+a_1^N)U_N=-\sum_{k=2}^{N-1}c_{N,k}^{(1)}\cdot U_k \qquad (7.93a)$$

$$(c_{N,2}^{(2)}+b_5^Nc_{N,2}^{(1)})W_2+(c_{N,N-1}^{(2)}+b_5^Nc_{N,N-1}^{(1)})W_{N-1}=-\sum_{k=3}^{N-2}(c_{N,k}^{(2)}+b_5^Nc_{N,k}^{(1)})W_k \qquad (7.93b)$$

Equations 7.92*a* and 7.93*a* can be coupled to give two solutions U_1 and U_N as

$$U_1=\frac{1}{AXN1}\cdot\sum_{k=2}^{N-1}[c_{1,N}^{(1)}\cdot c_{N,k}^{(1)}-(c_{N,N}^{(1)}+a_1^N)\cdot c_{1,k}^{(1)}]\cdot U_k \qquad (7.94a)$$

$$U_N=\frac{1}{AXN1}\cdot\sum_{k=2}^{N-1}[c_{N,1}^{(1)}\cdot c_{1,k}^{(1)}-(c_{1,1}^{(1)}+a_1^1)\cdot c_{N,k}^{(1)}]\cdot U_k \qquad (7.94b)$$

where

$$AXN1=(c_{1,1}^{(1)}+a_1^1)\cdot(c_{N,N}^{(1)}+a_1^N)-c_{1,N}^{(1)}\cdot c_{N,1}^{(1)}$$

In a similar way, Equations 7.92*b* and 7.93*b* can be coupled to provide two solutions, W_2 and W_{N-1}, as

$$W_2 = \frac{1}{AXN2} \cdot \sum_{k=3}^{N-2} AXK1 \cdot W_k \tag{7.95a}$$

$$W_{N-1} = \frac{1}{AXN2} \cdot \sum_{k=3}^{N-2} AXKN \cdot W_k \tag{7.95b}$$

where

$$AXN2 = (c_{1,2}^{(2)} + b_5^1 c_{1,2}^{(1)})(c_{N,N-1}^{(2)} + b_5^N c_{N,N-1}^{(1)}) - (c_{1,N-1}^{(2)} + b_5^1 c_{1,N-1}^{(1)})(c_{N,2}^{(2)} + b_5^N c_{N,2}^{(1)})$$

$$AXK1 = (c_{1,N-1}^{(2)} + b_5^1 c_{1,N-1}^{(1)})(c_{N,k}^{(2)} + b_5^N c_{N,k}^{(1)}) - (c_{N,N-1}^{(2)} + b_5^N c_{N,N-1}^{(1)})(c_{1,k}^{(2)} + b_5^1 c_{1,k}^{(1)})$$

$$AXKN = (c_{N,2}^{(2)} + b_5^N c_{N,2}^{(1)})(c_{1,k}^{(2)} + b_5^1 c_{1,k}^{(1)}) - (c_{1,2}^{(2)} + b_5^1 c_{1,2}^{(1)})(c_{N,k}^{(2)} + b_5^N c_{N,k}^{(1)})$$

Note that Equations 7.85*a*, 7.86*a*, 7.94 and 7.95 can be directly substituted into Equation 7.84. As a consequence, the following equation system can be obtained

$$[A] \cdot \{Z_I\} = -\rho h \omega^2 \cdot \{Z_I\} \tag{7.96}$$

where $[A]$ is a $(3N-8) \times (3N-8)$ dimension matrix.

7.4.4 Numerical Example: Free Vibration Analysis of A Composite Laminated Conical Shell

As an example, the free vibration of antisymmetric cross-ply conical shells is studied by the PDQ method. The simply supported condition is specified at both the small and large edges (SS-SS). In the following, the numerical results are displayed by the frequency parameter λ, defined as

$$\lambda = R_2 \omega \sqrt{\rho h / A_{11}} \tag{7.97}$$

where ω is the natural frequency. The elementary material parameters of each layer are given as

$$E_x / E_\theta = 15.0, \quad \nu_{x\theta} = 0.25, \quad G_{x\theta} / E_\theta = 0.5 \tag{7.98}$$

As a result, the coefficients in Equation 7.75 can be simplified as (Tong 1994)

$$A_{11} = A_{22} = \frac{h}{2}(Q_{11} + Q_{22}), \quad A_{12} = Q_{12}h, \quad A_{66} = Q_{66}h$$

$$B_{11} = -B_{22} = \pm\frac{h^2}{4N_l}(Q_{11} - Q_{22}), \quad B_{12} = 0, \quad B_{66} = 0 \tag{7.99}$$

$$D_{11} = D_{22} = \frac{h^3}{24}(Q_{11} + Q_{22}), \quad D_{12} = \frac{1}{12}Q_{12}h^3, \quad D_{66} = \frac{1}{12}Q_{66}h^3$$

where $Q_{ij}, i, j = 1, 2, 6$ are given from Equation 7.70. It is noted that the extension-bending coupling terms reach their maximum values for two plies (N_l = 2) and become zero for an infinite number of plies (N_l = ∞). Thus the case of N_l = ∞ is referred to as "without coupling" and the case of N_l = 2 is referred to as "maximum coupling". For the case of "maximum coupling", we have $B_{11} = -B_{22}$ $= (Q_{11} - Q_{22})h^2/8$, while for the case of "without coupling", we have $B_{11} = B_{22} = 0$. The convergence of the PDQ results for these two cases with SS-SS boundary condition, $h/R_2 = 0.01$, α = 30^0, $L\sin\alpha/R_2 = 0.25$, n = 0 is studied, and the numerical results provided by Shu (1996*a*) are listed in Table 7.3. From Table 7.3, it is apparent that the convergence of the PDQ results is very good. By comparing the DQ results with those of Tong (1994), it was found that the DQ results obtained using 13 grid points are very accurate. It can be seen from the table that when the number of grid points is larger than 13, the DQ results are independent of the grid. In other words, as the number of grid points is increased to above 13, the DQ results do not change. This demonstrates that 13 grid points are sufficient to obtain accurate numerical results. Furthermore, since accurate DQ results can be obtained using a considerably small number of grid points, the computational time required is very short.

Table 7.3 Convergence of frequency parameter λ for an antisymmetric cross-plied laminated conical shell under SS-SS condition (Shu 1996*a*)

$(h/R_2 = 0.01, \alpha = 30^0, L\sin\alpha/R_2 = 0.25, n = 0)$

	Without coupling ($N_l = \infty$)	*Maximum coupling* ($N_l = 2$)
Tong (1994)	0.1978	0.1769
PDQ, $N = 5$	0.3158	0.3155
PDQ, $N = 7$	0.2240	0.2226
PDQ, $N = 9$	0.1961	0.1820
PDQ, $N = 11$	0.1971	0.1787
PDQ, $N = 13$	0.1976	0.1799
PDQ, $N = 21$	0.1976	0.1799

Chapter 8

Miscellaneous Applications of Differential Quadrature Method

8.1 Introduction

In the previous chapters, we have described the applications of the DQ method to fluid mechanics and solid mechanics. Apart from these two areas, the DQ method has been applied in many other areas. For the solution of general partial differential equations, Mansell et al. (1993) compared the DQ method with conventional numerical approaches. Tomasiello (1998) also discussed the application of the DQ method for general initial-boundary-value problems. Another example is the application of the DQ method to solve ocean engineering problems by Gutierrez, Laura and Rossi (1994). In this chapter, we will briefly describe applications of the DQ method to heat transfer, chemical rector, lubrication, waveguide analysis problems, and to the solution of Helmholtz equation. The effect of mesh point distribution on the accuracy of the DQ results will also be discussed in this chapter. It should be indicated that some of these applications are still under active investigation, so the results reported here are very incomplete.

8.2 Application to Heat Transfer

The application of the DQ method to heat transfer was first reported by Shu and Richards (1990), who offered the solution of natural convection in a square cavity by the DQ method. Since then, the DQ method was further developed in this area for different applications including the analysis of heat conduction, convection, radiation and their combinations. To show this application, a simple case is cited from the work of Bert and Malik (1996*e*). The problem is a one-dimensional thin triangular fin, as shown in Figure 8.1. The heat is transmitted along its length by conduction and dissipated from its lateral surfaces to the surroundings by convection. The governing equation of the temperature in the fin can be written as

$$\xi \frac{d^2\theta}{d\xi^2} + \frac{d\theta}{d\xi} = m^2\theta \,, \qquad 0 \le \xi \le 1 \tag{8.1}$$

where θ is the non-dimensional temperature, ξ is the non-dimensional axial coordinate, m is a dimensionless parameter given by

$$m^2 = \frac{hL^2}{k\delta}$$

Here L and δ are the geometric parameters of the fin (see Figure 8.1), k and h are the thermal conductivity and fin-to-ambient heat transfer coefficient respectively.

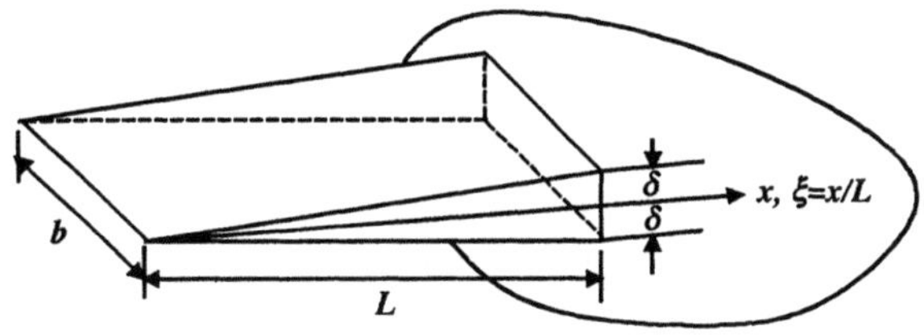

Figure 8.1 A triangular fin

The boundary conditions for Equation 8.1 are

$$\frac{d\theta}{d\xi} = m^2\theta \quad \text{at } \xi = 0 \tag{8.2}$$

and

$$\theta = 1 \quad \text{at } \xi = 1 \tag{8.3}$$

Applying the DQ method to Equation 8.1 leads to

$$\sum_{k=1}^{N} [\xi_i c_{i,k}^{(2)} + c_{i,k}^{(1)}]\theta_k - m^2\theta_i = 0, \text{ for } i = 2, 3, \ldots, N-1 \tag{8.4}$$

where N is the number of mesh points in the domain $0 \le \xi \le 1$, $c_{i,k}^{(1)}$ and $c_{i,k}^{(2)}$ are respectively the DQ weighting coefficients of the first and second order derivatives. Similarly, Equations 8.2 and 8.3 can be discretized as

$$\sum_{k=1}^{N} c_{1,k}^{(1)}\theta_k = m^2\theta_1 \tag{8.5}$$

$$\theta_N = 1 \tag{8.6}$$

Equations 8.4 to 8.6 form a set of N linear algebraic equations, which can be solved for the temperature values at N discrete points. However, since the temperature at

$\xi = 1$ is known, Equation 8.6 can be directly substituted into Equations 8.4 and 8.5. As a consequence, the following equations can be obtained

$$\sum_{k=1}^{N-1}[\xi_i c_{i,k}^{(2)} + c_{i,k}^{(1)}]\theta_k - m^2\theta_i = -\xi_i c_{i,N}^{(2)} - c_{i,N}^{(1)}, \quad \text{for } i = 2,3,\ldots,N-1 \tag{8.7}$$

$$\sum_{k=1}^{N-1} c_{1,k}^{(1)}\theta_k - m^2\theta_1 = -c_{1,N}^{(1)} \tag{8.8}$$

Equations 8.7 and 8.8 form a set of (*N*−1) linear algebraic equations which can yield temperature values at the discrete points ξ_i, $i = 1, 2, \ldots, N-1$. The solution of Equations 8.7 and 8.8 can be obtained using a direct method or using an iterative method.

By taking m = 1, Bert and Malik (1996*e*) reported some DQ results using the following two types of grid,

Type I: Uniform mesh

$$\xi_i = \frac{i-1}{N-1}, \qquad i = 1, 2, \ldots, N \tag{8.9a}$$

Type II: Non-uniform mesh

$$\xi_i = \frac{1}{2}\left(1 - \cos\frac{i-1}{N-1}\pi\right), \qquad i = 1, 2, \ldots, N \tag{8.9b}$$

The differential Equation 8.1 using the boundary conditions of Equations 8.2 and 8.3 has an exact solution, which can be expressed in terms of the Bessel function of the first kind with complex arguments (Lienhard 1987)

$$\theta = \frac{J_0(2m\sqrt{-\xi})}{J_0(2m\sqrt{-1})} \tag{8.10}$$

The solution given by Equation 8.10 will be used to validate the DQ results. The accuracy of the DQ results is measured by the percentage error, which is defined as

$$\varepsilon = \frac{\text{Exact solution - DQ solution}}{\text{Exact solution}} \times 100\% \tag{8.11}$$

Table 8.1 lists the percentage errors of the DQ results for the above-mentioned two types of grid with different number of mesh points. Also included in Table 8.1 is the exact solution given by Equation 8.10. It is noted that the results displayed in Table 8.1 are given at $\xi = 0.1$ intervals. If the mesh points for numerical computation do not coincide with the $\xi = 0.1$ intervals, the Lagrange interpolation scheme is utilized to obtain the temperature values at the $\xi = 0.1$ intervals over the length $0 \le \xi \le 1$.

Table 8.1 Comparison of exact and DQ solutions (Bert and Malik 1996*e*)

ξ	*Exact*	*Percent error, ε*				
		Number of mesh points, N				
	θ	11	20	31	40	51
Uniform mesh						
0.0	0.438676	-9.543	-5.165	-3.319	-2.581	-0.911
0.1	0.483653	-3.863	-1.693	-0.951	-0.694	-0.233
0.2	0.530897	-2.199	-0.976	-0.546	-0.399	-0.178
0.3	0.580485	-1.401	-0.629	-0.350	-0.257	-0.151
0.4	0.632494	-0.926	-0.422	-0.233	-0.172	-0.135
0.5	0.687003	-0.615	-0.287	-0.156	-0.116	-0.125
0.6	0.744096	-0.400	-0.193	-0.103	-0.078	-0.118
0.7	0.803855	-0.244	-0.050	-0.065	-0.050	-0.113
0.8	0.866367	-0.128	-0.074	-0.036	-0.029	-0.109
0.9	0.931718	-0.039	-0.036	-0.015	-0.013	-0.106
Non-uniform mesh						
0.0	0.438676	-3.023	-1.002	-0.451	-0.284	-0.183
0.1	0.483653	-0.635	-0.164	-0.071	-0.041	-0.025
0.2	0.530897	-0.388	-0.088	-0.041	-0.022	-0.015
0.3	0.580485	-0.184	-0.064	-0.025	-0.016	-0.010
0.4	0.632494	-0.194	-0.047	-0.017	-0.010	-0.006
0.5	0.687003	-0.110	-0.024	-0.012	-0.006	-0.004
0.6	0.744096	-0.048	-0.022	-0.008	-0.004	-0.003
0.7	0.803855	-0.060	-0.012	-0.005	-0.003	-0.002
0.8	0.866367	-0.026	-0.005	-0.003	-0.001	-0.001
0.9	0.931718	-0.010	-0.003	-0.001	-0.001	-0.000

It can be observed from Table 8.1 that the DQ results obtained using both two types of grid are very accurate. The maximum percentage error in the DQ solution occurs at $\xi = 0$, i.e., at the tip of the triangular fin. As the number of the grid points increases, the percentage error of the DQ results decreases. This implies that the accuracy of the DQ results can be improved by using a larger number of grid points. It is also interesting to note that the DQ results obtained from the non-uniform mesh are much more accurate than those obtained using uniform mesh. This example further confirms our recommendation of the use of a non-uniform grid in the application of the DQ method.

8.3 Application to Chemical Reactor

The application of the DQ method has also been extended to chemical engineering. In an earlier work, Wang (1982) employed the DQ method for the transient analysis of isothermal chemical reactors. The mathematical model used by Wang (1982) involves a single nonlinear initial-boundary-value equation in terms of partial pressure. This work was later extended by Naadimuthu et al. (1984) for adiabatic chemical reactors. Their mathematical model involves two coupled nonlinear initial-boundary-value equations in terms of partial pressure and temperature. It was shown in both of these works that the DQ results with only nine grid points are comparable in accuracy to the results of a finite difference scheme using 480 grid points. Chang, Tsai and Lin (1993) also applied the DQ method to chemical engineering problems with semi-infinite domains and steep gradients, and presented a few numerical examples. However, an examination of the pressure transients in the work of Wang (1982) and of the pressure and temperature transients in the work of Naadimuthu et al. (1984) shows that the DQ result actually degenerated as the steady state was approached. This drawback was overcome later by Civan (1994), who reworked the DQ computation for the isothermal reactor model of the earlier works (Wang 1982, Naadimuthu et al. 1984) and obtained a stable solution without any degeneration up to the steady state. In the following, the work of Civan (1994) is cited as an example to show the application of the DQ method to a chemical rector.

Consider the following dimensionless model equation given by Lee (1968)

$$\frac{\partial p}{\partial t}=\frac{1}{Pe}\frac{\partial^2 p}{\partial x^2}-\frac{\partial p}{\partial x}-rp^2, \quad 0\le x\le L, \quad 0<t<\infty \tag{8.12}$$

Equation 8.12 is subjected to the initial condition

$$p=p^0(x), \qquad 0\le x\le L, \qquad t=0 \tag{8.13}$$

and the Cauchy and Neumann boundary conditions

$$p-\frac{1}{Pe}\frac{\partial p}{\partial x}=p^*, \qquad x=0, \qquad 0<t<\infty \tag{8.14a}$$

$$\frac{\partial p}{\partial x}=0, \qquad x=L, \qquad 0<t<\infty \tag{8.14b}$$

where t denotes time, x denotes distance, L represents the reactor length, p^0, p^* and p are the reactor initial, entrance and instantaneous partial pressures of the reactant respectively, Pe is the Peclet number, and r is a reactor rate number.

For purposes of convenience during numerical computations, p, t and x are normalized by

$$P = (p - p^0)/\beta \tag{8.15}$$

$$T = 1 - \exp(-t/\alpha) \tag{8.16}$$

$$X = x/L \tag{8.17}$$

where β and α are the characteristic pressure and time respectively. Using Equations 8.15-8.17, Equation 8.12 can be transformed to

$$(1-T)\frac{\partial P}{\partial T} = \frac{\alpha}{PeL^2}\frac{\partial^2 P}{\partial X^2} - \frac{\alpha}{L}\frac{\partial P}{\partial X} - \alpha\beta r\left(\frac{p^0}{\beta} + P\right)^2 \tag{8.18}$$

Similarly, Equations 8.13 and 8.14 can be normalized to give

$$P = 0, \qquad 0 \le X \le 1,\ T = 0 \tag{8.19}$$

$$P - \frac{1}{PeL}\frac{\partial P}{\partial X} = \frac{p^* - p^0}{\beta}, \qquad X = 0, \quad 0 < T \le 1 \tag{8.20}$$

$$\frac{\partial P}{\partial X} = 0, \qquad X = 1, \quad 0 < T \le 1 \tag{8.21}$$

When the diffusion phenomenon is dominant in Equation 8.18, the coefficient of the $\partial^2 P/\partial X^2$ term is set equal to one to obtain the diffusion time scale as

$$\alpha = PeL^2 \tag{8.22}$$

When the convection phenomenon is dominant, then the coefficient of the $\partial P/\partial X$ term is set equal to one to obtain the convection time scale as

$$\alpha = L \tag{8.23}$$

When the reaction term is dominant, then the coefficient of $(p^0/\beta + P)^2$ is set equal to one to obtain the reaction time scale as

$$\alpha = (\beta r)^{-1} \tag{8.24}$$

Although not required, the characterization of the coefficient α by Equations 8.22-8.24, depending on whether the diffusion, convection or reaction may be dominant, can provide a better stability control in obtaining the numerical solutions. The coefficient β is characterized such that the right hand side of Equation 8.20 yields one. Thus

$$\beta = p^* - p^0 \tag{8.25}$$

For the present problem, the convection time scale is used. Using Equations 8.23 and 8.25, Equations 8.18–8.21 become

$$(1-T)\frac{\partial P}{\partial T} = \frac{1}{PeL}\frac{\partial^2 P}{\partial X^2} - \frac{\partial P}{\partial X} - rL(p^* - p^0)\left(\frac{p^0}{p^* - p^0} + P\right)^2 \tag{8.26}$$

$$P = 0, \qquad 0 \le X \le 1, \qquad T = 0 \tag{8.27}$$

$$P - \frac{1}{PeL}\frac{\partial P}{\partial X} = 1, \qquad X = 0, \qquad 0 < T \le 1 \tag{8.28}$$

$$\frac{\partial P}{\partial X} = 0, \qquad X = 1, \qquad 0 < T \le 1 \tag{8.29}$$

And accordingly, the normalized pressure, time and distance become

$$P = (p - p^0)/(p^* - p^0) \tag{8.30}$$

$$T = 1 - \exp(-t/L) \tag{8.31}$$

$$X = x/L \tag{8.32}$$

By taking $p^0 = 0$, $p^* = 0.07$, $Pe = 2$, $r = 1$ and $L = 48$ (Lee 1968), we get $\alpha = L = 48$, $\beta = p^* - p^0 = 0.07$ and $PeL = 96$. With these values, Equations 8.26-8.29 can be further reduced to

$$(1-T)\frac{\partial P}{\partial T} = \frac{1}{96}\frac{\partial^2 P}{\partial X^2} - \frac{\partial P}{\partial X} - 3.36P^2, \quad 0 \le X \le 1, \quad 0 < T \le 1 \tag{8.33}$$

$$P = 0, \qquad 0 \le X \le 1, T = 0 \tag{8.34}$$

$$\frac{\partial P}{\partial X} = 96(P-1), \qquad X = 0,\ 0 < T \le 1 \tag{8.35}$$

$$\frac{\partial P}{\partial X} = 0, \qquad X = 1,\ 0 < T \le 1 \tag{8.36}$$

Note that Equation 8.33 is the governing equation to be solved, which is subjected to the initial condition (Equation 8.34) and the boundary conditions (Equations 8.35 and 8.36). For the numerical computation, the physical domain $0 \le X \le 1$ is decomposed into (N–1) intervals with coordinates of mesh points given as $0 = X_1, X_2, ..., X_N = 1$. When the mesh point distribution is specified, the DQ weighting coefficients can be easily calculated. Then the spatial derivatives in Equation 8.33 can be discretized using the DQ method. Let $c_{i,j}^{(1)}$ be the DQ weighting coefficients of the first order derivative. The first and second order spatial derivatives in Equation 8.33 can be discretized as

$$\left(\frac{\partial P}{\partial X}\right)_i = \sum_{k=1}^{N} c_{i,k}^{(1)} \cdot P_k \tag{8.37}$$

$$\left(\frac{\partial^2 P}{\partial X^2}\right)_i = c_{i,1}^{(1)} \cdot \left(\frac{\partial P}{\partial X}\right)_1 + c_{i,N}^{(1)} \cdot \left(\frac{\partial P}{\partial X}\right)_N + \sum_{k1=2}^{N-1} c_{i,k1}^{(1)} \cdot \left(\frac{\partial P}{\partial X}\right)_{k1} \tag{8.38}$$

Substituting Equations 8.35 and 8.36 into Equation 8.38 and further discretizing the first order derivatives, we obtain

$$\left(\frac{\partial^2 P}{\partial X^2}\right)_i = 96 \cdot c_{i,1}^{(1)} \cdot (P_1 - 1) + \sum_{k=1}^{N} \tilde{c}_{i,k}^{(2)} \cdot P_k \tag{8.39}$$

where $\tilde{c}_{i,k}^{(2)} = \sum_{k1=2}^{N-1} c_{i,k1}^{(1)} \cdot c_{k1,k}^{(1)}$. With Equations 8.35, 8.36, 8.37 and 8.39, the discrete form of Equation 8.33 at each mesh point can be written as

$$\frac{dP_1}{dT} = \frac{1}{1-T}\left[c_{1,1}^{(1)}(P_1 - 1) + \frac{1}{96}\sum_{k=1}^{N} \tilde{c}_{1,k}^{(2)} \cdot P_k - 96(P_1 - 1) - 3.36 P_1^2\right] \tag{8.40a}$$

$$\frac{dP_i}{dT} = \frac{1}{1-T}\left[c_{i,1}^{(1)}(P_1 - 1) + \sum_{k=1}^{N}\left(\frac{\tilde{c}_{i,k}^{(2)}}{96} - c_{i,k}^{(1)}\right)P_k - 3.36 P_i^2\right], i = 2, ..., N-1 \tag{8.40b}$$

$$\frac{dP_N}{dT} = \frac{1}{1-T}\left[c_{N,1}^{(1)}(P_1 - 1) + \frac{1}{96}\sum_{k=1}^{N} \tilde{c}_{N,k}^{(2)} \cdot P_k - 3.36 P_N^2\right] \tag{8.40c}$$

Equation 8.40 is a set of ordinary differential equations which can be solved using explicit methods. Civan (1994) solved Equation 8.40 by using the Runge-Kutta-Fehlberg four method. He showed the DQ results using a uniform mesh with $N = 6$ and $N = 11$. Table 8.2 lists some results provided by Civan (1994). It can be seen from the table that the DQ results agree reasonably well with the approximate analytical solution (Grotch 1969) and the finite difference (FD) solution (Lee 1968) over the whole range of reactor length.

Table 8.2: Comparison of the steady-state solutions (Civan 1994)

Distance x	*Pressure, p*			
	Approximate solution (*Grotch* 1969)	*FD solution* (Lee 1968)	*DQ solution* $N = 6$	*DQ solution* $N = 11$
0.0	0.068	0.068	0.068	0.068
9.6	0.048	0.042	0.040	0.044
19.2	0.032	0.030	0.029	0.031
28.8	0.020	0.023	0.023	0.024
38.4	0.014	0.018	0.019	0.020
48.0	0.011	0.016	0.008	0.010

8.4 Application to Lubrication Problems

In fluid film lubrication, the Reynolds equation has been a research subject for more than one hundred years. In the past a few decades, numerical solutions of the full Reynolds equation are usually obtained using low order finite difference and finite element methods. In general, these low order methods can obtain accurate numerical results using a large number of mesh points and thus require a lot of computational effort. To reduce the computational effort, Malik and Bert (1994) applied the DQ method for the first time to solve the steady-state incompressible and compressible lubrication problems. Bert and Malik (1997*c*) also applied the DQ method for the transient response analysis of a plain gas-lubricated journal bearing. In the following, the work of Malik and Bert (1994) is taken as an example to show the application of the DQ method to the solution of steady-state incompressible lubrication problems.

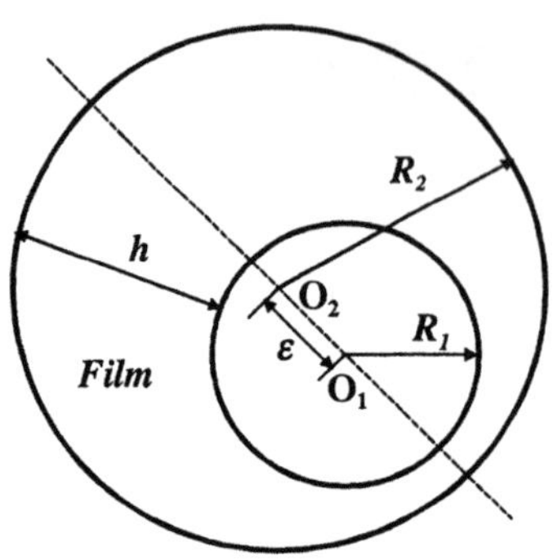

Figure 8.2 Configuration of a plain journal bearing

Figure 8.2 shows a plain journal bearing with an incompressible lubricant. The Reynolds equation for incompressible lubrication in non-dimensional form can be written as

$$\frac{\partial^2\psi}{\partial x^2}+\alpha^2\frac{\partial^2\psi}{\partial y^2}-\frac{3}{4}\frac{1}{h^2}\left[\left(\frac{dh}{dx}\right)^2+2h\frac{d^2h}{dx^2}\right]\psi-\frac{\beta}{h^{3/2}}\frac{dh}{dx}=0 \tag{8.41}$$

where

$$\alpha=\phi/(2\lambda),\quad \beta=6\phi \tag{8.42}$$

and

$$\psi=ph^{3/2} \tag{8.43}$$

ϕ is the extent of the active or positive pressure in the incompressible film of a journal bearing, λ is the bearing aspect ratio, ψ is a smoother function in comparison to the pressure (p) function since the pressure happens to be large for small film thickness h and vice versa. The boundary conditions are

$$\psi(0,y)=\psi(1,y)=\psi(x,0)=\psi(x,1)=0 \tag{8.44}$$

In addition, the boundary condition at the trailing edge of the active film in a journal bearing is assumed to be quasi-Reynolds, so that

$$\frac{\partial\psi}{\partial x}\bigg|_{x=1}=0 \tag{8.45}$$

Applying the DQ method to discretize the derivatives in Equation 8.41, we can obtain the following set of linear algebraic equations for ψ

$$\sum_{k=1}^{N}c_{i,k}^{(2)}\psi_{k,j}+\alpha^2\sum_{k=1}^{M}\bar{c}_{j,k}^{(2)}\psi_{i,k}-\frac{3}{4}\frac{1}{h_i^2}\left[\left(\frac{dh}{dx}\right)^2+2h\frac{d^2h}{dx^2}\right]_i\psi_{i,j}=\frac{\beta}{h_i^{3/2}}\left(\frac{dh}{dx}\right)_i \tag{8.46}$$

for $i=2,3,\ldots,N-1$; $j=2,3,\ldots,M-1$

where $c_{i,k}^{(2)}$ and $\bar{c}_{j,k}^{(2)}$ are respectively the DQ weighting coefficients of the second order derivatives in the x and y directions, N and M are the number of grid points in the x and y directions respectively.

With proper implementation of the boundary conditions, Equation 8.46 can be solved using a direct method or an iterative method. Once the ψ value is determined, the pressure p can be calculated from Equation 8.43. We can then compute the components of the load support along and perpendicular to the line of centers for a journal bearing

$$W_{\text{II}} = \phi \int_0^1 \int_0^1 p \cos\theta \, dxdy \tag{8.47}$$

$$W_{\perp} = -\phi \int_0^1 \int_0^1 p \sin\theta \, dxdy \tag{8.48}$$

and the attitude angle is defined as

$$\tan\theta_0 = W_{\text{II}} / W_{\perp} \tag{8.49}$$

The θ in Equations 8.47 and 8.48 is the angular coordinate measured from the line of centers in a journal bearing. In the numerical computation, the FDQ method is used in the circumferential direction while the PDQ method is applied in the other direction. The mesh point distribution for the PDQ method is given from the roots of the shifted Chebyshev polynomial. For the FDQ method, a uniform mesh is adopted.

For an infinitely long journal bearing with an incompressible lubricant, the problem becomes a one-dimensional plain journal bearing. The exact solution of this case exists, which can be employed to validate the DQ results. The comparison of attitude angle for a one-dimensional plain journal bearing between the DQ results and the exact solution (Szeri 1980) is shown in Table 8.3. The percentage error of the DQ results, defined by

$$\text{Percentage error} = \frac{\text{DQ result - Exact solution}}{\text{Exact solution}} \times 100\% \tag{8.50}$$

is also included in Table 8.3. It is to be noted that the Reynolds boundary condition at the trailing edge of the positive pressure film has been incorporated in the computation. It can be seen from Table 8.3 that the DQ results obtained from the use of 11 mesh points are in excellent agreement with the exact solution.

Table 8.3 Comparison of attitude angle for a one-dimensional plain journal bearing, $N = 11$ (Malik and Bert 1994)

ε	*Exact Solution*	*DQ Results*	*Percent error*
0.1	69.03	69.03	0.004
0.2	66.90	66.89	-0.007
0.3	64.46	64.45	-0.029
0.4	61.64	61.61	-0.041
0.5	58.30	58.28	-0.027
0.6	54.23	54.24	0.010
0.7	49.10	49.13	0.075
0.8	42.18	42.27	0.205
0.9	31.67	31.64	-0.073

To demonstrate the high accuracy and efficiency of the DQ method further, Malik and Bert (1994) also showed some results for plain journal bearings with finite lengths and incompressible lubricants. The attitude angle computed by the DQ method for this case is displayed in Table 8.4. The DQ results were obtained using a mesh size of 11×7. For purposes of comparison, the finite difference (FD) results given by Raimondi and Boyd (1958) are also included in Table 8.4. Obviously, the DQ results agree very well with the finite difference results.

Table 8.4 Comparison of attitude angle for finite length plain journal bearing, $N = 11$, $M = 7$ (Malik and Bert 1994)

λ	ε	*Attitude angle, θ_0*	
		FD results	*DQ results*
1.00	0.2	74.02	73.95
	0.4	63.10	62.68
	0.6	50.58	50.65
	0.8	36.24	36.33
	0.97	15.47	15.07
0.50	0.2	74.94	75.15
	0.4	61.45	61.83
	0.6	48.14	48.17
	0.8	33.31	32.97
	0.97	13.75	13.28
0.25	0.2	75.18	75.38
	0.4	60.68	61.21
	0.6	46.72	46.83
	0.8	31.04	31.17
	0.97	12.22	11.68

8.5 Application to Waveguide Analysis

The first application of the DQ method to waveguide analysis was given by Dong et al. (1997) for rectangular waveguides. Later, the present author successfully applied the DQ method to compute the cutoff wavelength of elliptical waveguides with arbitrary ellipticity. The DQ results obtained compare very favourably to the exact solution. In the following, the elliptical waveguide is selected as an example to show the DQ application to waveguide analysis.

Elliptical waveguides have been extensively used in many engineering problems such as radar feed lines, multi-channel communication and accelerator beam tubes. The determination of the cutoff wavelength of elliptical waveguides is one of the most important issues of consideration for designing a waveguide or analyzing the wave propagation in a waveguide. The governing equation for waveguide analysis is the Helmholtz equation, which has been presented in Chapter 5. For the two-dimensional case, the Helmholtz equation can be written in terms of Cartesian coordinates as

$$\frac{\partial^2\phi}{\partial x^2}+\frac{\partial^2\phi}{\partial y^2}+k_c^2\phi=0 \tag{8.51}$$

where k_c is the cutoff wavenumber. Equation 8.51 is subjected to two types of boundary conditions. The Neumann boundary condition

$$\left.\frac{\partial\phi}{\partial n}\right|_{\partial D}=0 \tag{8.52}$$

is for the *TE* waves whilst the Dirichlet boundary condition

$$\phi\big|_{\partial D}=0 \tag{8.53}$$

is for the *TM* waves. Here ∂D is the boundary of a waveguide. For a regular domain, Equation 8.51 can be solved directly (Gutierrez and Laura 1994). However, for an elliptical waveguide, a coordinate transformation is needed before the governing equation is solved. We set the following transformation for the elliptical waveguide

$$\begin{cases} x=ar\cdot\cos\theta \\ y=br\cdot\sin\theta \end{cases} \tag{8.54}$$

where a and b are the semi-major and semi-minor axes respectively. With Equation 8.54, the computational domain becomes a rectangle given by $0\le r\le 1$ and $0\le\theta\le 2\pi$. It is noted that the boundary of the elliptical cross section is now represented by $r = 1$. $r = 0$ represents the center of the cross section. With Equation 8.54, Equation 8.51 can be transformed accordingly into

$$A\frac{\partial^2\phi}{\partial r^2}+\frac{B}{r}\cdot\frac{\partial^2\phi}{\partial r\partial\theta}+\frac{C}{r^2}\cdot\frac{\partial^2\phi}{\partial\theta^2}+\frac{C}{r}\cdot\frac{\partial\phi}{\partial r}-\frac{B}{r^2}\cdot\frac{\partial\phi}{\partial\theta}+a^2k_c^2\phi=0 \tag{8.55}$$

where

$$A=\lambda^2\cdot\sin^2\theta+\cos^2\theta$$

$$B=(\lambda^2-1)\sin 2\theta$$

$$C=\lambda^2\cdot\cos^2\theta+\sin^2\theta$$

$\lambda = a / b$

Equation 8.55 is used to determine the wavenumber k_c. For the *TM* waves, the boundary condition at $r = 1$ is the same as that given by Equation 8.53, that is, $\phi = 0$, at $r = 1$. For the *TE* waves, the Neumann condition given by Equation 8.52 can be written in the (r,θ) coordinate system as

$$A\frac{\partial\phi}{\partial r} + \frac{B}{2}\frac{\partial\phi}{\partial\theta} = 0, \text{ at } r = 1 \tag{8.56}$$

In addition, Equation 8.55 needs a boundary condition for $r = 0$. This can be given from the consistency condition of Equation 8.55 at $r = 0$, which can be written as

$$\frac{\partial\phi}{\partial r} = 0, \text{ at } r = 0 \tag{8.57}$$

Equation 8.57 will be applied to both the *TM* and *TE* waves. In the θ direction, the periodic condition is applied.

For numerical computation, the derivatives in the r direction are discretized using the PDQ approach while the derivatives in the θ direction are approximated using the FDQ approach. It is supposed that there are N grid points in the r direction and M grid points in the θ direction. Using the DQ method, Equation 8.55 can be discretized at a mesh point (r_i, θ_j) as

$$\begin{aligned}&\sum_{k=1}^{N}\left(A_j \cdot c_{i,k}^{(2)} + \frac{C_j}{r_i}\cdot c_{i,k}^{(1)}\right)\cdot\phi_{k,j} + \frac{B_j}{r_i}\cdot\sum_{k1=1}^{N}\sum_{k2=1}^{M} c_{i,k1}^{(1)}\cdot\bar{c}_{j,k2}^{(1)}\cdot\phi_{k1,k2} + \\ &\sum_{k=1}^{M}\left(\frac{C_j}{r_i^2}\cdot\bar{c}_{j,k}^{(2)} - \frac{B_j}{r_i^2}\cdot\bar{c}_{j,k}^{(1)}\right)\cdot\phi_{i,k} + a^2\cdot k_c^2\cdot\phi_{i,j} = 0\end{aligned} \tag{8.58}$$

where $c_{i,k}^{(1)}$ and $c_{i,k}^{(2)}$ are respectively the weighting coefficients of the derivatives $\partial\phi/\partial r$ and $\partial^2\phi/\partial r^2$ while $\bar{c}_{j,k}^{(1)}$ and $\bar{c}_{j,k}^{(2)}$ are respectively the weighting coefficients of the derivatives $\partial\phi/\partial\theta$ and $\partial^2\phi/\partial\theta^2$. It should be indicated that Equation 8.58 has to be applied at the interior points, $2 \le i \le N-1$, $1 \le j \le M$. Similarly, the derivatives in the boundary conditions can be discretized using the DQ method. In the θ direction, no numerical condition is implemented since the periodic condition is automatically satisfied in the FDQ discretization. Suppose that the functional values at the interior points are represented by a vector $\{\phi_I\}$, and the remaining functional values at the boundary points are denoted by a vector $\{\phi_B\}$. With these definitions, Equation 8.58 can be written in the following matrix form

$$[A_{IB}]\cdot\{\phi_B\}+[A_{II}]\cdot\{\phi_I\}=-\Omega^2\cdot\{\phi_I\} \tag{8.59}$$

where $\Omega^2=a^2k_c^2$. Similarly, the discretized boundary conditions can be put in the following matrix form

$$[A_{BB}]\cdot\{\phi_B\}+[A_{BI}]\cdot\{\phi_I\}=0 \tag{8.60}$$

Substituting Equation 8.60 into Equation 8.59 gives the final eigenvalue equation system as

$$\left([A_{II}]-[A_{IB}]\cdot[A_{BB}]^{-1}\cdot[A_{BI}]\right)\cdot\{\phi_I\}=-\Omega^2\cdot\{\phi_I\} \tag{8.61}$$

Once the eigenvalue of the above system, Ω, is computed, the cutoff wavenumber k_c can be easily obtained. We can then compute the normalized cutoff wavelength defined by

$$\lambda_c=2\pi/k_c \tag{8.62}$$

Table 8.5 Normalized cutoff wavelengths of first nine *TM* modes (e = 0.9)

Modes	*Zhang and Shen* (1995)	*DQ Results*				
		9×19	13×19	14×19	17×19	14×17
TM_1	1.4906	1.4906	1.4906	1.4906	1.4906	1.4906
TM_2	1.1607	1.1610	1.1608	1.1607	1.1607	1.1607
TM_3	0.9375	0.9375	0.9375	0.9375	0.9375	0.9375
TM_4	0.8093	0.8096	0.8094	0.8093	0.8093	0.8093
TM_5	0.7803	0.7804	0.7803	0.7803	0.7803	0.7803
TM_6	0.7083	0.7083	0.7083	0.7083	0.7083	0.7083
TM_7	0.6651	0.6651	0.6651	0.6651	0.6651	0.6651
TM_8	0.6262	0.6262	0.6262	0.6262	0.6262	0.6262
TM_9	0.5780	0.5786	0.5780	0.5780	0.5780	0.5778

For the DQ computation, the following mesh point distribution is used for the elliptical waveguide,

$$r_i=\frac{1}{2}\left[1-\cos\left(\frac{i-1}{N-1}\cdot\pi\right)\right],\quad i=1,2,\ldots,N \tag{8.63a}$$

$$\theta_j = \frac{j-1}{M} 2\pi, \quad j = 1, 2, ..., M \tag{8.63b}$$

Table 8.5 shows the normalized wavelengths of the first nine *TM* modes for an eccentricity of 0.9 while Table 8.6 lists the normalized wavelengths of the first nine *TE* modes for an eccentricity of 0.9. Also included in these tables are the analytical solutions given by Zhang and Shen (1995). It is noted that for each mode, five mesh sizes were used to obtain the DQ results. It can be observed from the tables that for both the *TM* and *TE* modes, the convergence of the DQ results is evident. When the mesh size is slightly increased, the accuracy of the DQ results is greatly improved. It can also be seen from the tables that the convergence of the DQ results for the *TM* modes is faster than that for the *TE* modes. In other words, to obtain the same accuracy for the DQ results, the *TM* modes require fewer mesh points than the *TE* modes. It can be seen from Tables 8.5 and 8.6 that, for both the *TM* and *TE* modes, the DQ results obtained by using a few mesh points agree very well with the exact solutions.

Table 8.6 Normalized cutoff wavelengths of first nine *TE* modes ($e = 0.9$)

Modes	*Zhang and Shen* (1995)	*DQ Results*				
		9×19	13×19	15×19	31×19	15×17
TE_1	3.3482	3.3168	3.3340	3.3378	3.3459	3.3378
TE_2	1.8287	1.8287	1.8287	1.8287	1.8287	1.8287
TE_3	1.5650	1.5574	1.5615	1.5624	1.5644	1.5624
TE_4	1.2654	1.2687	1.2668	1.2664	1.2656	1.2664
TE_5	1.2292	1.2292	1.2292	1.2292	1.2292	1.2292
TE_6	0.9986	0.9996	0.9990	0.9989	0.9987	0.9989
TE_7	0.9698	0.9698	0.9698	0.9698	0.9698	0.9698
TE_8	0.8340	0.8340	0.8340	0.8340	0.8340	0.8342
TE_9	0.8177	0.8177	0.8177	0.8177	0.8177	0.8177

8.6 Solution of the Helmholtz Equation

To demonstrate the high efficiency and accuracy of the PDQ and FDQ methods, Shu and Xue (1999) studied the Helmholtz field problem on a rectangular domain.

The two-dimensional Helmholtz equation for a field problem can be written as

$$\frac{\partial^2 \phi}{\partial x^2} + \frac{\partial^2 \phi}{\partial y^2} + k^2 \phi = f, \qquad 0 \le x \le L_x, 0 \le y \le L_y \tag{8.64}$$

where ϕ and k are respectively the potential and the wave number defined in the rectangular domain, f is a source function, which was chosen by Shu and Xue (1999) to be

$$f(x,y)=k^2\left[1-\pi^2\left(\frac{1}{L_x^2}+\frac{1}{L_y^2}\right)\right]\sin\frac{k\pi x}{L_x}\cdot\sin\frac{k\pi y}{L_y} \tag{8.65}$$

When the Dirichlet boundary condition $\phi = 0$ is imposed on the boundary, Equation 8.64 has an exact solution given by

$$\phi=\sin\frac{k\pi x}{L_x}\cdot\sin\frac{k\pi y}{L_y} \tag{8.66}$$

For the DQ computation, Equation 8.64 can be discretized as

$$\sum_{k=1}^{N} c_{i,k}^{(2)}\cdot\phi_{k,j}+\sum_{k=1}^{N}\bar{c}_{j,k}^{(2)}\cdot\phi_{i,k}+k^2\cdot\phi_{i,j}=f_{i,j} \tag{8.67}$$

$$\text{for } i=2,3,\ldots,N-1;\ j=2,3,\ldots,M-1$$

where N and M are the number of grid points in the x and y directions, $c_{i,k}^{(2)}$ and $\bar{c}_{j,k}^{(2)}$ are respectively the weighting coefficients of the PDQ or FDQ method in the x and y directions, respectively. Equation 8.67 is a set of algebraic equations, which can be solved using a direct method of *LU* decomposition. The mesh point distribution for the PDQ and FDQ computation is taken to be the same, which is given by

$$x_i=\frac{1}{2}\left[1-\cos\left(\frac{i-1}{N-1}\cdot\pi\right)\right]L_x,\quad i=1,2,\ldots,N \tag{8.68a}$$

$$y_j=\frac{1}{2}\left[1-\cos\left(\frac{j-1}{M-1}\cdot\pi\right)\right]L_y,\quad j=1,2,\ldots,M \tag{8.68b}$$

Shu and Xue (1999) applied both the PDQ and FDQ methods to solve Equation 8.67. In their work, L_x and L_y were taken as π, and the number of grid points used in the x and y directions was taken to be the same as M. It was found that for this problem, the FDQ method behaves differently from the PDQ method. As the mesh size is increased, the accuracy of the PDQ results is gradually improved. However, the accuracy of the FDQ results seems to improve suddenly. When $M < 2k+1$, the FDQ results may not be sufficiently accurate. But when M is increased to $2k+1$, the FDQ results can match the exact solution up to 15 decimal places. As M is further increased to above $2k+1$, the accuracy of the FDQ results cannot be further improved. The good performance of the FDQ results is actually due to the feature of Fourier series approximation to the harmonic function given by Equation 8.66.

For the general case, when the same number of mesh points is used, the FDQ results are more accurate than the PDQ results. This can be observed from Figure 8.3, which compares the PDQ and FDQ results with the exact solution along the line of $y = \pi/2$ with $k = 7$.

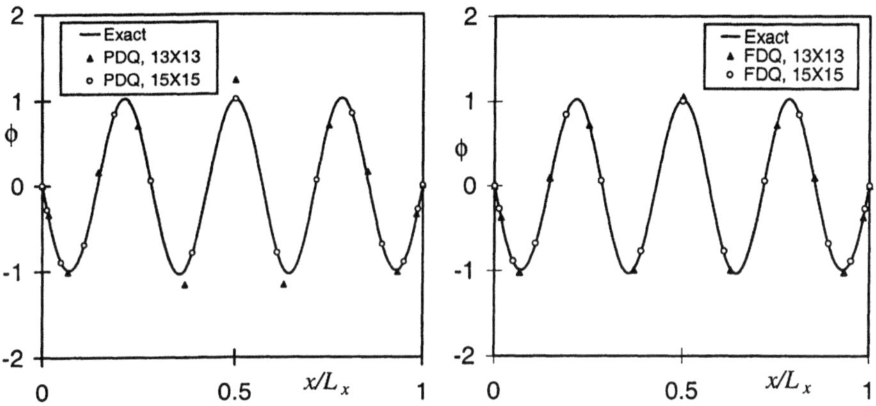

(a) PDQ results (b) FDQ results

Figure 8.3 Comparison of DQ results with exact solution along the line of $y = \pi/2$ for $k = 7$ (Shu and Xue 1999)

As discussed previously, for the general case, the FDQ results are very accurate when $M > 2k+1$. However, it was found that when $k = 5$, the FDQ results are quite erroneous for all mesh sizes. In contrast, the PDQ results of $k = 5$ for a mesh size of 15×15 are quite reasonable. This can be seen clearly in Figure 8.4, which shows the distribution of the PDQ and FDQ results along the line of $y = \pi/2$ for $k = 5$.

The irregular behavior of the FDQ results for $k = 5$ is due to the nature of the problem. It was found that $k = 5$ is the eigenvalue (natural wave number) of the Helmholtz eigenvalue equation

$$\nabla^2\phi + k^2\phi = 0 \tag{8.69}$$

with the same boundary condition and computational domain. So, theoretically, $k = 5$ is a singular case for Equation 8.64. When the FDQ method is used to solve Equation 8.69, the exact eigenvalue of 5 (error is in the order of 10^{-15}) can be obtained. Therefore, when $k = 5$ and the FDQ method is applied, the matrix of

Equation 8.67 is singular. As a consequence, no accurate results can be obtained. On the other hand, when the PDQ method is used to solve Equation 8.69, only an approximate eigenvalue to 5 can be obtained. So, when $k = 5$ and the PDQ method is applied, the matrix of Equation 8.67 is not singular, and accordingly, accurate numerical results can be obtained. As the number of the mesh points increases, the computed eigenvalue of Equation 8.69 by the PDQ method will be very close to 5. Then the matrix of Equation 8.67 will become highly ill-conditioned, and for this case, the PDQ results are not accurate. This has been confirmed by numerical experiments (Shu and Xue 1999).

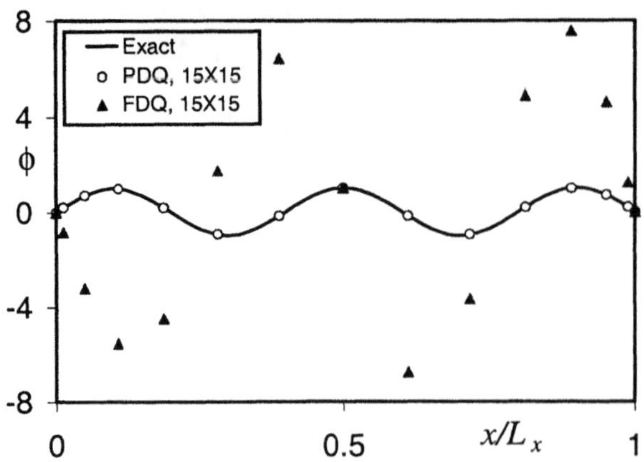

Figure 8.4 Comparison of DQ results with exact solution along the line of $y = \pi/2$ for $k = 5$ (Shu and Xue 1999)

8.7 Effect of Mesh Point Distribution on Accuracy of DQ Results

In Chapter 4, we have shown that the mesh point distribution has some effect on the stability of convection, diffusion and convection-diffusion equations. For the general case, the DQ computation requires the mesh to be non-uniform with stretching towards the boundary. Recently, the author has also studied the effect of the mesh point distribution on the accuracy of PDQ results. In the following, we will take the vibration of a uniform beam as an example to illustrate this effect.

The governing equation and DQ computation for the vibration of beams have been described in Section 7.2 of Chapter 7. To show the effect of mesh point distribution on the accuracy of the DQ results, we use the following grid

$$x_i = (1-\alpha)(3\xi_i^2 - 2\xi_i^3) + \alpha\xi_i, \quad i = 1, 2, ..., N; \quad \alpha \le 1 \tag{8.70}$$

where

$$\xi_i = \frac{1}{2}\left(1 - \cos\frac{i-1}{N-1}\pi\right), \quad i = 1, 2, \ldots, N \tag{8.71}$$

and α is a stretching parameter. Obviously, the smaller the value of α, the stronger the grid points are clustered towards the boundary. It is noted that when α is taken as a negative value, the x_2 obtained from Equation 8.70 may become a small negative value, which is outside of the domain [0, 1]. For this case, the coordinates of x_2 and x_{N-1} are given from the following formulations

$$x_2 = 0.1x_3, \quad x_{N-1} = 1 - x_2 \tag{8.72}$$

Note that the basic grid given by Equation 8.71 is the same as Grid I shown in Section 4.4.2 of Chapter 4. With Equation 8.70, the grid is changed through the use of different α-values. To show the accuracy of the DQ results, we define the error of the fundamental frequency as

$$\text{error} = \frac{\text{DQ result - Reference}}{\text{Reference}} \tag{8.73}$$

where the reference data is taken from the work of Blevins (1984). The vibration of beams with various combinations of SS, C and F boundary conditions was investigated using different grids. Figure 8.5 displays the α-error curves for various combinations of boundary conditions using 9 mesh points ($N = 9$). It is observed from the figure that the accuracy of the DQ results depends on the grid. There exists a best grid for each type of boundary conditions. The optimal stretching parameter depends on the boundary condition. As the order of the derivative term in the boundary conditions increases, the optimal stretching parameter α declines and the corresponding stretching towards the boundary is enhanced. To obtain accurate numerical solution, it was found that the C-F, SS-F and F-F beams require the strongest stretching towards the boundary (lowest value of α) among all test examples. The C-C beam, in which the boundary condition only involves the first order derivative, requires the least stretching.

It is interesting to see from Figure 8.5 that for the SS-F and F-F beams, the α-error curve is very sharp in the vicinity of the optimal stretching parameter. The accuracy of the DQ results can be greatly improved within a small range of α. It may be concluded that the DQ result is very sensitive to the grid size for these cases. In contrast, the remaining curves for C-C, C-SS, SS-SS and C-F beams are somewhat smooth.

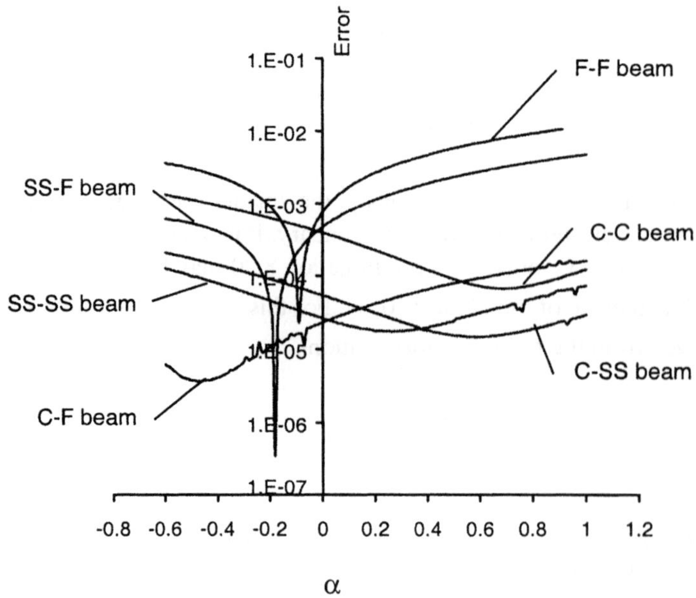

Figure 8.5 Effect of mesh point distribution of accuracy of DQ results for a beam

Mathematically, it is known that for a given boundary value problem, there exists an optimal polynomial which is the best approximation to the solution of the problem. However, it is very difficult to find such a polynomial by analytical tools. The results shown in Figure 8.5 confirm that the optimal polynomial does exist, and depends on the boundary condition. Figure 8.5 shows that for each case, there exists an optimal grid point distribution, which corresponds to the optimal stretching parameter. Since in the PDQ method, the coordinates of the mesh points are used to construct the approximated polynomial, the optimal mesh point distribution corresponds to the optimal polynomial approximation. The above study implies that, for a given boundary value problem, the optimal polynomial may be different for different types of boundary conditions.

Chapter 9

Application of Differential Quadrature to Complex Problems

9.1 Introduction

In the previous chapters, we have described applications of the DQ method for problems involving simple domains. These applications have proven to be very successful.

As shown in Chapters 2 and 3, similar to low order finite difference schemes, the DQ method discretizes the derivatives along a straight line. Thus, it requires the computational domain to be regular so that the physical boundary could be a mesh line. If the physical boundary is curved, the DQ method cannot be directly applied. There are two basic approaches to overcome this difficulty. The first is called the multi-domain DQ approach, in which the whole domain is decomposed into several subdomains and the DQ discretization is applied locally in each subdomain. The other is the coordinate transformation approach. In the latter approach, the irregular physical domain in the Cartesian coordinate system is transformed into a regular domain in the curvilinear coordinate system, and the governing equations in the Cartesian coordinate system are transformed into the relevant forms in the curvilinear coordinate system. All the numerical computations are then performed on the curvilinear coordinate system. In this chapter, some developments and applications of these two approaches will be reported.

9.2 Multi-domain DQ Method

The multi-domain DQ method was first proposed by Shu (1991) to simulate incompressible flows past a backward facing step and a square step. Shu and Richards (1992*b*) also applied the multi-domain DQ method to parallel simulation of driven cavity flows on a MIMD (multiple instruction, multiple data-stream) computer. Other applications of the multi-domain DQ method include the simulation of flows in the Czochralski crystal growth (Shu, Chew and Liu 1998), waveguide analysis with rectangular boundaries (Shu and Chew 1999), computation of Poisson and Laplace equations (Zhong and He 1998), structural and vibration analysis (Chen 1997, 1998*a*, *b*, 1999*a*, *b*, Liu and Liew 1998*a*, 1999, Striz, Chen and Bert 1994, 1997, Wang and Gu 1997, Wang, Wang and Chen 1998).

It is assumed that the physical domain of a problem can be represented by Ω and the boundary by Γ. The multi-domain technique, firstly, decomposes the domain Ω into several subdomains. In each subdomain, a local mesh is generated and the local DQ scheme is applied in the same fashion as the application of the scheme in a single domain. The solutions for the interior grid points are independent for each subdomain. Globally, the information exchange between subdomains is required. This can be done across the interface of subdomains. Since any complex geometry can be transformed into a rectangular domain or a combination of rectangular subdomains, a rectangular domain is chosen for demonstration without loss of generality. Basically, there are two kinds of interface topology: patched and overlapped.

9.2.1 Topology of Interface

The information exchange between neighbouring subdomains depends on the topology of the interface. In the following, the patched and overlapped interfaces will be described in detail.

9.2.1.1 Patched Interface

The topology of a patched interface is shown in Figure 9.1, where Γ_{ij} is the interface between two subdomains Ω_i and Ω_j. For the patched interface, the governing equation is not applied along the interface. Instead, the continuity condition is enforced. The common approach is to let the function and its normal derivative be continuous across the interface. In other words, the function is considered as C^1 continuous across the interface.

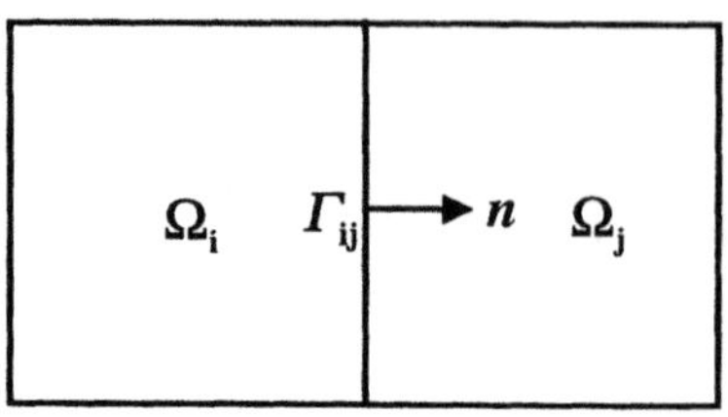

Figure 9.1 Topology of a patched interface

Mathematically, this continuity condition can be written as

$$f(x_N^i) = f(x_1^j) \quad \text{on } \Gamma_{ij} \tag{9.1}$$

$$f_n(x_N^i) = f_n(x_1^j) \quad \text{on } \Gamma_{ij} \tag{9.2}$$

where $f(x_N^i)$ and $f(x_1^j)$ represent respectively the values of the function f at the interface of the i-subdomain and the j-subdomain, $f_n(x_N^i)$ and $f_n(x_1^j)$ represent the values of the first-order derivative of f with respect to n at the interface. Equation 9.1 can be easily implemented. However, the implementation of Equation 9.2 involves the approximation of the first order derivative. Different approximations of the first order derivative would give different formulations to update the functional value at the interface. Basically, we have two methods to approximate the first order derivative in Equation 9.2.

DQ Approximation: For this approach, the DQ method is applied to discretize the first order derivatives in Equation 9.2. For the selected case, each subdomain is rectangular. Then the normal direction to the interface is parallel to one of the coordinate axes in the local coordinate system. For simplicity, this coordinate axis is assumed to be the x axis, and along this direction there are N grid points in the i-subdomain and M grid points in the j-subdomain. The local DQ weighting coefficients of the first-order derivative along the x direction are written as $a_{m,n}^i$ in the i-subdomain and as $a_{m,n}^j$ in the j-subdomain. Thus, using the DQ method, Equation 9.2 can be approximated by

$$\sum_{k=1}^{N} a_{N,k}^i \cdot f(x_k^i) = \sum_{k=1}^{M} a_{1,k}^j \cdot f(x_k^j) \tag{9.3}$$

Using Equation 9.3, and setting $f(x_N^i) = f(x_1^j) = \bar{f}$, we obtain

$$\bar{f} = \frac{\sum_{k=1}^{N-1} a_{N,k}^i \cdot f(x_k^i) - \sum_{k=2}^{M} a_{1,k}^j \cdot f(x_k^j)}{a_{1,1}^j - a_{N,N}^i} \tag{9.4}$$

where $\bar{f}$ is the value of the function f at the interface Γ_{ij}, which exchanges information between the two neighboring subdomains.

***First Order Finite Difference Approximation*:** For this approach, the first order finite difference scheme is used to discretize the first order derivatives in Equation 9.2. It is supposed that the spacing between the interface point and its adjacent point in the subdomain Ω_i is Δx_1, and the spacing between the interface point and its adjacent point in the subdomain Ω_j is Δx_2, that is,

$$\Delta x_1 = x_N^i - x_{N-1}^i, \quad \Delta x_2 = x_2^j - x_1^j$$

Applying the first order finite difference scheme to Equation 9.2 gives

$$\frac{\bar{f} - f(x_{N-1}^i)}{\Delta x_1} = \frac{f(x_2^j) - \bar{f}}{\Delta x_2} \tag{9.5}$$

which can be further reduced to

$$\bar{f} = \frac{\Delta x_2 \cdot f(x_{N-1}^i) + \Delta x_1 \cdot f(x_2^j)}{\Delta x_1 + \Delta x_2} \tag{9.6}$$

In the work of simulating Czochralski bulk flows, Shu, Chew and Liu (1998) compared the performance of Equations 9.4 and 9.6, and found that Equation 9.6 gives more accurate numerical results than Equation 9.4. As shown by Equations 9.1 and 9.2, the function is considered as C^1 continuous across the interface. For the first-order finite difference approximation, the function is assumed to be linear across the interface, which is coincident with the enforced condition of the patched interface. Thus, it gives more accurate results. In contrast, the DQ method is a global approach which is based on a high-order polynomial approximation. Therefore, it is overqualified when applied to a function with C^1 continuity, because the second- and higher-order terms may not fully cancel out during the numerical computation, and the resulting high-order errors may accumulate and spread into the whole computational domain. This is probably the reason why the DQ approximation (Equation 9.4) gives less accurate numerical results than the first order difference approximation (Equation 9.6).

9.2.1.2 Overlapped Interface

The topology of an overlapped interface is shown in Figure 9.2, where subdomain ABCD is overlapped with subdomain EFGH. It is noted that the right boundary of subdomain Ω_i, BC, is in the interior of subdomain Ω_j, and the left boundary of subdomain Ω_j, EH, is in the interior of subdomain Ω_i.

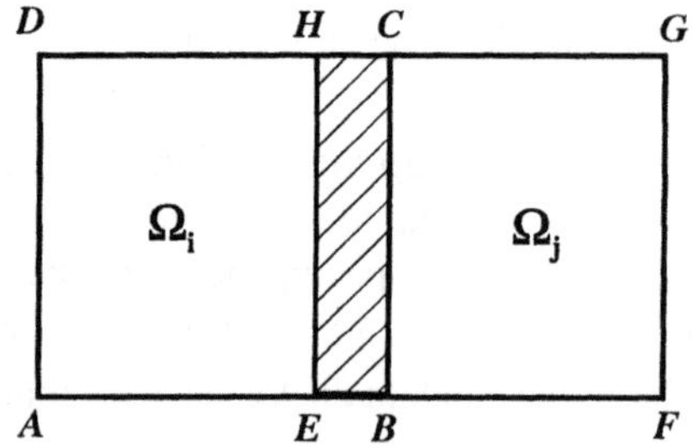

Figure 9.2 Topology of an overlapped interface

For the overlapped interface, the governing equations are also applied at the interface. It is supposed that the solution in the interior of subdomains is known at an iteration level. Then the functional values along the lines of BC and EH can be found out. These values are then transferred into the neighboring subdomains as new boundary conditions to get the solution in the interior of subdomains at the next iteration, i.e., the functional values along EH are transferred into subdomain Ω_j, and the functional values along BC are transferred into subdomain Ω_i. This process continues until converged solutions in all subdomains are obtained. For the overlapped topology, the functional values at the interface are obtained from the solution of the governing equations.

9.2.2 Multi-domain DQ Application in Fluid Mechanics

Shu and his collaborators have applied the multi-domain DQ method to simulate various incompressible flows by solving the vorticity-stream function formulation. Using a patched interface and Equation 9.4, Shu (1991) simulated the flow past a backward facing step by using 3 subdomains, and the flow past a square step by using 5 subdomains. Very accurate numerical results have been obtained. Some results for the backward facing step are described as follows. Figure 9.3 shows the problem definition and the computational domain for a backward facing step. For this problem, the expansion ratio is taken as 1:1.5. For the numerical simulation, all the lengths are normalized by the inlet width *D*, and the velocities by the maximum value of the longitudinal velocity at the inlet. The location of the inlet is taken as 6*D* upstream of the step, and the outlet is located at 12*D* downstream of the step.

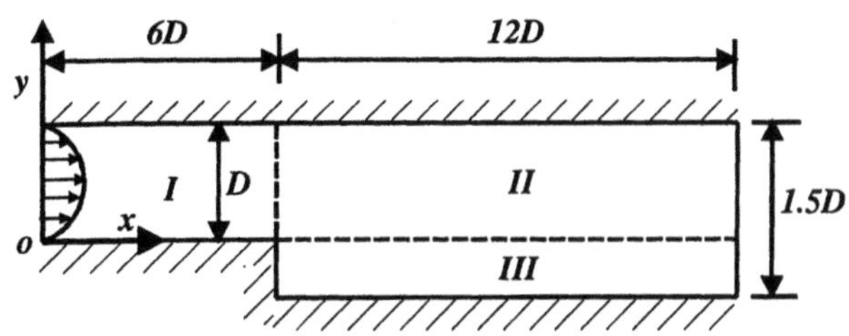

Figure 9.3 Computational domain of a backward facing step

As shown in Figure 9.3, a fully developed parabolic velocity profile is used as the boundary condition at the inlet. At the outlet, the natural boundary condition (zero gradient) is applied to both the vorticity and stream function. On the wall, the no-slip condition is implemented. After converting the boundary condition for velocity components into the condition for the stream function, we can obtain

$$\psi = 2y^2 - \frac{4}{3}y^3,\ \psi_x = 0 \qquad \text{at the inlet} \tag{9.7a}$$

$$\psi_x = 0, \ \omega_x = 0 \qquad \text{at the outlet} \tag{9.7b}$$

$$\psi = 0, \ \psi_n = 0 \qquad \text{on the bottom wall} \tag{9.7c}$$

$$\psi = \frac{2}{3}, \ \psi_n = 0 \qquad \text{on the top wall} \tag{9.7d}$$

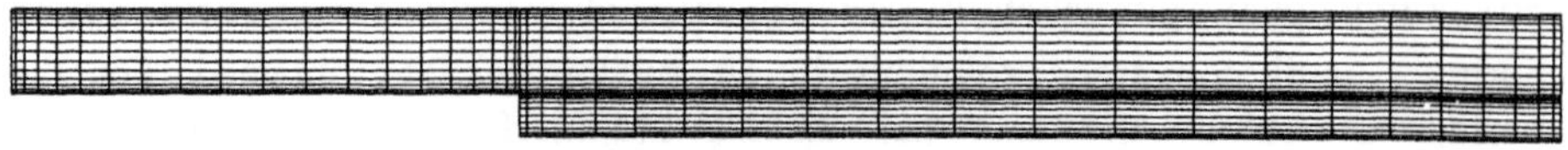

Figure 9.4 A computational mesh for flow past a backward facing step

For the DQ computation, the weighting coefficients are computed by the PDQ method. The mesh sizes used are 15×13 for domain I, 23×13 for domain II, and 23×11 for domain III. A computational mesh is shown in Figure 9.4. Figure 9.5 displays the streamlines of computed DQ results for Re = 500. The recirculation zone is clearly shown in the figure. The length of recirculation zone computed by multi-domain DQ method at various Reynolds numbers from Re = 50 to Re = 1000 is plotted in Figure 9.6. For comparison, the experimental results of Kueny and Binder (1984) and the finite element results of Bredif (1984) are also included in Figure 9.6. From the figure, it is obvious that the multi-domain DQ results agree well with the experimental data.

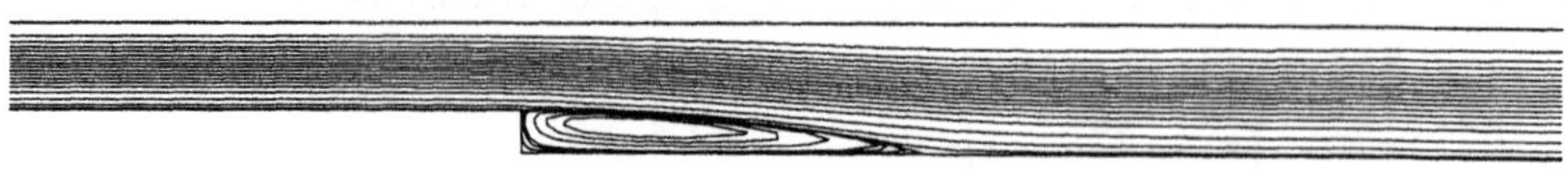

Figure 9.5 Streamlines of the flow past a backward facing step for Re = 500 (Shu 1991)

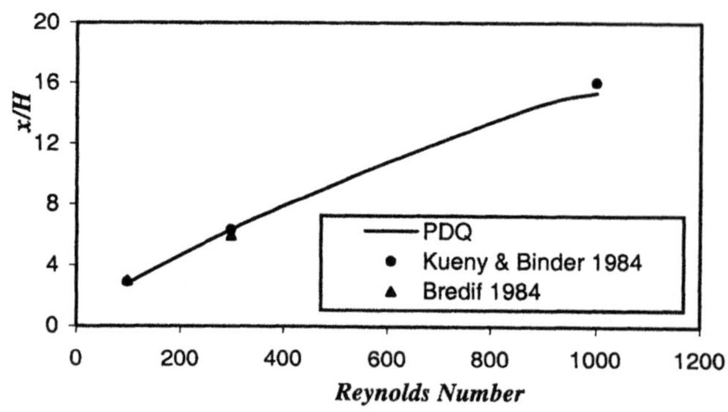

Figure 9.6 Length of recirculation zone vs Reynolds number (Shu 1991)

Comparison of the performances of the patched and overlapped interfaces has been conducted by Shu (1991) through the simulation of a driven cavity flow as well as Shu, Chew and Liu (1998) through the simulation of flows in Czochralski crystal growth. From these studies, it was found that the overlapped interface can generally yield better accuracy of numerical results than the patched interface. The reason may be that the governing equation is satisfied at the overlapped interface, but may not be satisfied at the patched interface.

9.2.3 Multi-domain DQ Application in Solid Mechanics

As shown above, many researchers have made contributions in applying the domain decomposition method to structural and vibration analysis, where the domain decomposition is based on the patched interface. Thus, the method is also called the DQ element approach. The DQ element approach consists of a collocation method in conjunction with a finite element technique to combine the high accuracy of the DQ method with the generality of the finite element formulation. As an example, Striz, Chen and Bert (1997) showed the free vibration analysis of a square isotropic plate with all edges simply supported using the DQ element approach. The element used in their work is a 25-node rectangular element, which is shown in Figure 9.7. The convergence of the frequency parameter λ ($\lambda = \rho h \omega^2 a^4 / D$) for the first eight modes was studied by using 3 element meshes of 1×1, 2×2, 3×3. The numerical results for this case are listed in Table 9.1. It can be observed from the table that the convergence trend for the DQ element results is very good.

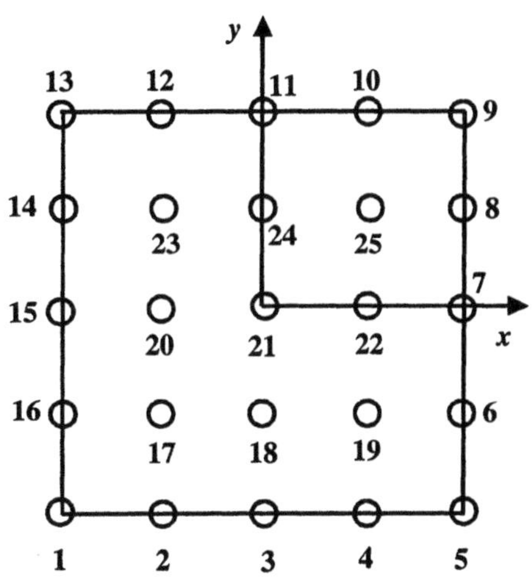

Figure 9.7 The nodal configuration of a rectangular element

Table 9.1 Convergence study of frequency parameter λ of a SS-SS-SS-SS square plate (Striz, Chen and Bert 1997)

Mode	*DQ element meshes*			*Exact*
	1×1	2×2	3×3	
1	19.7392	19.7392	19.7392	19.7392
2	49.4908	49.3480	49.3480	49.3480
3	79.1667	78.9568	78.9568	78.9568
4	100.117	98.7106	98.6961	98.6960
5	129.612	128.317	128.305	128.305
6	-	168.423	167.792	167.783
7	179.729	177.671	177.653	177.653
8	-	197.963	197.400	197.392

9.2.4 Multi-domain DQ Application in Waveguide Analysis

For analysis of various hollow conducting waveguides with rectangular boundaries, Shu and Chew (1999) applied the multi-domain DQ method with the overlapped interface to compute the cutoff wavenumbers of double-ridged, *L*-shaped, single-ridged, coaxial rectangular, and vaned rectangular waveguides. The multi-domain DQ results obtained are very accurate as compared with the results given by other numerical methods. Some results of *L*-shaped waveguides are shown below.

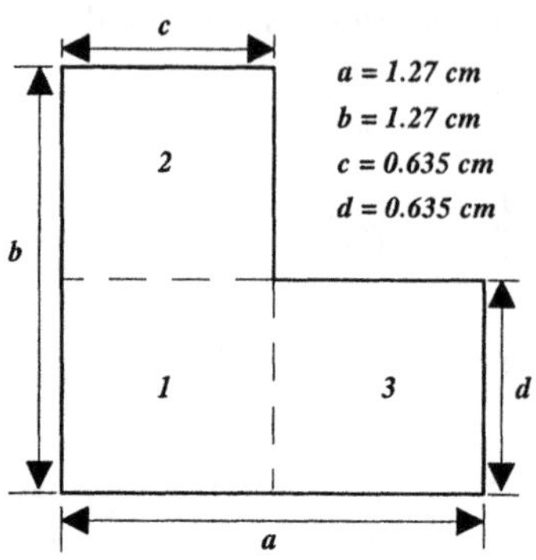

Figure 9.8 Configuration of a *L*-shaped waveguide and distribution of subdomains

The governing equation for the waveguide analysis is the Helmholtz equation, which has been described in Chapters 5 and 8. The configuration of a *L*-shaped waveguide with $a = b = 1.27$ cm, $c = d = a/2$ and the distribution of 3 subdomains are shown in Figure 9.8. In each subdomain, the same mesh size is used, and the PDQ method is applied to discretize the derivatives. Table 9.2 shows the computed cutoff wavenumbers of the first eight *TM* and *TE* modes by the multi-domain DQ approach.

Also included in this table are the results of the finite difference method with simultaneous iteration by Chebyshev acceleration (Guan and Su 1995). The multi-domain DQ results were obtained by using four different mesh sizes. It can be seen from the table that the convergence of the multi-domain DQ results is very good, and they agree very well with the results of Guan and Su (1995) which were based on a mesh size of 50×50. As shown by Guan and Su (1995), the analytical solutions of this problem for TM_3, TM_8, TE_3, TE_4, TE_7 modes are available. Their respective values are 6.9967, 11.0627, 4.9474, 4.9474, and 6.9967. Table 9.2 shows that when the mesh size in each subdomain is taken as 13×13, the multi-domain DQ results match the analytical solution of the above modes up to four decimal places. From the convergence trend, it is believed that the multi-domain DQ results with a mesh size of 9×9 are more accurate than the results of Guan and Su (1995).

Table 9.2 Cutoff wavenumbers of a *L*-shaped waveguide (Shu & Chew 1999)

	Multi-domain DQ				Guan & Su (1995)
Meshes	9×9	13×13	17×17	21×21	
			TM Modes		
TM_1	4.8976	4.8922	4.8907	4.8902	4.8949
TM_2	6.1390	6.1391	6.1392	6.1392	6.1350
TM_3	6.9967	6.9967	6.9967	6.9967	6.9921
TM_4	8.5565	8.5565	8.5565	8.5565	8.5458
TM_5	8.9073	8.9000	8.8980	8.8972	8.8940
TM_6	10.1493	10.1443	10.1430	10.1425	10.1262
TM_7	10.5577	10.5580	10.5580	10.5580	10.5318
TM_8	11.0628	11.0627	11.0627	11.0627	11.0380
			TE Modes		
TE_1	1.9053	1.9104	1.9118	1.9123	1.9111
TE_2	2.9604	2.9605	2.9605	2.9605	2.9600
TE_3	4.9474	4.9474	4.9474	4.9474	4.9452
TE_4	4.9474	4.9474	4.9474	4.9474	4.9452
TE_5	5.3147	5.3147	5.3147	5.3147	5.3128
TE_6	5.5752	5.5810	5.5825	5.5831	5.5799
TE_7	6.9967	6.9967	6.9967	6.9967	6.9937
TE_8	7.2736	7.2840	7.2868	7.2879	7.2784

A Fortran program for this problem is attached in Appendix *C*.

9.3 DQ Application in Curvilinear Coordinate System

The first application of the DQ method in the curvilinear coordinate system was given by Lam (1993) to study the heat conduction problem. Subsequently, the coordinate transformation approach was used by Shu, Chew, Khoo and Yeo (1995) to simulate the incompressible viscous flows around complex geometries, Bert and Malik (1996*a*), Han and Liew (1997*c*), for the structural and vibration analysis of irregular plates. This section will show in detail the application of the DQ method in the curvilinear coordinate system.

9.3.1 Coordinate Transformation

As shown in Figure 9.9, the physical domain in the (x, y) plane is irregular. However, this irregular domain can be transformed by the following formulation to a regular domain in the (ξ,η) plane,

$$\begin{cases} \xi = \xi(x, y) \\ \eta = \eta(x, y) \end{cases} \tag{9.8}$$

The regular domain in the (ξ,η) plane is also called the computational domain. In this approach, all the numerical computations are based on the computational domain.

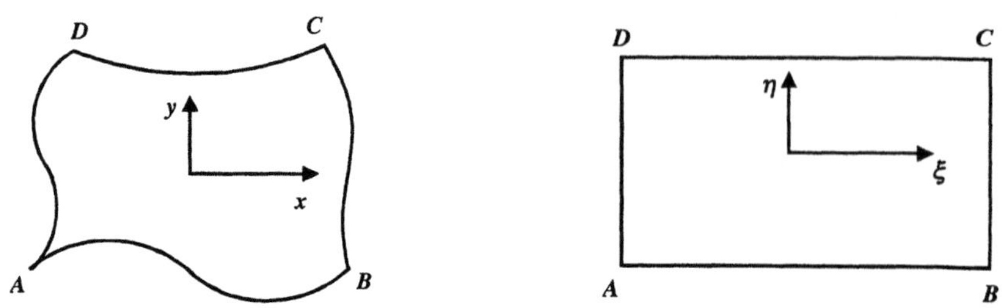

(*a*) Physical domain (*b*) Computational domain

Figure 9.9 Physical and computational domains

In the computational domain, the independent variables are ξ and η instead of x and y. Therefore, all the spatial derivatives with respect to x and y have to be transformed into derivatives with respect to ξ and η. The transformation can be made by invoking the differential chain rule. For example, the first order derivatives can be transformed by

$$\frac{\partial}{\partial x}=\frac{\partial \xi}{\partial x}\frac{\partial}{\partial \xi}+\frac{\partial \eta}{\partial x}\frac{\partial}{\partial \eta}=\xi_x\frac{\partial}{\partial \xi}+\eta_x\frac{\partial}{\partial \eta} \tag{9.9a}$$

$$\frac{\partial}{\partial y}=\frac{\partial \xi}{\partial y}\frac{\partial}{\partial \xi}+\frac{\partial \eta}{\partial y}\frac{\partial}{\partial \eta}=\xi_y\frac{\partial}{\partial \xi}+\eta_y\frac{\partial}{\partial \eta} \tag{9.9b}$$

The geometric parameters $x_\xi, x_\eta, y_\xi, y_\eta$ and $\xi_x, \xi_y, \eta_x, \eta_y$ have the following relationships

$$\xi_x = J^{-1} y_\eta,\ \eta_x = -J^{-1} y_\xi,\ \xi_y = -J^{-1} x_\eta,\ \eta_y = J^{-1} x_\xi \tag{9.10}$$

where

$$J = x_\xi y_\eta - x_\eta y_\xi \tag{9.11}$$

Let $f(x,y)$ be a differentiable scalar function of x and y. Then using Equations 9.9 and 9.10, we have

$$f_x = \frac{\partial f}{\partial x} = (y_\eta f_\xi - y_\xi f_\eta)/J \tag{9.12a}$$

$$f_y = \frac{\partial f}{\partial y} = (x_\xi f_\eta - x_\eta f_\xi)/J \tag{9.12b}$$

Repeated applications of Equation 9.12 give the following forms for the second order derivatives

$$\begin{aligned} f_{xx} &= \frac{\partial^2 f}{\partial x^2} = (y_\eta^2 f_{\xi\xi} - 2y_\xi y_\eta f_{\xi\eta} + y_\xi^2 f_{\eta\eta})/J^2 + [(y_\eta^2 y_{\xi\xi} - 2y_\xi y_\eta y_{\xi\eta} \\ &+ y_\xi^2 y_{\eta\eta})(x_\eta f_\xi - x_\xi f_\eta) + (y_\eta^2 x_{\xi\xi} - 2y_\xi y_\eta x_{\xi\eta} + y_\xi^2 x_{\eta\eta})(y_\xi f_\eta - y_\eta f_\xi)]/J^3 \end{aligned} \tag{9.13a}$$

$$\begin{aligned} f_{yy} &= \frac{\partial^2 f}{\partial y^2} = (x_\eta^2 f_{\xi\xi} - 2x_\xi x_\eta f_{\xi\eta} + x_\xi^2 f_{\eta\eta})/J^2 + [(x_\eta^2 y_{\xi\xi} - 2x_\xi x_\eta y_{\xi\eta} \\ &+ x_\xi^2 y_{\eta\eta})(x_\eta f_\xi - x_\xi f_\eta) + (x_\eta^2 x_{\xi\xi} - 2x_\xi x_\eta x_{\xi\eta} + x_\xi^2 x_{\eta\eta})(y_\xi f_\eta - y_\eta f_\xi)]/J^3 \end{aligned} \tag{9.13b}$$

$$\begin{aligned} f_{xy} &= \frac{\partial^2 f}{\partial x \partial y} = [(x_\xi y_\eta + x_\eta y_\xi) f_{\xi\eta} - x_\xi y_\xi f_{\eta\eta} - x_\eta y_\eta f_{\xi\xi}]/J^2 \\ &+ [x_\eta y_\eta x_{\xi\xi} - (x_\xi y_\eta + x_\eta y_\xi) x_{\xi\eta} + x_\xi y_\xi x_{\eta\eta}](y_\eta f_\xi - y_\xi f_\eta)/J^3 \\ &+ [x_\eta y_\eta y_{\xi\xi} - (x_\xi y_\eta + x_\eta y_\xi) y_{\xi\eta} + x_\xi y_\xi y_{\eta\eta}](x_\xi f_\eta - x_\eta f_\xi)/J^3 \end{aligned} \tag{9.13c}$$

In a similar manner, the Laplacian operator and the divergence formulation can be transformed to

$$\nabla^2 f = \frac{\partial^2 f}{\partial x^2} + \frac{\partial^2 f}{\partial y^2} = (\alpha f_{\xi\xi} - 2\beta f_{\xi\eta} + \gamma f_{\eta\eta} + \sigma f_{\eta} + \tau f_{\xi}) / J^2 \tag{9.14}$$

$$\nabla \cdot \vec{F} = \frac{\partial F_1}{\partial x} + \frac{\partial F_2}{\partial y} = \frac{1}{J}\left[\frac{\partial(y_\eta F_1 - x_\eta F_2)}{\partial \xi} + \frac{\partial(x_\xi F_2 - y_\xi F_1)}{\partial \eta}\right] \tag{9.15}$$

where

$$\alpha = x_\eta^2 + y_\eta^2 \tag{9.16a}$$

$$\beta = x_\xi x_\eta + y_\xi y_\eta \tag{9.16b}$$

$$\gamma = x_\xi^2 + y_\xi^2 \tag{9.16c}$$

$$\sigma = [y_\xi(\alpha x_{\xi\xi} - 2\beta x_{\xi\eta} + \gamma x_{\eta\eta}) - x_\xi(\alpha y_{\xi\xi} - 2\beta y_{\xi\eta} + \gamma y_{\eta\eta})] / J \tag{9.16d}$$

$$\tau = [x_\eta(\alpha y_{\xi\xi} - 2\beta y_{\xi\eta} + \gamma y_{\eta\eta}) - y_\eta(\alpha x_{\xi\xi} - 2\beta x_{\xi\eta} + \gamma x_{\eta\eta})] / J \tag{9.16e}$$

The above formulations will be used to transform the governing equations into the relevant forms in the curvilinear coordinate system. Note that the coordinate transformation (Equation 9.8) can be achieved by an algebraic expression or by the numerical grid generation method.

9.3.2 DQ Simulation of Incompressible Flows in Irregular Domains

For simulation of flows around a complex geometry, the calculation is based on the computational domain in the (ξ, η) plane. Thus, the incompressible Navier-Stokes equations should be transformed to the (ξ, η) system. Using the formulations given in the above section, the two-dimensional incompressible Navier-Stokes equations in the curvilinear coordinate system can be written as

continuity equation

$$\frac{1}{J}\left[y_\eta \frac{\partial u}{\partial \xi} - y_\xi \frac{\partial u}{\partial \eta} + x_\xi \frac{\partial v}{\partial \eta} - x_\eta \frac{\partial v}{\partial \xi}\right] = 0 \tag{9.17}$$

x-momentum equation

$$\frac{\partial u}{\partial t}+\frac{1}{J}\left[\frac{\partial(Uu)}{\partial\xi}+\frac{\partial(Vu)}{\partial\eta}\right]+\frac{1}{J}\left[y_\eta\frac{\partial p}{\partial\xi}-y_\xi\frac{\partial p}{\partial\eta}\right]$$
$$=\frac{1}{J}\left\{\frac{\partial}{\partial\xi}\left[A\frac{\partial u}{\partial\xi}+B\frac{\partial u}{\partial\eta}\right]+\frac{\partial}{\partial\eta}\left[C\frac{\partial u}{\partial\eta}+D\frac{\partial u}{\partial\xi}\right]\right\} \tag{9.18}$$

y-momentum equation

$$\frac{\partial v}{\partial t}+\frac{1}{J}\left[\frac{\partial(Uv)}{\partial\xi}+\frac{\partial(Vv)}{\partial\eta}\right]+\frac{1}{J}\left[-x_\eta\frac{\partial p}{\partial\xi}+x_\xi\frac{\partial p}{\partial\eta}\right]$$
$$=\frac{1}{J}\left\{\frac{\partial}{\partial\xi}\left[A\frac{\partial v}{\partial\xi}+B\frac{\partial v}{\partial\eta}\right]+\frac{\partial}{\partial\eta}\left[C\frac{\partial v}{\partial\eta}+D\frac{\partial v}{\partial\xi}\right]\right\} \tag{9.19}$$

where

$$U=uy_\eta-vx_\eta,\qquad V=vx_\xi-uy_\xi$$

$$A=\frac{\alpha}{\mathrm{Re}\cdot J},\qquad B=\frac{-\beta}{\mathrm{Re}\cdot J},\qquad C=\frac{\gamma}{\mathrm{Re}\cdot J},\qquad D=\frac{-\beta}{\mathrm{Re}\cdot J}$$

Similarly, the vorticity-stream function formulation can be transformed to

vorticity equation

$$\frac{\partial\omega}{\partial t}+\frac{1}{J}\left[\frac{\partial(U\omega)}{\partial\xi}+\frac{\partial(V\omega)}{\partial\eta}\right]=\frac{1}{J}\left\{\frac{\partial}{\partial\xi}\left[A\frac{\partial\omega}{\partial\xi}+B\frac{\partial\omega}{\partial\eta}\right]+\frac{\partial}{\partial\eta}\left[C\frac{\partial\omega}{\partial\eta}+D\frac{\partial\omega}{\partial\xi}\right]\right\} \tag{9.20}$$

stream function equation

$$\frac{\partial}{\partial\xi}\left[A_1\frac{\partial\psi}{\partial\xi}+B_1\frac{\partial\psi}{\partial\eta}\right]+\frac{\partial}{\partial\eta}\left[C_1\frac{\partial\psi}{\partial\eta}+D_1\frac{\partial\psi}{\partial\xi}\right]=J\omega \tag{9.21}$$

where

$$A_1=\mathrm{Re}\,A,\ B_1=\mathrm{Re}\,B,\ C_1=\mathrm{Re}\,C,\ D_1=\mathrm{Re}\,D$$

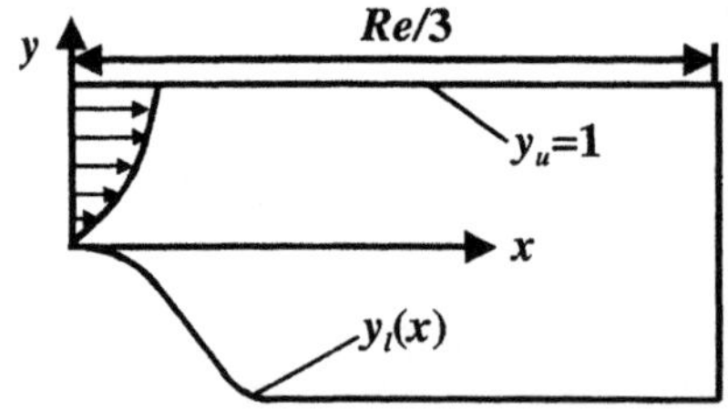

Figure 9.10 Geometry of an expansion channel

Using Equations 9.20 and 9.21, Shu, Chew, Khoo and Yeo (1995) successfully simulated a smooth expansion channel flow. The geometry of the problem is shown in Figure 9.10. This problem was initially proposed by Roache (1981), and then chosen by a workshop of International Association for Hydraulic Research (IAHR) working group (Napolitano and Orlandi 1985) as a suitable test case for assessing the capabilities of numerical methods to simulate laminar flows in complex geometries. This problem has a feature in that a quasi-self-similar solution can be obtained when the Reynolds number is much larger than 1. As shown in Figure 9.10, the total length of the channel is chosen to be Re/3. The lower boundary (solid wall) of the channel is given by the following expression

$$y_l(x) = \frac{1}{2}\left[\tanh\left(2 - \frac{30x}{\mathrm{Re}}\right) - \tanh(2)\right] \tag{9.22}$$

while the upper boundary (symmetry line) is located at $y_u = 1$. The boundary condition for ψ at the inlet is given by

$$\begin{cases} \psi = \dfrac{1}{2}(3y^2 - y^3) \\ \dfrac{\partial \psi}{\partial x} = 0 \end{cases} \qquad \text{for } x = 0,\ 0 \le y \le 1 \tag{9.23}$$

and the natural boundary condition

$$\psi_x = 0, \quad \omega_x = 0 \tag{9.24}$$

is used as the boundary condition at the outlet. The standard no-slip boundary condition is imposed at the wall while the symmetrical boundary condition is used along the symmetry line.

For the DQ application in the curvilinear coordinate system, the length scales in the ξ and η directions can be freely chosen by the user. In other words, one can choose the computational domain as $0 \le \xi \le 1$, $0 \le \eta \le 1$ or as $1 \le \xi \le N$, $1 \le \eta \le M$, where N and M are the number of grid points in the ξ and η direction respectively. When the coordinates of grid points in the (ξ, η) plane are specified, the weighting coefficients can be calculated using the formulations given in Chapters 2 and 3. For the expansion channel flow, Shu, Chew, Khoo and Yeo (1995) set up the coordinate transformation using a simple algebraic expression

$$\begin{cases} \xi = x / x_{out} = 3x/\mathrm{Re} \\ \eta = [y - y_l(x)]/[1 - y_l(x)] \end{cases} \tag{9.25}$$

With Equation 9.25, the geometric parameters x_ξ, x_η, y_ξ and y_η can be exactly computed. Equation 9.25 also implies that the computational domain in the (ξ,η) plane is $0 \le \xi \le 1$, $0 \le \eta \le 1$. Using the same mesh point distribution as given by Equation 6.43, the weighting coefficients are calculated by the PDQ formulations. Using a mesh size of 17×17, Shu, Chew, Khoo and Yeo (1995) obtained very accurate DQ results. Figure 9.11 shows the wall vorticity distribution for Re = 10 and Re = 100. The DQ results compare very well with the benchmark solution of the IAHR workshop given by Cliffe et al. using a finite element method with results being grid-independent. Note that the results of Cliffe et al. are shown in the paper of Napolitano and Orlandi (1985). It can be seen from the figure that, although a considerably small number of grid points and a non-orthogonal grid are used in the DQ computation, the DQ results are very accurate. Numerical computation also confirmed that as Reynolds number increases to a value of much larger than 1, the solution takes on a quasi-self-similar form, i.e., the wall vorticity becomes independent of Re when plotted against x / x_{out}.

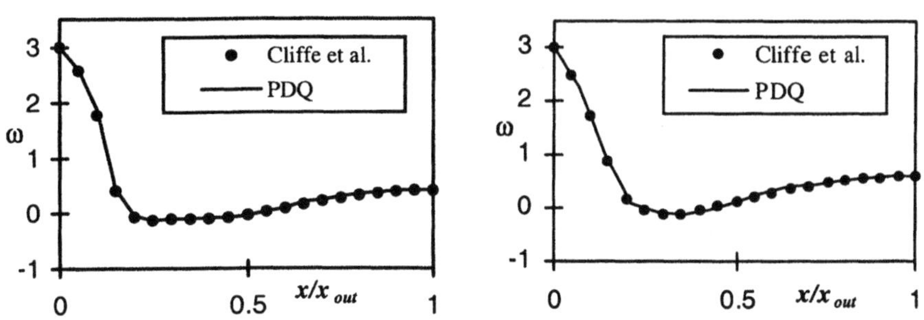

(a) Re = 10 (b) Re = 100

Figure 9.11 Wall vorticity distribution of an expansion channel flow (Shu, Chew, Khoo and Yeo 1995)

9.3.3 DQ Vibration Analysis of Irregular Plates

As shown in Chapter 7, the governing equation for vibration analysis of a thin plate involves fourth order partial derivatives with multiple boundary conditions. The application of the coordinate transformation technique for this case is much more complicated than for a flow problem. In this section, we will show two approaches in this application.

9.3.3.1 Partial Transformation

This approach was proposed by Bert and Malik in 1996. Having noted that the transformation forms of high order derivatives are very complicated, Bert and Malik (1996*a*) only performed coordinate transformation for the first order derivative, which is subsequently written in a matrix form. The discretizations of the second and higher order derivatives are made using the matrix multiplication technique.

Let $c_{i,j}^{(1)}$ and $\bar{c}_{i,j}^{(1)}$ be the weighting coefficients of the first order derivatives in the ξ and η direction respectively. Then, the discretization of Equation 9.12 in the curvilinear coordinate system gives

$$(f_x)_{i,j} = [(y_\eta)_{i,j} \sum_{k=1}^{N} c_{i,k}^{(1)} f_{k,j} - (y_\xi)_{i,j} \sum_{k=1}^{M} \bar{c}_{j,k}^{(1)} f_{i,k})] / J_{i,j} \tag{9.26a}$$

$$(f_y)_{i,j} = [(x_\xi)_{i,j} \sum_{k=1}^{M} \bar{c}_{j,k}^{(1)} f_{i,k} - (x_\eta)_{i,j} \sum_{k=1}^{N} c_{i,k}^{(1)} f_{k,j})] / J_{i,j} \tag{9.26b}$$

where N and M are the number of the mesh points in the ξ and η directions. Note that Equation 9.26 may involve all the functional values in the whole computational domain. Equation 9.26 can be written in a matrix form

$$\{f_x\} = [K^{(1)}]\{f\} \tag{9.27a}$$

$$\{f_y\} = [E^{(1)}]\{f\} \tag{9.27b}$$

where $\{f\}$ is a vector of functional values at all mesh points, $\{f_x\}$ and $\{f_y\}$ are vectors of the first order derivatives at all mesh points, the dimension of matrices $[K^{(1)}]$ and $[E^{(1)}]$ is NM by NM. Here $NM = N \times M$ is the total number of mesh points in the whole domain. In a similar manner, the discretization of high order derivatives can be put in the following matrix forms

$$\{f_{xx}\} = [K^{(2)}]\{f\} \tag{9.28a}$$

$$\{f_{yy}\} = [E^{(2)}]\{f\} \tag{9.28b}$$

$$\{f_{xxxx}\} = [K^{(4)}]\{f\} \tag{9.28c}$$

$$\{f_{yyyy}\} = [E^{(4)}]\{f\} \tag{9.28d}$$

$$\{f_{xxyy}\} = [K^{(2)}][E^{(2)}]\{f\} \tag{9.28e}$$

where the matrices $[K^{(2)}], [K^{(4)}], [E^{(2)}]$ and $[E^{(4)}]$ are determined by the matrix multiplication technique

$$[K^{(2)}] = [K^{(1)}][K^{(1)}] \tag{9.29a}$$

$$[K^{(4)}] = [K^{(2)}][K^{(2)}] \tag{9.29b}$$

$$[E^{(2)}] = [E^{(1)}][E^{(1)}] \tag{9.29c}$$

$$[E^{(4)}] = [E^{(2)}][E^{(2)}] \tag{9.29d}$$

Equation 9.28 can be directly substituted into Equation 7.34 to form an eigenvalue equation system. It is noted that the idea of this approach is very simple, and only the transformation of the first order derivatives is used in the curvilinear coordinate system. However, the computational effort for matrix multiplication by Equation 9.29 is very large, thus restricts the application of this approach, especially when a large number of mesh points is used. To reduce the computational effort, the present author proposed the complete transformation approach, which will be described below.

9.3.3.2 Complete Transformation

For this approach, all the spatial derivatives in the governing equation are transformed to the curvilinear coordinate system. Using the chain rule of differentiation, Equation 7.34 can be transformed into the following form in the (ξ, η) plane (curvilinear coordinate system),

$$\begin{aligned}
&\overline{D}^{(41)} w_{\xi\xi\xi\xi} + \overline{D}^{(42)} w_{\xi\xi\xi\eta} + \overline{D}^{(43)} w_{\xi\xi\eta\eta} + \overline{D}^{(44)} w_{\xi\eta\eta\eta} \\
&\quad + \overline{D}^{(45)} w_{\eta\eta\eta\eta} + \overline{D}^{(31)} w_{\xi\xi\xi} + \overline{D}^{(32)} w_{\xi\xi\eta} + \overline{D}^{(33)} w_{\xi\eta\eta} + \overline{D}^{(34)} w_{\eta\eta\eta} \\
&\quad + \overline{D}^{(21)} w_{\xi\xi} + \overline{D}^{(22)} w_{\xi\eta} + \overline{D}^{(23)} w_{\eta\eta} + \overline{D}^{(11)} w_{\xi} + \overline{D}^{(12)} w_{\eta} = \Omega^2 w
\end{aligned} \tag{9.30}$$

where

$$\overline{D}^{(41)} = a^2$$

$$\overline{D}^{(42)} = 2ab$$

$$\overline{D}^{(43)} = 2ac + b^2$$

$$\overline{D}^{(44)} = 2cb$$
$$\overline{D}^{(45)} = c^2$$
$$\overline{D}^{(31)} = 2da + 2aa_\xi + ba_\eta$$
$$\overline{D}^{(32)} = 2bd + 2ae + 2ab_\xi + 2ca_\eta + ba_\xi + bb_\eta$$
$$\overline{D}^{(33)} = 2dc + 2ac_\xi + 2be + 2cb_\eta + bb_\xi + bc_\eta$$
$$\overline{D}^{(34)} = 2ec + 2cc_\eta + bc_\xi$$
$$\overline{D}^{(21)} = d^2 + 2ad_\xi + da_\xi + ea_\eta + aa_{\xi\xi} + ba_{\xi\eta} + ca_{\eta\eta} + bd_\eta$$
$$\overline{D}^{(22)} = 2de + 2ae_\xi + 2cd_\eta + db_\xi + bd_\xi + eb_\eta + be_\eta + ab_{\xi\xi} + bb_{\xi\eta} + cb_{\eta\eta}$$
$$\overline{D}^{(23)} = e^2 + 2ce_\eta + dc_\xi + ec_\eta + ac_{\xi\xi} + bc_{\xi\eta} + cc_{\eta\eta} + be_\xi$$
$$\overline{D}^{(11)} = dd_\xi + ed_\eta + ad_{\xi\xi} + bd_{\xi\eta} + cd_{\eta\eta}$$
$$\overline{D}^{(12)} = de_\xi + ee_\eta + ae_{\xi\xi} + be_{\xi\eta} + ce_{\eta\eta}$$
$$a = A_1 / J \ , \ b = 2B_1 / J \ , \ c = C_1 / J \ , \ d = (A_{1\xi} + B_{1\eta}) / J \ , e = (B_{1\xi} + C_{1\eta}) / J$$
$$A_1 = \alpha / J \ , \ B_1 = -\beta / J \ , \ C_1 = \gamma / J$$

It is noted that the subscripts ξ and η in Equation 9.30 represent the partial derivatives in the respective directions. Equation 9.30 with varying coefficients is much more complicated in form than Equation 7.34. However, it should be indicated that, since the computational domain is regular, Equation 9.30 could be solved using exactly the same way as the application of the DQ method to solve regular domain problems. Thus, unlike the approach of partial transformation (Bert and Malik 1996*a*), the solution of Equation 9.30 does not involve the huge operation of matrix multiplication. As a consequence, the computational effort can be greatly reduced.

9.3.3.3 Implementation of Boundary Conditions

The simply supported (SS) and clamped (C) boundary conditions are considered in this section. For an irregular plate, the simply supported and clamped conditions can be written as

Clamped (C)

$$w = 0 \tag{9.31a}$$
$$w_x \cos\theta + w_y \sin\theta = 0 \tag{9.31b}$$

Simply-supported (SS)

$$w = 0 \tag{9.32a}$$

$$(\cos^2\theta + \nu\sin^2\theta)w_{xx} + (\sin^2\theta + \nu\cos^2\theta)w_{yy} + 2(1-\nu)\cos\theta\sin\theta w_{xy} = 0 \quad (9.32b)$$

where θ is the angle between the normal to the plate boundary and the x-axis. In the following, we will show that Equations 9.31*b* and 9.32*b* can be simplified along the ξ = constant and η = constant boundaries.

For the clamped and simply-supported edges, the deflection w is always zero. Thus, we have

$$\frac{\partial w}{\partial \eta} = 0, \quad \frac{\partial^2 w}{\partial \eta^2} = 0 \quad (9.33)$$

on the ξ = constant boundaries, and

$$\frac{\partial w}{\partial \xi} = 0, \quad \frac{\partial^2 w}{\partial \xi^2} = 0 \quad (9.34)$$

on the η = constant boundaries. On the other hand, we note that θ is the angle between the normal to the plate boundary and the x-axis. So, along the ξ = constant boundaries, we have

$$\cos\theta = y_\eta / \sqrt{\alpha}, \quad \sin\theta = -x_\eta / \sqrt{\alpha} \quad (9.35)$$

So, along the ξ = constant boundaries, Equation 9.31*b* can be simplified as

$$\frac{\partial w}{\partial \xi} = 0 \quad (9.36)$$

and Equation 9.32*b* can be reduced to

$$\frac{\partial^2 w}{\partial \xi^2} - \frac{2\beta}{\alpha}\frac{\partial^2 w}{\partial \xi \partial \eta} + s\frac{\partial w}{\partial \xi} = 0 \quad (9.37)$$

where

$$s = \frac{1}{J\alpha^2}\left[(\alpha^2 y_{\xi\xi} - 2\alpha\beta y_{\xi\eta} + \beta^2 y_{\eta\eta})x_\eta - (\alpha^2 x_{\xi\xi} - 2\alpha\beta x_{\xi\eta} + \beta^2 x_{\eta\eta})y_\eta\right] + \frac{\nu J}{\alpha^2}(y_{\eta\eta}x_\eta - x_{\eta\eta}y_\eta)$$

Similarly, along the η = constant boundaries, $\cos\theta$ and $\sin\theta$ can be expressed as

$$\cos\theta = y_\xi / \sqrt{\gamma}\ ,\ \sin\theta = -x_\xi / \sqrt{\gamma} \tag{9.38}$$

As a consequence, Equation 9.31*b* is reduced to

$$\frac{\partial w}{\partial \eta} = 0 \tag{9.39}$$

and Equation 9.32*b* is simplified to

$$\frac{\partial^2 w}{\partial \eta^2} - \frac{2\beta}{\gamma}\frac{\partial^2 w}{\partial \xi \partial \eta} + t\frac{\partial w}{\partial \eta} = 0 \tag{9.40}$$

where

$$t = \frac{1}{J\gamma^2}\left[(\beta^2 x_{\xi\xi} - 2\gamma\beta x_{\xi\eta} + \gamma^2 x_{\eta\eta})y_\xi - (\beta^2 y_{\xi\xi} - 2\gamma\beta y_{\xi\eta} + \gamma^2 y_{\eta\eta})x_\xi\right]$$
$$+ \frac{\nu J}{\gamma^2}(x_{\xi\xi} y_\xi - y_{\xi\xi} x_\xi)$$

9.3.3.4 Sample Application

As an example, we consider the free vibration of an eccentric sectorial plate. The geometry of the problem is shown in Figure 9.12. This problem was first studied by Bert and Malik (1996*a*). The coordinate transformation in their work was made by isoparametric shape functions which are often used in the finite element method. It should be indicated that the physical domain for DQ application is usually much larger than an element in the finite element application. Thus, for a geometry with high curvature, the selected shape function should be of high order. For the general case, it is suggested that coordinate transformation be performed by using mesh generation technique which is a popular tool in CFD.

For the sample problem, the coordinate transformation can be exactly set up by

$$x(\xi,\eta) = [r_d + (a - r_d)(\eta + 1)/2]\cos(\phi\xi/2) \tag{9.41a}$$

$$y(\xi,\eta) = [r_d + (a - r_d)(\eta + 1)/2]\sin(\phi\,\xi/2) \tag{9.41b}$$

where

$$r_d = \frac{\sin(\alpha + \phi\xi/2)}{\sin(\phi\xi/2)}b$$

$$\alpha = \arcsin\left[\frac{e}{b}\sin(\phi\xi/2)\right]$$

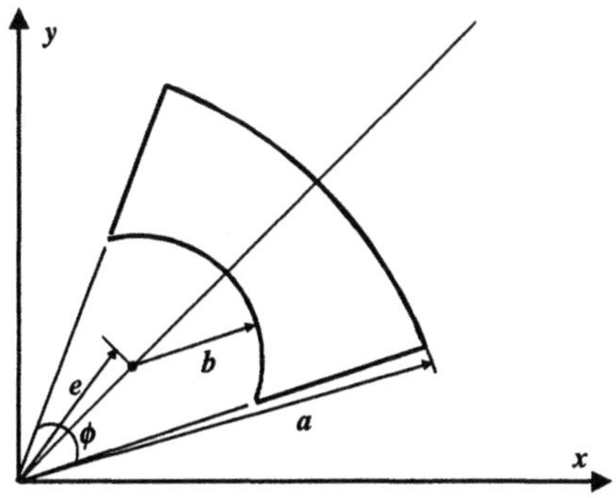

Figure 9.12 An eccentric sectorial domain

Note that the boundaries of ξ and η in Equation 9.41 define a square domain, $-1 \le \xi \le 1$, $-1 \le \eta \le 1$. The mesh point distribution in the (ξ, η) plane can be taken as

$$\xi_i = \cos\left(\frac{N-i}{N-1}\pi\right), \quad i = 1, 2, ..., N \tag{9.42a}$$

$$\eta_j = \cos\left(\frac{M-j}{M-1}\pi\right), \quad j = 1, 2, ..., M \tag{9.42b}$$

Table 9.3 First six frequencies of flexural vibration of eccentric sectorial plates ($a/b = 8/3$, $e/b = 1.0$, $\phi = 45°$, $\Omega = \omega a^2/\pi^2 \sqrt{\rho h/D}$)

Methods (N=M=15)	*Mode sequence*					
	1	2	3	4	5	6
SS-SS-SS-SS Plate						
Partial transformation	17.592	23.131	33.728	48.574	62.418	65.876
Complete transformation	17.717	23.168	33.745	48.557	62.545	65.854
C-C-C-C Plate						
Partial transformation	36.360	40.452	50.498	65.461	85.237	95.782
Complete transformation	36.405	40.417	50.488	65.443	85.306	95.718

The sample problem can be solved by both the partial transformation and complete transformation approaches. Table 9.3 displays the first six frequencies of flexural vibration of eccentric sectorial plates with SS-SS-SS-SS and C-C-C-C boundary conditions. The DQ results were obtained by two approaches using a mesh size of 15×15. Obviously, the two results agree very well. However, the CPU time required by the two approaches is quite different. It was found that for the same number of grid points, the DQ results using the complete transformation approach require much less CPU time than those using the partial transformation approach, especially for a large number of grid points. For example, for the results shown in Table 9.3, the partial transformation approach requires 35.72 seconds CPU time on HP C200 workstation while the complete transformation approach only needs 1.41 seconds CPU time on the same machine. It should be pointed out that the computational effort in the solution of a resultant eigenvalue equation system is the same for both approaches. The large computational effort of the partial transformation approach arises due to the huge operations involved in matrix multiplication.

9.4 Differential Cubature Method for Complex Problems

The differential cubature (DC) method has been introduced in Chapter 2. This method was proposed by Civan (1989) as an alternate technique for the numerical solution of multi-dimensional differential equations. By nature, the DC method has the ability to solve a problem with complex geometry since it discretizes the spatial derivative using the functional values at all mesh points in the whole domain. Obviously, there is no structure of mesh point distribution in applying the DC method. Civan (1989, 1991, 1994), Malik and Civan (1995) have made excellent contributions in the development and application of the DC method. Recently, in the application of the DC method to complex problems, Liew and Liu (1997), Liu and Liew (1998*b*) successfully performed the bending analysis of triangular, circular and skew plates. Very accurate numerical results have been obtained by using just a few mesh points. These applications showed that the DC method has a potential to become an efficient approach for solving complex problems.

On the other hand, we noticed that the number of mesh points used in the successful application of the DC method so far is still very small (usually less than 70). This is because the weighting coefficients in the DC method are determined by solving an algebraic equation system. When the number of mesh points is too large, this equation system will become highly ill-conditioned, leading to difficulties in computing the weighting coefficients. It should be indicated that the use of sufficient mesh points is necessary for the accurate numerical solution of some complex problems. Thus, there is a need to explore an efficient way to compute the weighting coefficients in the DC method before it can be extensively applied to engineering problems. Another concern regarding the application of the DC method is the selection of the set of test functions with respect to the completeness requirement. As pointed out by Malik and Civan (1995), the selection is not arbitrary.

To sum up, we can say that the DC method has the potential to solve a general multi-dimensional problem. However, much work is needed to further explore this method.

Chapter 10

Generalized Integral Quadrature (GIQ) and Its Application to Solve Boundary Layer Equations

10.1 Introduction

In practice, for some problems such as the area of a surface and the volume of a body, it is necessary to know the integration of a function over a domain. In most cases, it is difficult to obtain the value of the integration analytically. As a result, numerical integration is of great interest in engineering. The numerical integration of a function $f(x)$ over a domain $[a, b]$ can usually be written in the form

$$\int_a^b f(x) \cdot dx = \sum_{i=1}^{N} c_i \cdot f(x_i) \tag{10.1}$$

where c_i is the weighting coefficient. There are a lot of conventional rules to determine c_i. The low order Simpson's rule is the most frequently used approach in obtaining approximate integrals. To obtain accurate results, high order methods are preferable. Among them, the integration rules of Gaussian quadrature are extensively used, where both the weighting coefficients and the coordinates of the grid points are taken as unknowns. These unknowns are determined by $2N$ powers $1, x, \cdots, x^{2N-1}$. The major advantage of the Gaussian quadrature method is its high accuracy. But for the general case, it has a disadvantage in that the coordinates obtained may fall outside the region of interest. Thus it is necessary to check for the correct coordinates before numerical integration is performed. Some specific Gaussian quadrature techniques such as the Gauss-Legendre can overcome this drawback. Another drawback of Gaussian quadrature is that, since the coordinates are determined by the roots of some functions, they cannot be arbitrarily chosen. This may lead to difficulties in its use for some problems where the functional values are only known at some specific grid points which may not coincide with the coordinates obtained. In the case of solving a differential-integral equation, the grid points, where the unknowns are set, are usually generated in advance, and may not be the same as the coordinates required for Gaussian quadrature to be performed. Thus, some interpolation is necessary. This may introduce additional errors in the numerical computations.

In this chapter, we will show that c_i can be efficiently determined by the generalized integral quadrature method to a high order of accuracy.

10.2 Generalized Integral Quadrature (GIQ)

The generalized integral quadrature (GIQ) was developed following the same concept as PDQ. If a function is smooth in the whole domain, it can be approximated by a high order polynomial which would involve all the functional values in the whole domain. Then the integral of the function over a part of the whole domain can be computed by integrating the approximated polynomial. As a result, this approximation involves all the functional values in the whole domain with high order of accuracy even though the integral domain contains only two points. Obviously, the key procedure to this approach is to determine the weighting coefficients. We will show, in the following, that the weighting coefficients of GIQ can be easily obtained from those of the first order derivative in PDQ.

10.2.1 One dimensional Generalized Integral Quadrature

It is supposed that a function $f(x)$ is continuous in the overall domain $[a, b]$, which can be decomposed into N–1 intervals with grid points given as $a = x_1$, x_2, …, $x_N = b$. Since $f(x)$ is continuous in the whole domain, it can be approximated by an (N–1)th order polynomial. In particular, when the functional values at N grid points are known, $f(x)$ can be approximated by the Lagrange interpolation polynomials which are related to the functional values at all mesh points. As a result, the integral of this approximated polynomial over $[x_i, x_j]$ may involve the functional values outside the integral domain. As a general case, it is assumed that the integral of $f(x)$ over a part of the whole domain is approximated by a linear combination of all the functional values in the overall domain with the form

$$\int_{x_i}^{x_j} f(x)\cdot dx = \sum_{k=1}^{N} c_k^{ij}\cdot f(x_k) \tag{10.2}$$

where x_i and x_j are the coordinates that can be altered. When $x_i = a, x_j = b$, Equation 10.2 becomes a conventional integral. In a similar fashion to the analysis in PDQ, the (N–1)th order polynomial, which is an approximation to $f(x)$, constitutes an N–dimensional linear vector space. Thus if all the base polynomials satisfy Equation 10.2, so does any polynomial in the space. If the Lagrange interpolated polynomials, $r_k(x), k = 1, 2, \ldots, N$, are chosen as the base polynomials, c_k^{ij} can be determined by

$$c_k^{ij} = \int_{x_i}^{x_j} r_k(x) \cdot dx \tag{10.3}$$

The expression for c_k^{ij} is very complicated. Therefore, it is difficult to calculate c_k^{ij} accurately using Equation 10.3. We will adopt another approach to determine c_k^{ij}. Setting

$$f(x) = \frac{du(x)}{dx} \tag{10.4}$$

we can clearly see that if $f(x)$ is an (N−1)th order polynomial, $u(x)$ should be an Nth order polynomial. It is supposed that $f(x)$ is approximated by an (N−1)th order polynomial with the following form

$$f(x) = a_0 + a_1 \cdot x + \cdots + a_{N-1} \cdot x^{N-1} \tag{10.5}$$

where $a_0, a_1, \ldots, a_{N-1}$ are constants. Integrating Equation 10.5 from a constant c to the variable x gives

$$u(x) = \int_c^x f(t) \cdot dt + u(c) = F(x,c) + u(c) \tag{10.6}$$

where

$$F(x,c) = x\left(a_0 + \frac{a_1}{2}x + \cdots + \frac{a_{N-1}}{N}x^{N-1}\right) - c\left(a_0 + \frac{a_1}{2}c + \cdots + \frac{a_{N-1}}{N}c^{N-1}\right)$$

Clearly, $F(x,c)$ depends on N constants. It can be shown that $F(x,c)$ constitutes an N dimensional linear polynomial vector space. One set of its base polynomials can be chosen as

$$p_k(x) = (x-c) \cdot r_k(x) \quad , \quad k = 1, 2, \ldots, N \tag{10.7}$$

where $r_k(x)$ is the Lagrange interpolated polynomial. Similar to the PDQ, we can set

$$F_x(x_i, c) = \sum_{j=1}^{N} \underline{a}_{ij} \cdot F(x_j, c), \quad \text{for } i = 1, 2, \ldots, N \tag{10.8}$$

where $\underline{a}_{ij}$ are the weighting coefficients. The value of $\underline{a}_{ij}$ can be determined using the same fashion as that used for PDQ. Substituting Equation 10.7 into Equation 10.8 gives the weighting coefficients $\underline{a}_{ij}$ as

$$\underline{a}_{ij} = \frac{x_i - c}{x_j - c} \cdot w_{ij}^{(1)} \quad , \quad \text{when } j \neq i \tag{10.9a}$$

$$\underline{a}_{ii} = w_{ii}^{(1)} + \frac{1}{x_i - c} \tag{10.9b}$$

where $w_{ij}^{(1)}$ is the weighting coefficient of the first order derivative in PDQ. From Equation 10.9, c cannot be chosen to be the coordinates of grid points, x_i. This can be seen from Equation 10.6 that, when $c = x$, the integral domain is zero. Thus it is not necessary to approximate the integral for $c = x$.

On the other hand, Equation 10.8 can be written as a vector form

$$\{F_x\} = [\underline{A}]\{F\} \tag{10.10}$$

where

$$\{F\} = \{F(x_1,c), F(x_2,c), \cdots, F(x_N,c)\}^T$$

$$\{F_x\} = \{f\} = \{F_x(x_1,c), F_x(x_2,c), \cdots, F_x(x_N,c)\}^T$$

Substituting Equations 10.4 and 10.6 into Equation 10.10 and setting

$$\{f^I\} = \{F\} = \{\int_c^{x_1} f(x)dx, \int_c^{x_2} f(x)dx, \cdots, \int_c^{x_N} f(x)dx\}^T \tag{10.11}$$

we obtain

$$\{f\} = [\underline{A}] \cdot \{f^I\} \tag{10.12}$$

Setting $[W^I] = [\underline{A}]^{-1}$, Equation 10.12 can be further reduced to

$$\int_c^{x_i} f(x) \cdot dx = \sum_{k=1}^{N} w_{ik}^I \cdot f(x_k), \quad \text{for } i = 1, 2, \ldots, N \tag{10.13}$$

Thus

$$\int_{x_i}^{x_j} f(x) \cdot dx = \sum_{k=1}^{N} (w_{jk}^I - w_{ik}^I) \cdot f(x_k) \tag{10.14}$$

By comparing Equation 10.14 with Equation 10.2, we obtain

$$c_k^{ij} = w_{jk}^I - w_{ik}^I \tag{10.15}$$

Obviously, the weighting coefficients in GIQ are computed from those of the first order derivative discretization in PDQ.

10.2.2 Error Analysis

The error of the numerical integral of $f(x)$ in the domain $[x_i, x_j]$ is defined as

$$E_I(f, x_i, x_j) = \int_{x_i}^{x_j} f(x)dx - \sum_{k=1}^{N}(w_{jk}^I - w_{ik}^I)f(x_k) \tag{10.16}$$

where w_{ij}^I is the weighting coefficient in GIQ. Note that $f(x)$ in GIQ is approximated by an (N−1)th order polynomial. So, the error estimation for $f(x)$ is given by Equation 2.58. Thus, for the general case, using Equation 2.58, we can get

$$E_I(f, x_i, x_j) = \frac{1}{N!} \cdot \int_{x_i}^{x_j} f^{(N)}(\xi) \cdot M(x) \cdot dx \tag{10.17}$$

If the integral domain is $[x_i, x_{i+1}]$, then $M(x)$ does not change its sign in $[x_i, x_{i+1}]$. By using the second mean-value theorem, Equation 10.17 can be reduced to

$$E_I(f, x_i, x_{i+1}) = \frac{f^{(N)}(\eta)}{N!} \cdot \int_{x_i}^{x_{i+1}} M(x) \cdot dx \tag{10.18}$$

In general, $M(x)$ may change its sign in the domain $[x_i, x_j]$, but $|M(x)|$ is always positive in the domain. If it is assumed that $\left| f^{(N)}(\xi) \right| \le C$, then Equations 10.17, 10.18 can be written as

$$E_I(f, x_i, x_j) \le \frac{C}{N!} \cdot \int_{x_i}^{x_j} |M(x)| \cdot dx \tag{10.19}$$

$$E_I(f, x_i, x_{i+1}) \le \frac{C}{N!} \cdot \left| \int_{x_i}^{x_{i+1}} M(x) \cdot dx \right| \tag{10.20}$$

10.2.3 Extension to Multi-dimensional Cases

First, we consider the two-dimensional case. The whole domain is $a \le x \le b$, $c \le y \le d$, which can be decomposed into $(N-1)\times(M-1)$ intervals with N grid points in the x direction $a = x_1,\ x_2, \cdots,\ x_N = b$, and M grid points in the y direction $c = y_1,\ y_2, \cdots,\ y_M = d$. Suppose that a function $f(x, y)$ is continuous in the whole domain and we just consider an integral of $f(x, y)$ over a part of the whole domain as shown by the shaded area in Figure 10.1.

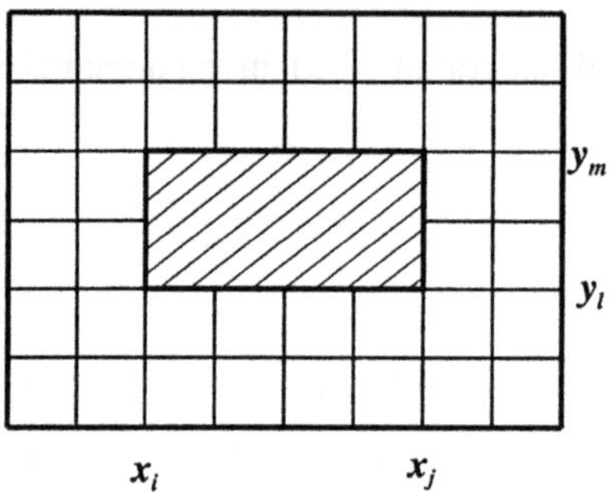

Figure 10.1 Integral region in a two-dimensional domain

Using the same technique as in the one-dimensional case and in the PDQ analysis, the integral of $f(x, y)$ over a part of the whole domain can be approximated by all the functional values in the overall domain. As a result, the formulation of the two-dimensional integral of $f(x, y)$ over part of the domain $x_i \le x \le x_j$, $y_l \le y \le y_m$ can be written as

$$\int_{x_i}^{x_j}\int_{y_l}^{y_m} f(x, y)dxdy = \sum_{n=1}^{N}\sum_{k=1}^{M}(w_{jn}^{I} - w_{in}^{I})(\overline{w}_{mk}^{I} - \overline{w}_{lk}^{I})f(x_n, y_k) \qquad (10.21)$$

where w_{jn}^{I} and $\overline{w}_{mk}^{I}$ are the weighting coefficients of the one-dimensional integral in the x and y directions respectively, and are given by

$$w_{jn}^{I} - w_{in}^{I} = \int_{x_i}^{x_j} r_n(x)\cdot dx, \qquad i, j, n = 1, 2, \cdots, N \qquad (10.22)$$

$$\overline{w}_{mk}^{I} - \overline{w}_{lk}^{I} = \int_{y_l}^{y_m} s_k(y)\cdot dy, \qquad l, m, k = 1, 2, \cdots, M \qquad (10.23)$$

Here $r_n(x)$ and $s_k(y)$ are the Lagrange interpolated polynomials. As shown in the last section, w_{jn}^I and $\overline{w}_{mk}^I$ can be easily determined by inversion of the PDQ weighting coefficient matrices of the first order derivative. For the three-dimensional integral, it is supposed that the function $f(x, y, z)$ is continuous in the overall domain $a \le x \le b$, $c \le y \le d$, $g \le z \le h$ which can be divided into $(N-1)\times(M-1)\times(L-1)$ intervals with N grid points in the x direction $a = x_1$, $x_2, \cdots$, $x_N = b$, M grid points in the y direction $c = y_1$, $y_2, \cdots$, $y_M = d$ and L grid points in the z direction $g = z_1$, $z_2, \cdots$, $z_L = h$. It is shown that the integral of $f(x, y, z)$ over any interval in the domain can be approximated by the linear combination of all the functional values in the whole domain. The GIQ formulation for this case can be written as

$$\int_{x_i}^{x_j}\int_{y_l}^{y_m}\int_{z_e}^{z_q} f(x, y, z)dxdydz =$$

$$\sum_{n=1}^{N}\sum_{k=1}^{M}\sum_{p=1}^{L}(w_{jn}^I - w_{in}^I)(\overline{w}_{mk}^I - \overline{w}_{lk}^I)(\overline{\overline{w}}_{qp}^I - \overline{\overline{w}}_{ep}^I) f(x_n, y_k, z_p) \tag{10.24}$$

where $\overline{\overline{w}}_{qp}^I$ is the weighting coefficient of the one-dimensional integral in the z direction, and given by

$$\overline{\overline{w}}_{qp}^I - \overline{\overline{w}}_{ep}^I = \int_{z_e}^{z_q} t_p(z)\cdot dz \qquad e,\ q,\ p = 1, 2, \cdots, L \tag{10.25}$$

Here $t_p(z)$ is the Lagrange interpolated polynomial in the z direction. $\overline{\overline{w}}_{qp}^I$ can also be obtained by the inversion of the PDQ weighting coefficient matrix of the first order derivative in the z direction.

10.2.4 Sample Application of GIQ Method

As an example, the GIQ method is applied to solve the following model integral equation

$$y(x) = 0.5x(1-x) + \int_0^1 K(x,\xi)y(\xi)\cdot d\xi \tag{10.26}$$

with a symmetrical kernel

$$K(x,\xi) = \begin{bmatrix} \xi(1-x) & \text{when} \quad 0 \le \xi \le x \\ x(1-\xi) & \text{when} \quad x \le \xi \le 1 \end{bmatrix}$$

Equation 10.26 has an exact solution

$$y = A\sin(x) + B\cos(x) - 1 \tag{10.27}$$

where $A = \tan(0.5)$ and $B = 1$. For numerical computation of Equation 10.26, the mesh point distribution is selected by Shu (1991) to be

$$x_i = -\cos\left[\frac{(2i-1)\pi}{2N}\right] / [2\cos(\pi / 2N)] + \frac{1}{2}, \quad \text{for } i = 1, 2, \ldots, N \tag{10.28}$$

where N is the number of mesh points. With Equation 10.28, we can easily compute the PDQ weighting coefficients of the first order derivative. Then by inversion, the weighting coefficients of GIQ can be obtained from the PDQ weighting coefficient matrix. Using the GIQ method, Equation 10.26 can be discretized as

$$y_i = 0.5x_i(1-x_i) + \sum_{k=1}^{N} (w_{N,k}^I - w_{1,k}^I) K_{i,k} y_k \tag{10.29}$$

where $w_{N,k}^I$ and $w_{1,k}^I$ are the GIQ weighting coefficients, $K_{i,k} = K(x_i, x_k)$, $y_i = y(x_i)$. Applying Equation 10.29 at all mesh points would result in an algebraic equation system, which can be solved by a direct method of *LU* decomposition. Table 10.1 shows the GIQ results provided by Shu (1991) and exact solution of the model integral equation. The GIQ results are obtained by 11 mesh points (N = 11). Since $y(x)$ is symmetrical with respect to x = 0.5, only the results in [0, 0.5] are shown in the table. From this table, it can be seen that the GIQ results agree very well with the exact solution although very few mesh points are used.

Table 10.1 Results of the model integral equation (Shu 1991)

x	0.040507	0.118239	0.226900	0.357685	0.500000
GIQ	0.021312	0.057509	0.097403	0.128106	0.139771
Exact	0.021303	0.057462	0.097264	0.127974	0.139494

10.3 DQ-GIQ Algorithm to Solve Boundary Layer Equations

The boundary layer approximation is still an efficient approach in CFD because it greatly reduces the computational effort as compared to a Navier-Stokes solver. The boundary layer equations can be solved when the dependent variables are the primitive variables or the stream function. For the case of stream function as the dependent variable, the continuity equation need not be considered in the solution procedure. Accordingly, the order of the differential equation is increased by one, which may create difficulties in dealing with the boundary conditions. For this case, the DQ method can be used to solve the boundary layer equations. On the other hand, the use of primitive variables as the dependent variables is attractive since it can be directly extended to three-dimensional cases. The difficulty then is the coupling of the continuity equation with the momentum and energy equations. Conventionally, no work has been done in using the integral form of the continuity equation to obtain the normal velocity. The reason is that the normal velocity obtained by integrating the continuity equation along the normal coordinate is less accurate because some integral domains do not contain sufficient grid points. As will be shown in this section, the GIQ technique can provide a promising way to obtain the normal velocity accurately by an explicit formulation derived from the integration of the continuity equation. The determination of the normal velocity at any mesh point in the normal coordinate direction can be accomplished with the same order of accuracy.

On the other hand, it has been shown that the DQ method is a robust, efficient technique to discretize spatial derivatives, which can achieve high order of accuracy by using just a few grid points. In this section, we will describe two algorithms to solve the boundary layer equations. In these two algorithms, the integrals and derivatives in the normal direction are approximated by GIQ and DQ methods, while the derivatives in the streamwise or the crossflow direction can be discretized by the DQ method or low order finite difference schemes. Although two specific problems are chosen for purposes of demonstration, the DQ-GIQ algorithm can be used for general boundary layer problems.

10.3.1 Stream Function as Dependent Variable

For simplicity, we consider the two-dimensional Howarth boundary layer problem. The governing equation (Keller and Cebeci 1970) is

$$\frac{\partial^3 f}{\partial \eta^3} + f \cdot \frac{\partial^2 f}{\partial \eta^2} + \beta(\xi) \cdot \left[1 - \left(\frac{\partial f}{\partial \eta}\right)^2\right] = 2\xi \left[\frac{\partial f}{\partial \eta} \cdot \frac{\partial^2 f}{\partial \xi \partial \eta} - \frac{\partial^2 f}{\partial \eta^2} \cdot \frac{\partial f}{\partial \xi}\right] \tag{10.30}$$

where $f(\xi,\eta)$ is the dimensionless stream function and is subjected to the boundary conditions

$$f(\xi,0)=\frac{\partial f}{\partial \eta}(\xi,0)=0 \tag{10.31}$$

$$\frac{\partial f}{\partial \eta}(\xi,\eta)=1, \quad \text{when } \eta\to\infty \tag{10.32}$$

Here (ξ,η) are the Levy-Lees coordinates; ξ increases in the free stream direction and η increases away from the wall. $\beta(\xi)$ is given by

$$\beta(\xi)=\frac{\xi}{\xi-4} \tag{10.33}$$

The third order differential Equation 10.30 can be directly solved by the DQ method. On the other hand, it can be alternatively solved by the DQ-GIQ algorithm. Setting $u=\partial f/\partial\eta$, Equation 10.30 can be reduced to

$$\frac{\partial^2 u}{\partial \eta^2}+f\cdot\frac{\partial u}{\partial \eta}+\beta(\xi)\cdot(1-u^2)=2\xi\left(u\cdot\frac{\partial u}{\partial \xi}-\frac{\partial u}{\partial \eta}\cdot\frac{\partial f}{\partial \xi}\right) \tag{10.34}$$

$$f=\int_0^\eta u\cdot d\eta+f(0) \tag{10.35}$$

Note that Equation 10.34 is a second order partial differential equation. For numerical computation, the infinite domain in the η direction can be truncated to a finite domain $[0,\eta_\infty]$. Using the DQ and GIQ methods in the η direction, we have

$$\left(\frac{\partial^2 u}{\partial \eta^2}\right)_{i,j}=\sum_{k=1}^{M}\bar{c}_{j,k}^{(2)}\cdot u_{i,k} \tag{10.36a}$$

$$\left(\frac{\partial u}{\partial \eta}\right)_{i,j}=\sum_{k=1}^{M}\bar{c}_{j,k}^{(1)}\cdot u_{i,k} \tag{10.36b}$$

$$f_{i,j}=\sum_{k=1}^{M}(w_{j,k}^{I}-w_{1,k}^{I})\cdot u_{i,k}+f_{i,1} \tag{10.36c}$$

where $\bar{c}_{i,j}^{(m)}$ are the DQ weighting coefficients of the mth order derivative in the η direction, $w_{i,j}^{I}$ are the GIQ weighting coefficients of the integral along the η

direction, and M is the number of mesh points in the η direction. With the new variable u, the boundary conditions 10.31-10.32 become

$$u_{i,1} = 0, \quad u_{i,M} = 1, \quad f_{i,1} = 0 \tag{10.37}$$

which can be easily implemented in the solution procedure. It is clear that there are two boundary conditions for u in the η direction. The derivatives in the ξ direction can be discretized using the DQ method or a low order finite difference scheme.

When the second order finite difference scheme is applied at ($\xi_{i-1/2}, \eta_j$) for the discretization of derivatives in the ξ direction, after linearization of the non-linear terms, the resultant algebraic equations can be written as

$$[A]\{U\} = \{b\} \tag{10.38}$$

where $\{U\} = \{u_{i,2}, u_{i,3}, \cdots, u_{i,M-1}\}^T$ is a vector of unknowns, $[A]$ is a matrix and $\{b\}$ is a known vector. Using Equation 10.38, the boundary layer solution in the whole domain can be obtained by a marching technique along the ξ direction.

When the DQ method is used in the ξ direction, after linearization of the non-linear terms, we can get a similar algebraic equation system like Equation 10.38, but the vector $\{U\}$ includes all the interior functional values in the whole domain. Thus the marching technique cannot be used in this case. If only the steady state solution is of interest, we can use another method. Introducing an unsteady term in Equation 10.34, we can get

$$\frac{\partial^2 u}{\partial \eta^2} + f\frac{\partial u}{\partial \eta} + \beta(\xi)(1-u^2) = \frac{\partial u}{\partial t} + 2\xi\left(u\frac{\partial u}{\partial \xi} - \frac{\partial u}{\partial \eta}\frac{\partial f}{\partial \xi}\right) \tag{10.39}$$

The spatial discretization of Equation 10.39 leads to a set of ordinary differential equations which can be solved by the 4-stage Runge-Kutta scheme. Equation10.39 is not a true unsteady boundary layer equation, and is only used for the steady state solution.

10.3.2 Primitive Variables as Dependent Variables

For demonstration, we consider the two-dimensional unsteady flow past a circular cylinder started impulsively from rest. The non-dimensional form of the governing equations (Cebeci 1979, Liakopoulos 1988) is written as

$$u_x + v_y = 0 \tag{10.40}$$

$$u_t + u \cdot u_x + v \cdot u_y = Ue \cdot \frac{dUe}{dx} + u_{yy} \tag{10.41}$$

with initial condition

$$u(x, y, 0) = Ue(x) = \sin x \quad (y \neq 0) \tag{10.42}$$

and boundary conditions

$$u(x, 0, t) = v(x, 0, t) = 0 \tag{10.43a}$$

$$u(x, \infty, t) = Ue(x) = \sin x \tag{10.43b}$$

$$u(0, y, t) = u(\pi, y, t) = 0 \tag{10.43c}$$

Note that the subscripts x and y in Equations 10.40 and 10.41 represent the partial derivatives in the respective directions. The computational domain in the y direction can be obtained by truncating the infinite domain to $[0, y_\infty]$. Similar to Equations 10.36*a* and 10.36*b*, u_y and u_{yy} can be discretized using the DQ method. The velocity v can be obtained by applying the GIQ method to Equation 10.40

$$v_{i,j} = -\sum_{k=1}^{M} (w_{j,k}^I - w_{1,k}^I) \cdot (u_x)_{i,k} + v_{i,1} \tag{10.44}$$

It is noted that Equation 10.44 is different from Equation 10.36*c*. In a similar fashion to the algorithm presented in Section 10.3.1, when the derivatives in the x direction are discretized by a second order finite difference scheme at $(x_{i-1/2}, y_j)$, the numerical solution can be obtained by marching along the x direction. However, when the DQ method is also used in the x direction, the marching technique is invalid. For this case, the 4-stage Runge-Kutta scheme can be used to solve the resultant ordinary differential equations.

The use of the DQ method in the x direction is attractive although it may increase the amount of virtual storage required. Since the DQ method can achieve the same order of accuracy using a few grid points as the second order finite difference scheme using a large number of grid points, the amount of virtual storage and computational effort required can be greatly reduced if the DQ and GIQ methods are used in all coordinate directions. In the following examples, we will use the DQ and GIQ methods to discretize the spatial derivatives and the integral in all directions, and the 4-stage Runge-Kutta scheme to solve the resultant ordinary differential equations.

10.4 Steady Boundary Layer Solutions

In this section, we will demonstrate the application of the DQ-GIQ algorithm to solve the one-dimensional, two-dimensional and three-dimensional steady boundary layer equations. One test problem will be shown in each case.

10.4.1 One-dimensional Case

The Blasius boundary layer is an example of one-dimensional boundary layer problems. This problem was first solved by Shu, Chew and Richards (1995) using the DQ-GIQ algorithm. The governing equation of the problem is written as

$$\frac{d^3 f}{d\eta^3} + f\frac{d^2 f}{d\eta^2} = 0 \tag{10.45}$$

Here f is the stream function. The boundary conditions for Equation 10.45 are

$$f = 0, \; f_\eta = 0, \quad \text{when } \eta = 0 \tag{10.46a}$$

$$f_\eta = 2, \quad \text{when } \eta \to \infty \tag{10.46b}$$

Setting $u = \partial f / \partial \eta$ and introducing an unsteady term, Equation 10.45 can be rewritten as

$$\frac{\partial u}{\partial t} = \frac{\partial^2 u}{\partial \eta^2} + f \cdot \frac{\partial u}{\partial \eta} \tag{10.47}$$

$$f = \int_0^\eta u \cdot d\eta + f(0) \tag{10.48}$$

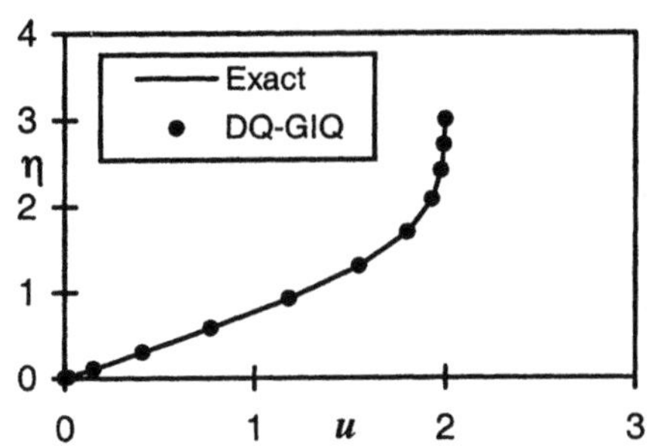

Figure 10.2 Velocity profiles of Blasius boundary layer (Shu, Chew and Richards 1995)

For the numerical computation, the computational domain is truncated to [0, 3], and the grid is stretched near $\eta = 0$. After spatial discretization by the PDQ and GIQ methods, the resultant ordinary differential equations are solved by the 4-stage Runge-Kutta scheme. For the sample problem, it was found that the numerical results are very accurate. Using 12 grid points and a time step size of 1.3×10^{-2}, accurate numerical results can be obtained by using only 385 time steps. The obtained wall shear stress is 1.3286 while the exact value is 1.3284. Figure 10.2 shows the computed and the exact velocity profiles of the Blasius boundary layer. Good agreement between the computed and exact solutions can be clearly observed.

10.4.2 Two-dimensional Case

As an example, we consider the Howarth boundary layer which was solved by Shu, Chew and Richards (1995). The governing equations of the problem have been shown in Section 10.3.1. This problem has a separation point at $\xi \cong 0.901$. So, the computational domain in the ξ direction should be [0, b] where $b < 0.901$ because of the Goldstein singularity. In the η direction, the infinite domain is truncated to [0, 5.0]. The PDQ and GIQ methods are used in both the ξ and η directions, and the resultant ordinary differential equations are solved by the 4-stage Runge-Kutta scheme. It was found that the DQ-GIQ approach is very sensitive to the choice of b when b is close to the separation point. Actually, when b is taken as 0.90, the computation will diverge quickly after a few time steps, but when b is chosen as less than or equal to 0.894, the steady state solution can be obtained quickly and accurately. The rate of convergence is very fast when b is far from the separation point, and is slightly slower when b is very close to the separation point. Figure 10.3 displays the wall shear stress distribution of the Howarth boundary layer computed by the DQ-GIQ algorithm using a mesh size of 11×12 and b = 0.894. Clearly, the DQ-GIQ results are very close to those given by the Keller-box finite difference scheme using a mesh size of 51×121 (Keller et al. 1970).

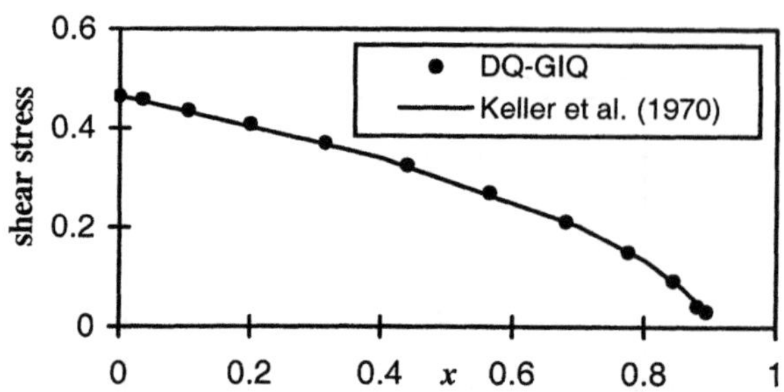

Figure 10.3 Wall shear stress distribution of the Howarth boundary layer (Shu, Chew and Richards 1995)

10.4.3 Three-dimensional Case

For three-dimensional boundary layer simulations, we choose the laminar flow past a flat plate with an attached cylinder as a test example. The problem definition is shown in Figure 10.4. This problem has been studied by Cebeci (1975) using the Keller-box finite difference scheme, and solved by Shu, Chew, Khoo and Yeo (1996) using the DQ-GIQ algorithm.

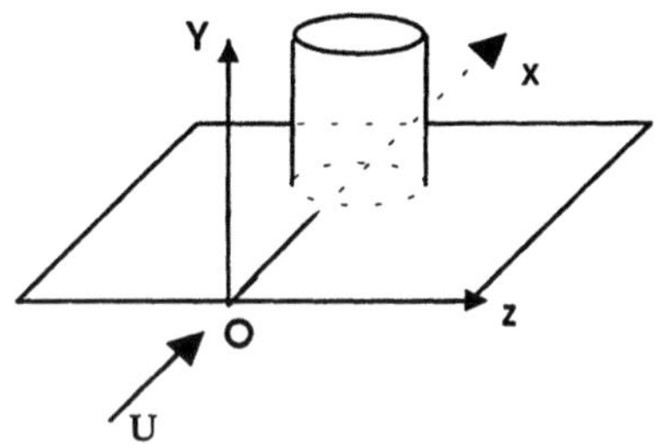

Figure 10.4 Geometry of the flow past a flat plate with attached cylinder

The governing equations can be written as

$$\frac{\partial u}{\partial x}+\frac{\partial v}{\partial y}+\frac{\partial w}{\partial z}=0 \tag{10.49}$$

$$\frac{\partial u}{\partial t}+u\frac{\partial u}{\partial x}+v\frac{\partial u}{\partial y}+w\frac{\partial u}{\partial z}=-\frac{1}{\rho}\frac{\partial p}{\partial x}+\nu\frac{\partial^2 u}{\partial y^2} \tag{10.50}$$

$$\frac{\partial w}{\partial t}+u\frac{\partial w}{\partial x}+v\frac{\partial w}{\partial y}+w\frac{\partial w}{\partial z}=-\frac{1}{\rho}\frac{\partial p}{\partial z}+\nu\frac{\partial^2 w}{\partial y^2} \tag{10.51}$$

with

$$-\frac{1}{\rho}\frac{\partial p}{\partial x}=\frac{\partial Ue}{\partial t}+Ue\frac{\partial Ue}{\partial x}+We\frac{\partial Ue}{\partial z} \tag{10.52}$$

$$-\frac{1}{\rho}\frac{\partial p}{\partial z}=\frac{\partial We}{\partial t}+Ue\frac{\partial We}{\partial x}+We\frac{\partial We}{\partial z} \tag{10.53}$$

where Ue and We are the components of the inviscid flow velocity in the x and z directions respectively. At the leading edge of the flat plate, the boundary layer thickness is zero. Thus, Equations 10.49-10.51 encounter a singularity at $x=0$. To remove this singularity, we perform the following transformation

$$\begin{cases} T = Ue \cdot t / x, \quad \xi = x, \quad \zeta = z, \quad \eta = \sqrt{Ue/(\nu x)} \cdot y \\ F = u / Ue, \quad G = w / We \end{cases} \tag{10.54}$$

With Equation 10.54, Equations 10.49-10.51 can be transformed to

$$\xi \frac{\partial F}{\partial \xi} + \frac{\partial V}{\partial \eta} + \frac{F}{2}(1 + K_1) + \frac{\xi We}{Ue}\frac{\partial G}{\partial \zeta} + G(K_3 - 0.5K_2) = 0 \tag{10.55}$$

$$\frac{\partial F}{\partial T} + \xi F \frac{\partial F}{\partial \xi} + V \frac{\partial F}{\partial \eta} + \xi G \frac{We}{Ue}\frac{\partial F}{\partial \zeta} + K_1(F^2 - 1) + K_2(FG - 1) = \frac{\partial^2 F}{\partial \eta^2} \tag{10.56}$$

$$\frac{\partial G}{\partial T} + \xi F \frac{\partial G}{\partial \xi} + V \frac{\partial G}{\partial \eta} + \xi G \frac{We}{Ue}\frac{\partial G}{\partial \zeta} + K_3(G^2 - 1) + K_4(FG - 1) = \frac{\partial^2 G}{\partial \eta^2} \tag{10.57}$$

where

$$V = \sqrt{x/(\nu Ue)}v + xF\frac{\partial \eta}{\partial x} + xG\frac{We}{Ue}\frac{\partial \eta}{\partial z}$$

$$K_1 = \frac{x}{Ue}\frac{\partial Ue}{\partial x}, \qquad K_2 = \frac{x}{Ue}\frac{We}{Ue}\frac{\partial Ue}{\partial z}$$

$$K_3 = \frac{x}{Ue}\frac{\partial We}{\partial z}, \qquad K_4 = \frac{x}{We}\frac{\partial We}{\partial x}$$

The inviscid velocity distribution of the problem is given by

$$Ue = u_\infty (1 + a^2 \cdot \Delta_2 / \Delta_1^2) \tag{10.58}$$

$$We = -2u_\infty a^2 \cdot \Delta_3 / \Delta_1^2 \tag{10.59}$$

where

$$\Delta_1 = (x - x_0)^2 + z^2$$

$$\Delta_2 = -(x - x_0)^2 + z^2$$

$$\Delta_3 = (x - x_0) \cdot z$$

Here u_∞ is the reference velocity, a is the radius of the cylinder, and x_0 denotes the distance of the cylinder axis from the leading edge, $x = 0$. To make a direct comparison with the results obtained by Cebeci (1975), we choose $u_\infty = 30.50$ m/sec, $a = 0.061$ m, $x_0 = 0.457$ m, $\eta_\infty = 8.0$.

As we know, the boundary layer equations pose a combined initial and boundary value problem. The boundary conditions are usually obtained from the independent inviscid flow equations, whereas the initial conditions must be somehow obtained from the boundary layer equations themselves. For the sample problem, since the primitive variables are chosen as the independent variables, then for simplicity, only two boundary conditions in the η direction are implemented for F and G, that is

$$F = 0, \quad G = 0, \quad \text{at } \eta = 0 \tag{10.60}$$

$$F = 1, \quad G = 1, \quad \text{at } \eta = \eta_\infty \tag{10.61}$$

Using the GIQ method, V can be computed by

$$V_{i,j,k} = -\sum_{m=1}^{M} (w_{j,m}^I - w_{1,m}^I) \cdot \Phi_{i,m,k} + V_{i,1,k} \tag{10.62}$$

where

$$\Phi_{i,m,k} = \left[\xi \frac{\partial F}{\partial \xi} + \frac{F}{2}(1 + K_1) + \frac{\xi We}{Ue} \frac{\partial G}{\partial \zeta} + G(K_3 - 0.5K_2) \right]_{i,m,k}$$

Along the $\xi = 0$ line of the flat plate, the governing equation is reduced to the Blasius equation for both F and G. Thus the initial condition at $\xi = 0$ can be obtained from the Blasius boundary layer solutions. In addition, the initial condition along a $\zeta =$ constant line is needed. For the problem considered, the initial condition along the line of symmetry ($\zeta = 0$) can be obtained. Since, along the line of symmetry, w and We are zero, we define a variable G_z as

$$G_z = \frac{\partial w / \partial \zeta}{\partial (We) / \partial \zeta} \tag{10.63}$$

Thus the governing equations along the line of symmetry are reduced to

$$\xi \frac{\partial F}{\partial \xi} + \frac{\partial V}{\partial \eta} + 0.5F(1 + K_1) + G_z K_3 = 0 \tag{10.64}$$

$$\frac{\partial F}{\partial T} + \xi F \frac{\partial F}{\partial \xi} + V \frac{\partial F}{\partial \eta} = \frac{\partial^2 F}{\partial \eta^2} + K_1(1 - F^2) \tag{10.65}$$

$$\frac{\partial G_z}{\partial T} + \xi F \frac{\partial G_z}{\partial \xi} + V \frac{\partial G_z}{\partial \eta} = \frac{\partial^2 G_z}{\partial \eta^2} + K_3(1 - G_z^2) + K_5(1 - FG_z) \tag{10.66}$$

where $K_5 = \dfrac{\xi}{(We)_z} \cdot \dfrac{\partial (We)_z}{\partial \xi}$, $(We)_z = \partial (We) / \partial \zeta$. After obtaining the initial values along the line of $\xi = 0$, the initial values along the line of symmetry can be found by solving Equations 10.64-10.66. With the boundary and initial values, the full boundary layer solutions can be obtained by solving Equations 10.55-10.57, using the approach as shown in Section 10.3.2.

Table 10.2 Comparison of DQ-GIQ results with those of Cebeci (1975) (Shu, Chew, Khoo and Yeo 1996)

$z = 0$				
x	$(\partial F / \partial \eta)_w$		$(\partial G / \partial \eta)_w$	
	Cebeci (1975)	*DQ-GIQ*	*Cebeci* (1975)	*DQ-GIQ*
0	0.332066	0.331898	0.332066	0.331898
0.0488	0.324951	0.325169	0.702488	0.702674
0.0976	0.312821	0.312574	1.124300	1.124564
0.1464	0.290184	0.290472	1.624250	1.623227
0.1952	0.243524	0.243612	2.250740	2.247190
0.2440	0.125972	0.123533	3.126830	3.119115
$z = 0.0488$				
x	$(\partial F / \partial \eta)_w$		$(\partial G / \partial \eta)_w$	
	Cebeci (1975)	*DQ-GIQ*	*Cebeci* (1975)	*DQ-GIQ*
0	0.332066	0.331898	0.332066	0.331898
0.0488	0.325184	0.325653	0.717599	0.695967
0.0976	0.314423	0.314859	1.124210	1.106794
0.1464	0.295233	0.295767	1.605920	1.585971
0.1952	0.259005	0.259305	2.191930	2.171027
0.2440	0.181979	0.177060	2.959350	2.940921

For the numerical computations, the PDQ and GIQ methods are applied to discretize all the spatial derivatives and integrals, and the resultant equations are solved by the 4-stage Runge-Kutta method. The computational domain in the z direction is chosen to be between $z = 0$ and $z = 0.20$. For the problem, the presence of the cylinder results in an adverse pressure gradient in front of itself. Thus, it is expected that the streamwise velocity will reverse in direction along a line in front of the cylinder. Since the adverse pressure gradient is maximum along the line of symmetry, the flow reversal in the boundary layer will first occur along the line of symmetry. Thus the computational domain in the x direction should be before the separation point in the line of symmetry because of the Goldstein singularity. It was found that the DQ-GIQ approach is very sensitive to the Goldstein singularity. When the computational domain in the x direction is taken as [0.0, 0.25970], the steady state solution can be obtained very quickly and accurately. But when the computational domain is taken as [0.0, 0.25975], the computation will diverge after a few time steps. Thus it appears that the separation point in the line of symmetry lies between 0.25970 and 0.25975, which agrees well with other researchers' results. Table 10.2 lists the computed values of $(\partial F/\partial \eta)_w$ and $(\partial G/\partial \eta)_w$ at some specific points. The DQ-GIQ results were obtained by using 11 grid points in the x and z directions, and 13 grid points in the η direction. Also included in Table 10.2 are the results of Cebeci (1975) obtained by the Keller-box difference scheme with Richardson extrapolation. Clearly, the DQ-GIQ results agree very well with those of Cebeci (1975).

10.5 Unsteady Boundary Layer Solutions

The unsteady boundary layer simulation is an area of great interest in CFD. This is because many practical problems are unsteady. On the other hand, there are still some controversies in the literature concerning the occurrence of singularities in the solution of unsteady boundary layers. For steady boundary layer problems, it is well known that there may exist a Goldstein singularity in the boundary layer solution. But for unsteady boundary layer problems, there are different opinions as to whether there exists a finite time singularity in the solution of the unsteady boundary layer equations. For the unsteady, two-dimensional laminar flow past a circular cylinder started impulsively from rest, some researchers (e.g. Bodonyi and Stewartson 1977, Van Dommelen and Shen 1980) claimed that there exists a finite time singularity in the solution procedure, while others (e.g. Cebeci 1979, 1986) suggested the absence of a finite time singularity. The governing equation for this problem has been shown in Section 10.3.2.

Shu, Khoo, Chew and Yeo (1996) also studied this problem by using the DQ-GIQ algorithm. In their work, two approaches were used. The first approach applies the PDQ method to discretize the spatial derivatives in both the x and y directions. The approach is termed as the DQ-GIQ approach. The x coordinates for this approach are chosen as

$$x_i = \frac{1}{2}\cdot\left[1-\cos\left(\frac{i-1}{N-1}\cdot\pi\right)\right]\cdot\pi\ , \qquad i=1,2,\cdots,N \tag{10.67}$$

The second approach applies the PDQ method to discretize the spatial derivatives in the y direction and a second order finite difference scheme to discretize the spatial derivatives in the x direction. This approach is known as the DQ-GIQ-FD approach. A uniform mesh is used in the x direction for this approach. After spatial discretization of Equation 10.41, the resultant ordinary differential equations for u are solved by the 4-stage Runge-Kutta scheme. The velocity v is computed from Equation 10.44 by the GIQ scheme.

For the numerical simulation, y_∞ is chosen as 30, and the computational domain in the x direction is taken from 0 to π. First, the DQ-GIQ approach is applied to simulate the problem with a mesh size of 21×31. It was found that the reverse flow first starts at $\theta = 180^0$ and time, $t = 0.644$. This is in good agreement with the results of Cebeci (1979). As time increases, the point of zero wall shear stress moves along the surface of the cylinder towards the steady state value $\theta_G = 104.5^0$ (the position associated with Goldstein singularity). However, the computation fails to yield the steady state solution because numerical instability breaks down the calculation at $t \approx 3.0$. It was found that, up to time $t = 2.5$, the numerical solution proceeds very smooth. As time passes, a large disturbance develops in the vicinity of the separation point. This disturbance subsequently spreads into the whole computational field and grows very rapidly. The large amounts of disturbances distort the numerical results and finally cause the numerical computation to break down. This can be clearly seen in Figure 10.5. Some wiggles are seen to occur in the streamlines. The reason is that PDQ is a global method with very small numerical damping. When the singular point is approached, some disturbances could develop in the vicinity of the separation point, which may not be damped due to small numerical error of the method. These disturbances would subsequently spread throughout the whole flow field and eventually cause the numerical computation to break down.

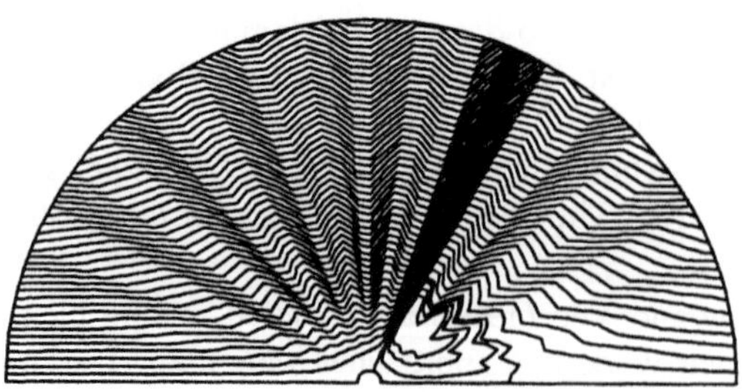

Figure 10.5 Streamlines past a circular cylinder computed by DQ-GIQ approach, $t = 2.7$ (Shu, Khoo, Chew and Yeo 1996)

To study this problem further, the DQ method is replaced by a second order finite difference scheme to discretize the spatial derivatives in the x direction. That is, the DQ-GIQ-FD approach is used to simulate the same problem. For this case, 81 grid points are used in the x direction. It was found that, up to time $t = 2.0$, the DQ-GIQ-FD results agree very well with the DQ-GIQ results. After time $t = 2.0$, the numerical results are significantly improved by the DQ-GIQ-FD approach. The reason for such an improvement by the DQ-GIQ-FD approach is that, the second order finite difference scheme is a low order method, which provides a much higher numerical damping than the DQ method. Hence when the singularity point is approached, the initial small disturbances could be damped locally, leading to more plausible results than the DQ-GIQ approach when $t > 2.0$. However, the singularity arises due to the nature of the boundary layer equations. Any disturbance caused by the singularity would grow quickly and finally cause the numerical computation to break down. It is interesting to see from Figure 10.6 that the solution seems to reveal a similar phenomenon to the formation of wave steepening (shock wave) in compressible flows. This result is in agreement with the work of Van Dommelen and Shen (1980), Liakopoulos (1988). Table 10.3 compares the results for the position and time of zero wall shear stress. It should be noted that the results of Bar-Lev and Yang (1975) were obtained by solving the full Navier-Stokes equations with the use of a matched asymptotic method. The results of Cebeci (1979) were given from the solution of the boundary layer equations with the use of the Keller-box finite difference scheme. It can be seen that the DQ-GIQ and DQ-GIQ-FD results agree well with other researchers' results.

Table 10.3 Comparison of the time and zero wall shear stress (Shu, Khoo, Chew and Yeo 1996)

References	180^0	166^0	146^0	138^0	124^0	110^0
Bar-Lev & Yang (1975)	0.644	0.660	0.778	0.876	1.204	2.188
Cebeci (1979)	0.640	0.660	0.780	0.872	1.192	2.200
DQ-GIQ	0.644	0.664	0.790	0.874	1.193	2.098
DQ-GIQ-FD	0.644	0.668	0.791	0.878	1.196	2.204

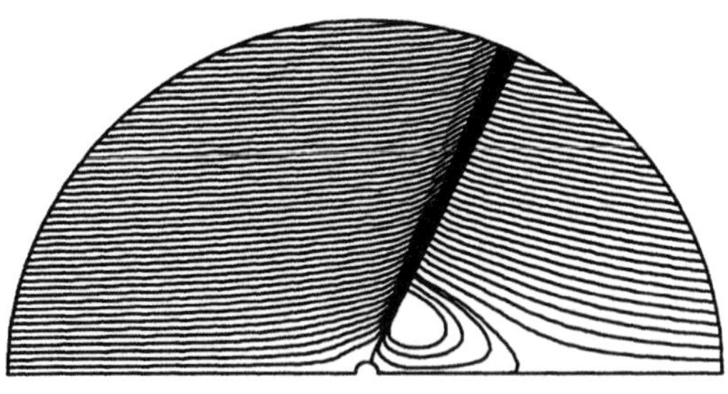

Figure 10.6 Streamlines past a circular cylinder computed by DQ-GIQ-FD approach, $t = 2.8$ (Shu, Khoo, Chew and Yeo 1996)

Appendix A

A Fortran Program for Simulation of Natural Convection in A Square Cavity

This program is based on the algorithms shown in Section 6.3. The spatial derivatives in the governing differential equations and the boundary conditions are discretized by the PDQ method. After spatial discretization, the resultant ordinary differential equations for vorticity (Equation 6.22) and temperature (Equation 6.23) are solved by the 4-stage Runge-Kutta method, and the resultant algebraic equations for stream function (Equation 6.21) are solved by a direct method of the *LU* decomposition.

The two-layer approach is used in this program to implement the boundary condition for the stream function. The Dirichlet condition is implemented directly, while the discrete Neumann condition is directly substituted into Equation 6.21.

The feature of the DQ method is that it can obtain very accurate numerical results by using just a few grid points. Although the DQ solution at the coarse mesh point is very accurate, it may not give the accurate maximum and minimum values of some variables. In addition, the coarse mesh may not be good enough for contour plotting. To improve these, an interpolation process is introduced in the program. When the converged solution on a coarse mesh is obtained, the functional values on a fine mesh are given from interpolation by using Equations 2.128-2.130. It is noted that the use of Equations 2.128-2.130 will not affect the accuracy of the numerical solution since the PDQ discretization is also based on the Lagrange interpolated polynomials. The DQ results on the fine mesh can be used for contour plotting, and can also be used to compute the maximum values of stream function, velocity u, v and their respective positions.

The program also provides the computation of Nusselt numbers. Through any line parallel to the y-axis, the average Nusselt number is defined as

$$Nu_x = \int_0^1 Q(x, y)dy \tag{A.1}$$

where

$$Q(x, y) = uT - \partial T / \partial x \tag{A.2}$$

Clearly, Nu_x is a function of x. The average Nusselt number through out the whole cavity is given by

$$\overline{Nu} = \int_0^1 Nu_x dx \tag{A.3}$$

In the program, the average Nusselt numbers are computed from a fine mesh by using Lagrange interpolation, and the integrals in Equations A.1 and A.3 are approximated by the trapezoidal rule.

The convergence of computation greatly depends on the time step size *H*. The optimal value of *H* can be obtained from a trial-and-error process.

Description of Variables

N — Number of grid points in the x direction

M — Number of grid points in the y direction

Pr — Prandtl number

Ra — Rayleigh number

H — Time step size

NTT — Maximum time steps to control the computation

EPS1 — Convergence criterion for temperature Equation 6.23

EPS2 — Convergence criterion for vorticity Equation 6.22

XX(I) — x-coordinate of a mesh point, x_i

YY(J) — y-coordinate of a mesh point, y_j

AX(I,J) — Weighting coefficients of the first order derivative in the x direction, $c_{i,j}^{(1)}$

BX(I,J) — Weighting coefficients of the second order derivative in the x direction, $c_{i,j}^{(2)}$

AY(I,J) — Weighting coefficients of the first order derivative in the y direction, $\bar{c}_{i,j}^{(1)}$

BY(I,J) — Weighting coefficients of the second order derivative in the y direction, $\bar{c}_{i,j}^{(2)}$

U(I,J) — Velocity component in the x direction at a mesh point (x_i, y_j), $u_{i,j}$

V(I,J) — Velocity component in the y direction at a mesh point (x_i, y_j), $v_{i,j}$

P(I,J) — Stream function value at a mesh point (x_i, y_j), $\psi_{i,j}$

W(I,J) — Vorticity value at a mesh point (x_i, y_j), $\omega_{i,j}$

T(I,J) — Temperature value at a mesh point (x_i, y_j), $T_{i,j}$

DW(I,J) — Residual of vorticity Equation 6.22 at a mesh point (x_i, y_j), $d\omega_{i,j}/dt$

DT(I,J) — Residual of temperature Equation 6.23 at a mesh point (x_i, y_j), $dT_{i,j}/dt$

PD(I,J) — Matrix of algebraic equation system 6.21

CNU(I) — Average Nusselt number at x_i

CNUAVG — Average Nusselt number through the whole cavity

```
C  *******************************************************************
C  *                         PROGRAM PDQNC                           *
C  *                                                                 *
C  *     THIS IS A TWO-DIMENSIONAL SOLVER FOR NUMERICAL SIMULATION   *
C  *  OF NATURAL CONVECTION IN A SQUARE CAVITY. THE SOLVER USES      *
C  *  THE GLOBAL METHOD OF POLYNOMIAL-BASED DIFFERENTIAL QURATURE TO *
C  *  DISCRETIZE THE SPATIAL DERIVATIVES. AS A RESULT OF SPATIAL     *
C  *  DISCRETIZATION, THE RESULTANT ORDINARY DIFFERENTIAL EQUATIONS  *
C  *  FOR VORTICITY (W) AND TEMPERATURE (T) ARE SOLVED BY THE 4-STAGE*
C  *  RUNGE-KUTTA METHOD AND THE RESULTANT ALGEBRAIC EQUATIONS FOR   *
C  *  STREAM FUNCTION ARE SOLVED BY A DIRECT METHOD OF LU            *
C  *  DECOMPOSITION.                                                 *
C  *                                                                 *
C  *******************************************************************
C  *                                                                 *
      PROGRAM MAIN
      IMPLICIT REAL*8(A-H,O-Z)
      DIMENSION AX(50,50),BX(50,50),XX(50),AY(50,50),BY(50,50),YY(50)
      DIMENSION U(50,50),V(50,50),PD(401,401),P(50,50)
      DIMENSION T(50,50),W(50,50),DT(50,50),DW(50,50)
C
C INPUT NUMBER OF MESH POINTS, MAXIMUM ITERATION NUMBER, TIME STEP
C SIZE, CONVERGENCE CRITERIA, PRANDTL NUMBER AND RAYLEIGH NUMBER
C
      WRITE(6,*) 'N,M,NTT,H,EPS1,EPS2,PR,RA'
      READ(5,*) N,M,NTT,H,EPS1,EPS2,PR,RA
C
C MESH GENERATION AND SET-UP OF INITIAL FLOW FIELD
C
      CALL GRID(N,M,XX,YY,U,V,W,T,DT,DW)
C
C COMPUTATION OF WEIGHTING COEFFICIENTS OF THE FIRST AND SECOND ORDER
C DERIVATIVES IN THE X AND Y DIRECTIONS
C
      CALL WCOEF(1,N,XX,AX,BX)
      CALL WCOEF(1,M,YY,AY,BY)
C
C FORM THE MATRIX OF EQUATION SYSTEM 6.21. THE OBTAINED MATRIX IS
C THEN DECOMPOSED BY LU DECOMPOSITION (SEE EQUATION 5.27)
C
      CALL PDCO(N,M,AX,BX,AY,BY,PD)
```

```
      NT=1
C
C START ITERATION
C
 88   CONTINUE
C
C IMPLEMENT BOUNDARY CONDITIONS FOR VORTICITY AND TEMPERATURE
C
      CALL BOCO(N,M,U,V,W,T,AX,AY)
C
C 4-STAGE RUNGE-KUTTA METHOD TO SOLVE RESULTANT ORDINARY DIFFERENTIAL
C EQUATIONS FOR VORTICITY AND TEMPERATURE
C
      CALL RUNGU(N,M,H,RA,PR,AX,BX,AY,BY,U,V,T,W,DT,DW)
C
C COMPUTE STREAM FUNCTION
C
      CALL COREP(N,M,W,P,PD)
C
C UPDATE STREAM FUNCTION ON THE BOUNDARY
C
      CALL BPCO(N,M,AX,AY,P)
C
C COMPUTE VELOCITY U, V
C
      CALL CORUV(N,M,AX,AY,U,V,P)
C
C FIND OUT MAXIMUM RESIDUALS FOR EQUATIONS 6.22 AND 6.23
C
      ERR1=DT(1,1)**2
      ERR2=DW(1,1)**2
      DO 91 I=1,N
      DO 91 J=1,M
      ERR1=DT(I,J)**2+ERR1
      ERR2=DW(I,J)**2+ERR2
   91 CONTINUE
      NM=N*M
      ERR1=SQRT(ERR1/NM)
      ERR2=SQRT(ERR2/NM)
      WRITE(6,113) NT,ERR1,ERR2
  113 FORMAT(1X,I10,5X,'    ERR1=    ',E10.5,5X,'    ERR2=    ',E10.5)
C
C CHECK CONVERGENCE CRITERIA
C
      IF(ERR1.LE.EPS1.AND.ERR2.LE.EPS2) GO TO 92
      IF(NT.GE.NTT) GO TO 92
      NT=NT+1
      GO TO 88
   92 CONTINUE
C
C COMPUTE NUSSELT NUMBERS
C
```

```
      CALL NUSEC(N,M,XX,YY,AX,U,T)
C
C INTERPOLATE NUMERICAL RESULTS FROM COARSE MESH TO FINE MESH FOR
C PLOTTING AND OTHER PURPOSE
C
      CALL OUTPLT(N,M,XX,YY,U,V,W,T,P)
      STOP
      END
C
C THIS SUBROUTINE IS TO COMPUTE THE PDQ WEIGHTING COEFFICIENTS OF
C THE FIRST AND SECOND ORDER DERIVATIVES WHEN THE COORDINATES OF
C MESH POINTS ARE INPUT
C
      SUBROUTINE WCOEF(N0,NN,XX,AX,BX)
      IMPLICIT REAL*8(A-H,O-Z)
      DIMENSION AX(50,50),BX(50,50),XX(50)
      DIMENSION AIX(50)
      NC=NN-N0+1
C
C COMPUTE THE WEIGHTING COEFFICIENTS OF THE FIRST ORDER DERIVATIVE
C
      DO 72 I=N0,NN
      II=I-N0+1
      AX1=1.
      DO 73 J=N0,NN
      IF(J.EQ.I) GO TO 73
      AX1=AX1*(XX(I)-XX(J))
   73 CONTINUE
   72 AIX(II)=AX1
      DO 74 I=1,NC
      II=I+N0-1
      DO 74 J=1,NC
      JJ=J+N0-1
      IF(J.EQ.I) GO TO 74
      AX(I,J)=AIX(I)/((XX(II)-XX(JJ))*AIX(J))
   74 CONTINUE
      DO 77 I=1,NC
      AXI=0.
      DO 78 J=1,NC
      IF(J.EQ.I) GO TO 78
      AXI=AXI+AX(I,J)
   78 CONTINUE
      AX(I,I)=-AXI
   77 CONTINUE
C
C COMPUTE THE WEIGHTING COEFFICIENTS OF THE SECOND ORDER DERIVATIVE
C
      DO 300 I=1,NC
      II=I+N0-1
      DO 300 J=1,NC
      JJ=J+N0-1
```

```
      IF(J.EQ.I) GO TO 300
      BX(I,J)=2.*AX(I,J)*(AX(I,I)-1./(XX(II)-XX(JJ)))
  300 CONTINUE
      DO 400 I=1,NC
      BXI=0.
      DO 410 J=1,NC
      IF(J.EQ.I) GO TO 410
      BXI=BXI+BX(I,J)
  410 CONTINUE
      BX(I,I)=-BXI
  400 CONTINUE
      RETURN
      END
C
C THIS SUBROUTINE IS TO GENERATE THE COORDINATES OF MESH POINTS
C
      SUBROUTINE GRID(N,M,XX,YY,U,V,W,T,DT,DW)
      IMPLICIT REAL*8(A-H,O-Z)
      DIMENSION XX(50),YY(50),U(50,50),V(50,50)
      DIMENSION T(50,50),W(50,50),DT(50,50),DW(50,50)
      WRITE(6,*) 'ALAX,ALAY'
      READ(5,*) ALAX,ALAY
      PI=4.*ATAN(1.)
C
      DO 71 I=1,N
      ALFA=PI*FLOAT(I-1)/FLOAT(N-1)
  71  XX(I)=0.5*(1.-COS(ALFA))
      X11=XX(1)
C
C STRETCH THE MESH TOWARDS THE BOUNDARY, ALAX<1
C
      DO 76 I=1,N
  76  XX(I)=(1.-ALAX)*(3.*XX(I)**2-2.*XX(I)**3)+ALAX*XX(I)
C
      DO 81 J=1,M
      ALFA=PI*FLOAT(J-1)/FLOAT(M-1)
  81  YY(J)=0.5*(1.-COS(ALFA))
C
C STRETCH THE MESH TOWARDS THE BOUNDARY, ALAY<1
C
      DO 86 J=1,M
  86  YY(J)=(1.-ALAY)*(3.*YY(J)**2-2.*YY(J)**3)+ALAY*YY(J)
      YY(MH)=0.5
C
C SET UP INITIAL FLOW FIELD
C
      DO 18 J=1,M
      DO 10 I=1,N
      U(I,J)=0.
      V(I,J)=0.
      T(I,J)=0.
      W(I,J)=0.
```

```
      DT(I,J)=0.
      DW(I,J)=0.
  10  CONTINUE
      T(1,J)=1.0
  18  CONTINUE
      RETURN
      END
C
C IMPLEMENT BOUNDARY CONDITIONS FOR VORTICITY AND TEMPERATURE
C
      SUBROUTINE BOCO(N,M,U,V,W,T,AX,AY)
      IMPLICIT REAL*8(A-H,O-Z)
      DIMENSION AX(50,50),AY(50,50),U(50,50),V(50,50)
      DIMENSION T(50,50),W(50,50)
C
C UPDATE VORTICITY ON THE BOTTOM AND TOP BOUNDARIES
C
      DO 72 I=1,N
      SUM1=0.
      SUM2=0.
      DO 10 J=1,M
      SUM1=SUM1+AY(1,J)*U(I,J)
  10  SUM2=SUM2+AY(M,J)*U(I,J)
      W(I,1)=SUM1
  72  W(I,M)=SUM2
C
C UPDATE VORTICITY ON THE LEFT AND RIGHT BOUNDARIES
C
      DO 73 J=1,M
      SUM3=0.
      SUM4=0.
      DO 20 I=1,N
      SUM3=SUM3+AX(1,I)*V(I,J)
  20  SUM4=SUM4+AX(N,I)*V(I,J)
      W(1,J)=-SUM3
      W(N,J)=-SUM4
C
C TEMPERATURE CONDITION AT LEFT AND RIGHT BOUNDARIES
C
      T(1,J)=1.
      T(N,J)=0.
  73   CONTINUE
      RETURN
      END
C
C 4-STAGE RUNGE-KUTTA METHOD TO SOLVE ORDINARY DIFFERENTIAL EQUATIONS
C FOR VORTICITY AND TEMPERATURE.THE SCHEME IS SHOWN IN EQUATION 5.15
C
      SUBROUTINE RUNGU(N,M,H,RA,PR,AX,BX,AY,BY,U,V,T,W,DT,DW)
      IMPLICIT REAL*8(A-H,O-Z)
      DIMENSION AX(50,50),BX(50,50),AY(50,50),BY(50,50)
```

```
      DIMENSION U(50,50),V(50,50),T(50,50),W(50,50)
      DIMENSION CT(50,50),CW(50,50),DT(50,50),DW(50,50)
C STAGE 1
      CALL TRFW(N,M,RA,PR,AX,BX,AY,BY,U,V,T,W,DW)
      CALL TRFT(N,M,AX,BX,AY,BY,U,V,T,DT)
      DO 3 I=1,N
      DO 3 J=1,M
      CW(I,J)=W(I,J)+0.25*H*DW(I,J)
  3   CT(I,J)=T(I,J)+0.25*H*DT(I,J)
C STAGE 2
      CALL TRFW(N,M,RA,PR,AX,BX,AY,BY,U,V,CT,CW,DW)
      CALL TRFT(N,M,AX,BX,AY,BY,U,V,CT,DT)
      DO 5 I=1,N
      DO 5 J=1,M
      CW(I,J)=W(I,J)+H*DW(I,J)/3.
  5   CT(I,J)=T(I,J)+H*DT(I,J)/3.
C STAGE 3
      CALL TRFW(N,M,RA,PR,AX,BX,AY,BY,U,V,CT,CW,DW)
      CALL TRFT(N,M,AX,BX,AY,BY,U,V,CT,DT)
      DO 7 I=1,N
      DO 7 J=1,M
      CW(I,J)=W(I,J)+H*DW(I,J)/2.
  7   CT(I,J)=T(I,J)+H*DT(I,J)/2.
C STAGE 4
      CALL TRFW(N,M,RA,PR,AX,BX,AY,BY,U,V,CT,CW,DW)
      CALL TRFT(N,M,AX,BX,AY,BY,U,V,CT,DT)
      DO 9 I=2,N-1
      DO 9 J=2,M-1
      W(I,J)=W(I,J)+H*DW(I,J)
  9   T(I,J)=T(I,J)+H*DT(I,J)
C
      CALL TRFW(N,M,RA,PR,AX,BX,AY,BY,U,V,T,W,DW)
      CALL TRFT(N,M,AX,BX,AY,BY,U,V,T,DT)
      RETURN
      END
C
C COMPUTE THE RESIDUAL OF VORTICITY EQUATION 6.22
C
      SUBROUTINE TRFW(N,M,RA,PR,AX,BX,AY,BY,U,V,CT,CW,DW)
      IMPLICIT REAL*8(A-H,O-Z)
      DIMENSION AX(50,50),BX(50,50),AY(50,50),BY(50,50)
      DIMENSION U(50,50),V(50,50),CT(50,50),CW(50,50),DW(50,50)
      DO 10 I=2,N-1
      DO 10 J=2,M-1
      DWI=0.
      DTI=0.
      DDWI=0.
      DO 20 K=1,N
      DDWI=DDWI+BX(I,K)*CW(K,J)
      DTI=DTI+AX(I,K)*CT(K,J)
 20   DWI=DWI+AX(I,K)*CW(K,J)
      DWJ=0.
```

```
      DDWJ=0.
      DO 30 K=1,M
      DWJ=DWJ+AY(J,K)*CW(I,K)
   30 DDWJ=DDWJ+BY(J,K)*CW(I,K)
      DW(I,J)=(DDWI+DDWJ)*PR-U(I,J)*DWI-V(I,J)*DWJ-DTI*PR*RA
   10 CONTINUE
      RETURN
      END
C
C COMPUTE THE RESIDUAL OF TEMPERATURE EQUATION 6.23
C
      SUBROUTINE TRFT(N,M,AX,BX,AY,BY,U,V,CT,DT)
      IMPLICIT REAL*8(A-H,O-Z)
      DIMENSION AX(50,50),BX(50,50),AY(50,50),BY(50,50)
      DIMENSION U(50,50),V(50,50),CT(50,50),DT(50,50)
C
C UPDATE THE TEMPERATURE ON THE TOP AND BOTTOM BOUNDARIES USING
C EQUATIONS 6.42B AND 6.42C
C
      DO 70 I=2,N-1
      C2=0.
      C3=0.
      DO 80 J=2,M-1
      C2=C2+(AY(1,J)*AY(M,M)-AY(M,J)*AY(1,M))*CT(I,J)
      C3=C3+(AY(1,J)*AY(M,1)-AY(M,J)*AY(1,1))*CT(I,J)
   80 CONTINUE
      CT(I,1)=-C2/(AY(1,1)*AY(M,M)-AY(M,1)*AY(1,M))
   70 CT(I,M)=-C3/(AY(1,M)*AY(M,1)-AY(M,M)*AY(1,1))
C
C COMPUTE RESIDUALS OF TEMPERATURE EQUATION 6.23 FOR ALL INTERIOR
C POINTS
C
      DO 10 I=2,N-1
      DO 10 J=2,M-1
      DTI=0.
      DDTI=0.
      DO 20 K=1,N
      DDTI=DDTI+BX(I,K)*CT(K,J)
   20 DTI=DTI+AX(I,K)*CT(K,J)
      DTJ=0.
      DDTJ=0.
      DO 30 K=1,M
      DTJ=DTJ+AY(J,K)*CT(I,K)
   30 DDTJ=DDTJ+BY(J,K)*CT(I,K)
      DT(I,J)=DDTI+DDTJ-U(I,J)*DTI-V(I,J)*DTJ
   10 CONTINUE
      RETURN
      END
C
C FORM THE MATRIX OF EQUATION SYSTEM 6.21. THE TWO-LAYER APPROACH
C IS USED TO IMPLEMENT THE BOUNDARY CONDITION FOR STREAM FUNCTION.
```

```
C THE OBTAINED MATRIX IS DECOMPOSED BY LU DECOMPOSITION
C
      SUBROUTINE PDCO(N,M,AX,BX,AY,BY,PD)
      IMPLICIT REAL*8(A-H,O-Z)
      DIMENSION AX(50,50),BX(50,50),AY(50,50),BY(50,50)
      DIMENSION PD(401,401),PL(401,401),PU(401,401)
      NN=N-4
      MM=M-4
      NM=NN*MM
      N1=N-1
      M1=M-1
C
C AXN AND AYM ARE USED IN EQUATIONS 6.30 AND 6.31 FOR TWO-LAYER
C APPROACH
C
      AXN=AX(1,2)*AX(N,N1)-AX(N,2)*AX(1,N1)
      AYM=AY(1,2)*AY(M,M1)-AY(M,2)*AY(1,M1)
      DO 10 I=3,N-2
      DO 10 J=3,M-2
      IJ=MM*(I-3)+J-2
      DO 20 K=1,NM
   20 PD(IJ,K)=0.
      DO 30 K1=3,N-2
      KJ=MM*(K1-3)+J-2
C
C EQUATION 6.30 IS DIRECTLY SUBSTITUTED INTO EQUATION 6.21
C
      AXK1=AX(1,K1)*AX(N,N1)-AX(N,K1)*AX(1,N1)
      AXKN=AX(N,K1)*AX(1,2)-AX(1,K1)*AX(N,2)
   30 PD(IJ,KJ)=BX(I,K1)-(AXK1*BX(I,2)+AXKN*BX(I,N1))/AXN
      DO 40 K2=3,M-2
      IK=MM*(I-3)+K2-2
C
C EQUATION 6.31 IS DIRECTLY SUBSTITUTED INTO EQUATION 6.21
C
      AYK1=AY(1,K2)*AY(M,M1)-AY(M,K2)*AY(1,M1)
      AYKM=AY(M,K2)*AY(1,2)-AY(1,K2)*AY(M,2)
   40 PD(IJ,IK)=PD(IJ,IK)+BY(J,K2)-(AYK1*BY(J,2)+AYKM*BY(J,M1))/AYM
   10 CONTINUE
C
C LU DECOMPOSITION USING EQUATION 5.27
C
      DO 50 I=1,NM
      DO 50 J=1,NM
      IF(J.GE.I) GO TO 51
      J1=J-1
      SUM1=0.
      IF(J1.EQ.0) GO TO 57
      DO 55 JJ=1,J1
   55 SUM1=SUM1+PL(I,JJ)*PU(JJ,J)
   57 PL(I,J)=(PD(I,J)-SUM1)/PU(J,J)
      GO TO 50
```

```
   51 IF(I.EQ.1) GO TO 52
      I1=I-1
      SUM2=0.
      DO 53 II=1,I1
   53 SUM2=SUM2+PL(I,II)*PU(II,J)
      PU(I,J)=PD(I,J)-SUM2
      GO TO 50
   52 PU(I,J)=PD(I,J)
   50 CONTINUE
C
C THE TRIANGULAR MATRICES [PL] AND [PU] ARE STORED IN [PD]
C
      DO 60 I=1,NM
      DO 60 J=1,NM
      IF(J.GE.I) GO TO 61
      PD(I,J)=PL(I,J)
      GO TO 60
   61 PD(I,J)=PU(I,J)
   60 CONTINUE
      RETURN
      END
C
C THE ALGEBRAIC EQUATION SYSTEM FOR STREAM FUNCTION IS SOLVED BY
C LU DECOMPOSITION.
C
      SUBROUTINE COREP(N,M,W,P,PD)
      IMPLICIT REAL*8(A-H,O-Z)
      DIMENSION PD(401,401),P(50,50),W(50,50)
      DIMENSION B(401),BI(401),BJ(401)
      NN=N-4
      MM=M-4
      NM=NN*MM
      DO 10 I=3,N-2
      DO 10 J=3,M-2
      IJ=MM*(I-3)+J-2
      B(IJ)=W(I,J)
  10  CONTINUE
C
C EQUATION 5.26A IS USED TO GET THE SOLUTION AT THE FIRST STEP
C
      DO 40 I=1,NM
      IF(I.EQ.1) GO TO 41
      I1=I-1
      SUM1=0.
      DO 42 J=1,I1
   42 SUM1=SUM1+PD(I,J)*BI(J)
      BI(I)=B(I)-SUM1
      GO TO 40
   41 BI(1)=B(1)
   40 CONTINUE
C
```

```
C EQUATION 5.26B IS USED TO GET THE SOLUTION AT THE SECOND STEP
C
      DO 50 I=1,NM
      IV=NM-I+1
      IF(I.EQ.1) GO TO 51
      I2=I-1
      SUM2=0.
      DO 52 J=1,I2
      IVJ=NM-J+1
   52 SUM2=SUM2+PD(IV,IVJ)*BJ(IVJ)
      BJ(IV)=(BI(IV)-SUM2)/PD(IV,IV)
      GO TO 50
   51 BJ(IV)=BI(IV)/PD(IV,IV)
   50 CONTINUE
C
      DO 60 I=3,N-2
      DO 60 J=3,M-2
      IJ=MM*(I-3)+J-2
   60 P(I,J)=BJ(IJ)
      RETURN
      END
C
C IMPLEMENT BOUNDARY CONDITION FOR STREAM FUNCTION. THE TWO-LAYER
C APPROACH IS USED
C
      SUBROUTINE BPCO(N,M,AX,AY,P)
      IMPLICIT REAL*8(A-H,O-Z)
      DIMENSION AX(50,50),AY(50,50),P(50,50)
      N1=N-1
      M1=M-1
C
C AXN AND AYM ARE USED IN EQUATIONS 6.30 AND 6.31 FOR TWO-LAYER
C APPROACH
C
      AXN=AX(1,2)*AX(N,N1)-AX(N,2)*AX(1,N1)
      AYM=AY(1,2)*AY(M,M1)-AY(M,2)*AY(1,M1)
C
C EQUATION 6.31 IS USED TO UPDATE THE STREAM FUNCTION
C
      DO 20 I=3,N-2
      SUM3=0.
      SUM4=0.
      DO 21 J=3,M-2
      SUM3=SUM3-(AY(1,J)*AY(M,M1)-AY(M,J)*AY(1,M1))*P(I,J)
   21 SUM4=SUM4-(AY(M,J)*AY(1,2)-AY(1,J)*AY(M,2))*P(I,J)
      P(I,2)=SUM3/AYM
   20 P(I,M1)=SUM4/AYM
C
C EQUATION 6.30 IS USED TO UPDATE THE STREAM FUNCTION
C
      DO 10 J=2,M-1
      SUM1=0.
```

```
      SUM2=0.
      DO 11 I=3,N-2
      SUM1=SUM1-(AX(1,I)*AX(N,N1)-AX(N,I)*AX(1,N1))*P(I,J)
   11 SUM2=SUM2-(AX(N,I)*AX(1,2)-AX(1,I)*AX(N,2))*P(I,J)
      P(2,J)=SUM1/AXN
   10 P(N1,J)=SUM2/AXN
C
      DO 30 J=1,M
      P(1,J)=0.
   30 P(N,J)=0.
      DO 40 I=2,N1
      P(I,1)=0.
   40 P(I,M)=0.
      RETURN
      END
C
C COMPUTE VELOCITY U, V FROM STREAM FUNCTION
C
      SUBROUTINE CORUV(N,M,AX,AY,U,V,P)
      IMPLICIT REAL*8(A-H,O-Z)
      DIMENSION AX(50,50),AY(50,50),U(50,50),V(50,50),P(50,50)
      DO 10 I=2,N-1
      DO 10 J=2,M-1
      DPI=0.
C COMPUTE V VELOCITY FOR INTERIOR POINTS
      DO 20 K=1,N
   20 DPI=DPI+AX(I,K)*P(K,J)
      V(I,J)=-DPI
      DPJ=0.
C COMPUTE U VELOCITY FOR INTERIOR POINTS
      DO 30 K=1,M
   30 DPJ=DPJ+AY(J,K)*P(I,K)
   10 U(I,J)=DPJ
      DO 40 I=1,N
      U(I,1)=0.
      V(I,1)=0.
      U(I,M)=0.
  40  V(I,M)=0.
      DO 50 J=2,M-1
      U(1,J)=0.
      V(1,J)=0.
      U(N,J)=0.
  50  V(N,J)=0.
      RETURN
      END
C
C COMPUTE NUSSELT NUMBERS
C
      SUBROUTINE NUSEC(N,M,XX,YY,AX,U,T)
      IMPLICIT REAL*8(A-H,O-Z)
      DIMENSION XX(50),YY(50),X2(201),Y2(201),AX(50,50),U(50,50)
```

```
      DIMENSION T(50,50),QQ(50,50),QI(50),QT(50,201),CNU(50),CNUT(201)
C
C COMPUTE LOCAL HEAT FLUX IN A HORIZONTAL DIRECTION Q(X,Y)
C
      DO 10 I=1,N
      DO 10 J=1,M
      DTI=0.
      DO 11 K=1,N
   11 DTI=DTI+AX(I,K)*T(K,J)
   10 QQ(I,J)=U(I,J)*T(I,J)-DTI
C
C SET UP A FINE MESH
C
      NN=101
      MM=101
      TDX=(XX(N)-XX(1))/FLOAT(NN-1)
      TDY=(YY(M)-YY(1))/FLOAT(MM-1)
      DO 20 I=1,NN
  20  X2(I)=FLOAT(I-1)*TDX
      DO 30 J=1,MM
  30  Y2(J)=FLOAT(J-1)*TDY
C
C INTERPOLATE Q(X,Y) TO THE FINE MESH IN THE Y DIRECTION
C
      DO 40 I=1,N
      DO 41 J=1,M
      QI(J)=QQ(I,J)
   41 CONTINUE
      DO 42 J=1,MM
      YJ=Y2(J)
      CALL CZ(M,YJ,YY,QI,QOUT)
      QT(I,J)=QOUT
   42 CONTINUE
  40  CONTINUE
C
C COMPUTE AVERAGE NUSSELT NUMBER ON THE VERTICAL LINE
C
      DO 188 I=1,N
      SUM=0.0
      DO 198 J=2,MM
      SUM=SUM+0.5*(QT(I,J)+QT(I,J-1))*TDY
  198 CONTINUE
      CNU(I)=SUM
  188 CONTINUE

      DO 52 I=1,NN
      XI=X2(I)
      CALL CZ(N,XI,XX,CNU,CNUOUT)
      CNUT(I)=CNUOUT
  52  CONTINUE
C
C COMPUTE AVERAGE NUSSELT NUMBER THROUGHOUT THE CAVITY
```

```
C
      CNUAVG=0.0
      DO 918 I=2,NN
      CNUAVG=CNUAVG+0.5*(CNUT(I)+CNUT(I-1))*TDX
  918 CONTINUE
C
C FIND OUT MAXIMUM AND MINIMUM LOCAL NUSSELT NUMBERS AT X=0 TOGETHER
C WITH THEIR POSITIONS
C
      AMIN=QT(1,1)
      AMAX=QT(1,1)
      DO 518 J=1,MM
      IF(QT(1,J).LE.AMAX) GO TO 888
      AMAX=QT(1,J)
      YMAX=Y2(J)
  888 IF(QT(1,J).GE.AMIN) GO TO 518
      AMIN=QT(1,J)
      YMIN=Y2(J)
  518 CONTINUE
      NH=(N+1)/2
C
C OUTPUT THE NUSSELT NUMBERS
C
      WRITE(6,70) CNU(1)
   70 FORMAT(2X,'AVERAGE NUSSELT NUMBER AT X=0',5X,F15.4)
      WRITE(6,72) CNU(NH)
   72 FORMAT(2X,'AVERAGE NUSSELT NUMBER AT X=0.5',3X,F15.4)
      WRITE(6,74) CNUAVG
   74 FORMAT(2X,'AVERAGE NUSSELT NUMBER OVER CAVITY',F15.4)
      WRITE(6,76) AMAX,YMAX
   76 FORMAT(2X,'NUMAX AND POSITION AT X=0',5X,F10.4,5X,F10.4)
      WRITE(6,78) AMIN,YMIN
   78 FORMAT(2X,'NUMIN AND POSITION AT X=0',5X,F10.4,5X,F10.4)
      RETURN
      END
C
C LAGRANGE INTERPOLATION
C
      SUBROUTINE CZ(N,XX,X,F,FV)
      IMPLICIT REAL*8(A-H,O-Z)
      DIMENSION X(50),F(50)
      FV=0.
      DO 10 I=1,N
      AD=1.
      DO 20 J=1,N
      IF(J.EQ.I) GO TO 21
      AD1=(XX-X(J))/(X(I)-X(J))
      GO TO 22
   21 AD1=1.
   22 AD=AD*AD1
   20 CONTINUE
```

```
      FV=FV+AD*F(I)
   10 CONTINUE
      RETURN
      END
C
C INTERPOLATE THE NUMERICAL RESULTS FROM A COARSE MESH TO A FINE MESH
C
      SUBROUTINE OUTPLT(N,M,X1,Y1,U,V,W,G,P)
      IMPLICIT REAL*8(A-H,O-Z)
      DIMENSION X1(50),Y1(50),X2(101),Y2(101),P(50,50),W(50,50),
     1           U(50,50),V(50,50),G(50,50),UT(101,101),VT(101,101),
     2           GT(101,101),PT(101,101),WT(101,101),UI(51),UJ(51),
     3          VI(51),VJ(51),WI(51),WJ(51),PI(51),PJ(51),GI(50),GJ(50)
      X2(1)=X1(1)
      Y2(1)=Y1(1)
      WRITE(6,*) 'READ THE VALUES OF NN, MM FOR INTERPOLATION'
      READ(5,*) NN,MM
      TDX=(X1(N)-X1(1))/FLOAT(NN-1)
      TDY=(Y1(M)-Y1(1))/FLOAT(MM-1)
C GENERATE FINE MESH
      DO 20 I=2,NN
   20 X2(I)=X2(I-1)+TDX
      DO 30 J=2,MM
   30 Y2(J)=Y2(J-1)+TDY
C
C INTERPOLATION IN THE X DIRECTION
C
      DO 40 J=1,M
      DO 41 I=1,N
      UI(I)=U(I,J)
      VI(I)=V(I,J)
      WI(I)=W(I,J)
      GI(I)=G(I,J)
      PI(I)=P(I,J)
   41 CONTINUE
      DO 42 I=1,NN
      XI=X2(I)
      CALL CZ(N,XI,X1,UI,UOUT)
      CALL CZ(N,XI,X1,VI,VOUT)
      CALL CZ(N,XI,X1,WI,WOUT)
      CALL CZ(N,XI,X1,GI,GOUT)
      CALL CZ(N,XI,X1,PI,POUT)
      UT(I,J)=UOUT
      VT(I,J)=VOUT
      WT(I,J)=WOUT
      GT(I,J)=GOUT
      PT(I,J)=POUT
   42 CONTINUE
   40 CONTINUE
C
C INTERPOLATION IN THE Y DIRECTION
C
```

```
      DO 50 I=1,NN
      DO 51 J=1,M
      UJ(J)=UT(I,J)
      VJ(J)=VT(I,J)
      WJ(J)=WT(I,J)
      GJ(J)=GT(I,J)
      PJ(J)=PT(I,J)
   51 CONTINUE
      DO 52 J=1,MM
      YJ=Y2(J)
      CALL CZ(M,YJ,Y1,UJ,UOUT)
      CALL CZ(M,YJ,Y1,VJ,VOUT)
      CALL CZ(M,YJ,Y1,WJ,WOUT)
      CALL CZ(M,YJ,Y1,GJ,GOUT)
      CALL CZ(M,YJ,Y1,PJ,POUT)
      UT(I,J)=UOUT
      VT(I,J)=VOUT
      WT(I,J)=WOUT
      GT(I,J)=GOUT
      PT(I,J)=POUT
  52  CONTINUE
  50  CONTINUE
C
C OUTPUT DATA FOR CONTOUR PLOTTING BY TECPLOT (A SOFTWARE)
C
      OPEN(7,FILE='NSTO.DAT')
      WRITE(7,*) 'TITLE= "  " '
      WRITE(7,*) 'VARIABLES= X,Y,U,V,T,P,W'
      WRITE(7,*) 'ZONE I=',NN,', J=',MM,', F=POINT'
      DO 850 J=1,MM
      DO 850 I=1,NN
      WRITE(7,950) X2(I),Y2(J),UT(I,J),VT(I,J),GT(I,J),PT(I,J),WT(I,J)
 850  CONTINUE
 950  FORMAT(F7.4,1X,F7.4,1X,E11.5,2X,E11.5,2X,E11.5,2X,E11.5,
     1       2X,E11.5)
      CLOSE(7)
      RETURN
      END
```

Appendix B

A Fortran Program for Vibration Analysis of Rectangular Plates

This program is based on the algorithm shown in the Section 7.3.3.3. The spatial derivatives in the governing differential equations and the boundary conditions are discretized by the PDQ method. The discrete boundary conditions are directly substituted into the discrete governing equations. The standard QR algorithm is used to compute the eigenvalues of the resultant eigenvalue equation system 7.50. For details of the algorithm, the reader is advised to refer to the Section 7.3.3.3.

Description of Variables

N — Number of grid points in the x direction

M — Number of grid points in the y direction

RABDA — Aspect ratio, $\lambda = a/b$

CMU — Poisson's ratio, ν

ISTAX1 — Type of boundary condition for $x = 0$, which can be taken as 1 for simply supported condition, or 2 for clamped condition

ISTAXN — Type of boundary condition for $x = 1$, which can be taken as 1 for simply supported condition, or 2 for clamped condition

ISTAY1 — Type of boundary condition for $y = 0$, which can be taken as 1 for simply supported condition, or 2 for clamped condition

ISTAYM — Type of boundary condition for $y = 1$, which can be taken as 1 for simply supported condition, or 2 for clamped condition

XX(I) — x-coordinate of a mesh point, x_i

YY(J) — y-coordinate of a mesh point, y_j

AX(I,J) — Weighting coefficients of the first order derivative in the x direction, $c_{i,j}^{(1)}$

BX(I,J) — Weighting coefficients of the second order derivative in the x direction, $c_{i,j}^{(2)}$

CX(I,J) — Weighting coefficients of the third order derivative in the x direction, $c_{i,j}^{(3)}$

DX(I,J) — Weighting coefficients of the fourth order derivative in the x direction, $c_{i,j}^{(4)}$

AY(I,J) — Weighting coefficients of the first order derivative in the y direction, $\bar{c}_{i,j}^{(1)}$

BY(I,J) — Weighting coefficients of the second order derivative in the y direction, $\bar{c}_{i,j}^{(2)}$

CY(I,J) — Weighting coefficients of the third order derivative in the y direction, $\bar{c}_{i,j}^{(3)}$

DY(I,J) — Weighting coefficients of the fourth order derivative in the y direction, $\bar{c}_{i,j}^{(4)}$

W(I,J) — Dimensionless mode shape function

PDD(I,J) — Matrix of the eigenvalue equation system 7.50

U(I) — Real part of eigenvalues given by Equation 7.50, which is the dimensionless frequency for the problem

V(I) — Imaginary part of eigenvalues given by Equation 7.50, which is zero for the problem

EIGOT(I) — Distribute the computed eigenvalues in sequence from low to high

```
C VIBRATION ANALYSIS OF RECTANGULAR PLATES BY PDQ METHOD
C THE PROGRAM IS BASED ON THE ALGORITHM SHOWN IN THE SECTION 7.3.3.3
C ANY COMBINATION OF SIMPLY SUPPORTED AND CLAMPED BOUNDARY CONDITIONS
C CAN BE SOLVED BY THIS PROGRAM
C
      PROGRAM MAIN
      IMPLICIT REAL*8(A-H,O-Z)
      DIMENSION AX(50,50),BX(50,50),CX(50,50),DX(50,50),XX(50)
      DIMENSION AY(50,50),BY(50,50),CY(50,50),DY(50,50),YY(50)
      DIMENSION W(50,50),PDD(1401,1401),U(1401),V(1401),
     1          EIGOT(1401),IEIG(1401)
      WRITE(6,*) 'READ THE VALUES OF N,M,RABDA,CMU'
      READ(5,*) N,M,RABDA,CMU
      WRITE(6,*) 'READ THE VALUES OF ISTAX1,ISTAXN,ISTAY1,ISTAYM'
      READ(5,*) ISTAX1,ISTAY1,ISTAXN,ISTAYM
C
C MESH GENERATION
C
      CALL GRID(N,M,XX,YY)
C
C COMPUTATION OF WEIGHTING COEFFICIENTS OF THE FIRST AND SECOND ORDER
C DERIVATIVES IN THE X AND Y DIRECTIONS BY PDQ METHOD. THE SUBROUTINE
C WCOEF(1,N,XX,AX,BX) HAS BEEN SHOWN IN APPENDIX A
C
      CALL WCOEF(1,N,XX,AX,BX)
```

```
      CALL WCOEF(1,M,YY,AY,BY)
C
C COMPUTATION OF WEIGHTING COEFFICIENTS OF THE THIRD AND FOURTH ORDER
C DERIVATIVES IN THE X DIRECTION BY MATRIX MULTIPLICATION TECHNIQUE
C
      DO 555 I=1,N
      DO 555 J=1,N
      SUM1=0.0
      SUM2=0.0
      DO 556 K=1,N
      SUM1=SUM1+BX(I,K)*AX(K,J)
      SUM2=SUM2+BX(I,K)*BX(K,J)
 556  CONTINUE
      CX(I,J)=SUM1
      DX(I,J)=SUM2
 555  CONTINUE
C
C COMPUTATION OF WEIGHTING COEFFICIENTS OF THE THIRD AND FOURTH ORDER
C DERIVATIVES IN THE Y DIRECTION BY MATRIX MULTIPLICATION TECHNIQUE
C
      DO 557 I=1,M
      DO 557 J=1,M
      SUM1=0.0
      SUM2=0.0
      DO 558 K=1,M
      SUM1=SUM1+BY(I,K)*AY(K,J)
      SUM2=SUM2+BY(I,K)*BY(K,J)
 558  CONTINUE
      CY(I,J)=SUM1
      DY(I,J)=SUM2
 557  CONTINUE
C
      NMD=(N-4)*(M-4)
      DO 604 I=1,NMD
      DO 604 J=1,NMD
      PDD(I,J)=0.0
 604  CONTINUE
C
C FORM THE MATRIX OF THE EIGENVALUE EQUATION SYSTEM IN EQUATION 7.50
C
      CALL PDCO(N,M,ISTAX1,ISTAY1,ISTAXN,ISTAYM,RABDA,CMU,
     1          AX,BX,DX,AY,BY,DY,PDD)
C
C COMPUTE EIGENVALUES OF THE MATRIX [PDD] BY USING QR ALGORITHM
C
      CALL HBG(PDD,NMD)
      CALL HQR(PDD,NMD,U,V)
      DO 909 I=1,NMD
      U(I)=SQRT(ABS(U(I)))
 909  CONTINUE
C
C RE-DISTRIBUTE THE EIGENVALUES IN SEQUENCE
```

```
C
      DO 905 I=1,NMD
      CALL EIGVMX(NMD,U,EIGOT(I),IEIG(I))
 905  CONTINUE
C
C OUTPUT EIGENVALUES IN SEQUENCE
C
      WRITE(6,*) 'READ THE NUMBER FOR OUTPUT EIGENVALUES,NEIG'
      READ(5,*) NEIG
      WRITE(6,902) (EIGOT(I),I=1,NEIG)
 902  FORMAT(4F20.6)
      STOP
      END
C
C THIS SUBROUTINE IS TO GENERATE THE COORDINATES OF MESH POINTS
C
      SUBROUTINE GRID(N,M,XX,YY)
      IMPLICIT REAL*8(A-H,O-Z)
      DIMENSION XX(50),YY(50)
C
      PI=4.*ATAN(1.)
C
C GENERATE THE X-COORDINATES USING EQUATION 7.39A
C
      DO 71 I=1,N
      ALFA=PI*FLOAT(I-1)/FLOAT(N-1)
      XX(I)=0.5*(1.-COS(ALFA))
  71  CONTINUE
      WRITE(6,102)
 102  FORMAT(15X,'          HORIZONTAL COORDINATE                ')
      WRITE(6,100) (XX(J),J=1,N)
 100  FORMAT(1X,6F12.6)
C
C GENERATE THE Y-COORDINATES USING EQUATION 7.39B
C
      DO 81 J=1,M
      ALFA=PI*FLOAT(J-1)/FLOAT(M-1)
      YY(J)=0.5*(1.-COS(ALFA))
  81  CONTINUE
      WRITE(6,103)
 103  FORMAT(15X,'          LONGITUDINAL COORDINATE              ')
      WRITE(6,100) (YY(J),J=1,M)
      RETURN
      END
C
C THIS SUBROUTINE IS TO FORM THE MATRIX OF THE EIGENVALUE EQUATION
C SYSTEM 7.50
C
      SUBROUTINE PDCO(N,M,ISTAX1,ISTAY1,ISTAXN,ISTAYM,RABDA,CMU,
     1          AX,BX,DX,AY,BY,DY,PDD)
      IMPLICIT REAL*8(A-H,O-Z)
```

```
      DIMENSION AX(50,50),BX(50,50),DX(50,50),BY(50,50),DY(50,50)
      DIMENSION AY(50,50),PDD(1401,1401)
      NN=N-4
      MM=M-4
      NM=NN*MM
C
C COMPUTE AXN AND AYM USED IN EQUATION 7.50
C
      CALL AXYNM(N,ISTAX1,ISTAXN,AX,BX,AXN)
      CALL AXYNM(M,ISTAY1,ISTAYM,AY,BY,AYM)
C
      DO 10 I=3,N-2
      DO 10 J=3,M-2
      IJ=MM*(I-3)+J-3+1
C
C COMPUTE C1 USED IN EQUATION 7.50 AND ITS CONTRIBUTION TO THE
C MATRIX OF EQUATION SYSTEM 7.50
C
      DO 30 K1=3,N-2
      KJ=MM*(K1-3)+J-3+1
      CALL AXYK(N,K1,ISTAX1,ISTAXN,AX,BX,AXK1,AXKN)
      PDD(IJ,KJ)=DX(I,K1)-(AXK1*DX(I,2)+AXKN*DX(I,N-1))/AXN
   30 CONTINUE
C
C COMPUTE C3 USED IN EQUATION 7.50 AND ITS CONTRIBUTION TO THE
C MATRIX OF EQUATION SYSTEM 7.50
C
      DO 40 K2=3,M-2
      IK=MM*(I-3)+K2-3+1
      CALL AXYK(M,K2,ISTAY1,ISTAYM,AY,BY,AYK1,AYKM)
      PDD(IJ,IK)=PDD(IJ,IK)+RABDA**4*(DY(J,K2)-(AYK1*DY(J,2)+
     1           AYKM*DY(J,M-1))/AYM)
   40 CONTINUE
C
C COMPUTE C2 USED IN EQUATION 7.50 AND ITS CONTRIBUTION TO THE
C MATRIX OF EQUATION SYSTEM 7.50
C
      DO 50 K1=3,N-2
      DO 50 K2=3,M-2
      IK=MM*(K1-3)+K2-3+1
      CALL AXYK(N,K1,ISTAX1,ISTAXN,AX,BX,AXK1,AXKN)
      CALL AXYK(M,K2,ISTAY1,ISTAYM,AY,BY,AYK1,AYKM)
      PDD(IJ,IK)=PDD(IJ,IK)+2.0*RABDA**2*(BX(I,K1)*BY(J,K2)-
     1           (AXK1*BX(I,2)*(BY(J,K2)-BY(J,2)*AYK1/AYM)+
     2            AXKN*BX(I,N-1)*(BY(J,K2)-BY(J,2)*AYK1/AYM))/AXN-
     3           (AYK1*BY(J,2)*BX(I,K1)+
     4            AYKM*BY(J,M-1)*(BX(I,K1)-(BX(I,2)*AXK1+
     5            BX(I,N-1)*AXKN)/AXN))/AYM)
   50 CONTINUE
   10 CONTINUE
      RETURN
      END
```

```
C
C THIS SUBROUTINE IS TO COMPUTE AXN USED IN EQUATION 7.47 AND
C AYM USED IN EQUATION 7.48
C
      SUBROUTINE AXYNM(N,ISTAX1,ISTAXN,AX,BX,AXN)
      IMPLICIT REAL*8 (A-H,O-Z)
      DIMENSION AX(50,50),BX(50,50)
      IF(ISTAX1.EQ.2) THEN
        IF(ISTAXN.EQ.2) THEN
          AXN=AX(1,2)*AX(N,N-1)-AX(N,2)*AX(1,N-1)
        ELSE
          AXN=AX(1,2)*BX(N,N-1)-BX(N,2)*AX(1,N-1)
        END IF
      ELSE
        IF(ISTAXN.EQ.2) THEN
          AXN=BX(1,2)*AX(N,N-1)-AX(N,2)*BX(1,N-1)
        ELSE
          AXN=BX(1,2)*BX(N,N-1)-BX(N,2)*BX(1,N-1)
        END IF
      END IF
      RETURN
      END
C
C THIS SUBROUTINE IS TO COMPUTE AXK1 AND AXKN USED IN EQUATION
C 7.47 AND AYK1 AND AYKM USED IN EQUATION 7.48
C
      SUBROUTINE AXYK(N,K1,ISTAX1,ISTAXN,AX,BX,AXK1,AXKN)
      IMPLICIT REAL*8 (A-H,O-Z)
      DIMENSION AX(50,50),BX(50,50)
      IF(ISTAX1.EQ.2) THEN
        IF(ISTAXN.EQ.2) THEN
          AXK1=AX(1,K1)*AX(N,N-1)-AX(N,K1)*AX(1,N-1)
          AXKN=AX(N,K1)*AX(1,2)-AX(1,K1)*AX(N,2)
        ELSE
          AXK1=AX(1,K1)*BX(N,N-1)-BX(N,K1)*AX(1,N-1)
          AXKN=BX(N,K1)*AX(1,2)-AX(1,K1)*BX(N,2)
        END IF
      ELSE
        IF(ISTAXN.EQ.2) THEN
          AXK1=BX(1,K1)*AX(N,N-1)-AX(N,K1)*BX(1,N-1)
          AXKN=AX(N,K1)*BX(1,2)-BX(1,K1)*AX(N,2)
        ELSE
          AXK1=BX(1,K1)*BX(N,N-1)-BX(N,K1)*BX(1,N-1)
          AXKN=BX(N,K1)*BX(1,2)-BX(1,K1)*BX(N,2)
        END IF
      END IF
      RETURN
      END
C
C THIS SUBROUTINE IS TO RE-DISTRIBUTE THE EIGENVALUES FROM LOW
C TO HIGH. THE SUBROUTINE IS USED FOR OUTPUT
```

```
C
      SUBROUTINE EIGVMX(NMD,U,EIG,J)
      IMPLICIT REAL*8(A-H,O-Z)
      DIMENSION U(1401)
      EIG=U(1)
      DO 10 I=1,NMD
      IF(U(I).GE.EIG) GO TO 10
      EIG=U(I)
      J=I
   10 CONTINUE
      U(J)=1.0E30
      RETURN
      END
C
C CONVERT THE MATRIX [A] INTO A HESSENBERG MATRIX
C
      SUBROUTINE HBG(A,N)
      IMPLICIT REAL*8 (A-H,O-Z)
      DIMENSION A(1401,1401)
      DO 100 K=2,N-1
      D=0.0
      DO 10 J=K,N
      IF(ABS(A(J,K-1)).GT.ABS(D)) THEN
      D=A(J,K-1)
      I=J
      END IF
  10  CONTINUE
      IF(ABS(D)+1.0.NE.1.0) THEN
      IF(I.NE.K) THEN
      DO 20 J=K-1,N
      T=A(I,J)
      A(I,J)=A(K,J)
  20  A(K,J)=T
      DO 30 J=1,N
      T=A(J,I)
      A(J,I)=A(J,K)
  30  A(J,K)=T
      END IF
      DO 90 I=K+1,N
      T=A(I,K-1)/D
      A(I,K-1)=0.0
      DO 40 J=K,N
  40  A(I,J)=A(I,J)-T*A(K,J)
      DO 50 J=1,N
  50  A(J,K)=A(J,K)+T*A(J,I)
  90  CONTINUE
      END IF
 100  CONTINUE
      RETURN
      END
C
C USE STANDARD QR ALGORITHM TO COMPUTE THE EIGENVALUES
```

```
C OF A HESSENBERG MATRIX
C
      SUBROUTINE HQR(A,N,WR,WI)
      IMPLICIT REAL*8 (A-H,O-Z)
      DIMENSION A(1401,1401),WR(1401),WI(1401)
      ANORM=ABS(A(1,1))
      DO 12 I=2,N
      DO 11 J=I-1,N
      ANORM=ANORM+ABS(A(I,J))
11    CONTINUE
12    CONTINUE
      NN=N
      T=0.
1     IF(NN.GE.1)THEN
      ITS=0
2     DO 13 L=NN,2,-1
      S=ABS(A(L-1,L-1))+ABS(A(L,L))
      IF(S.EQ.0.)S=ANORM
      IF(ABS(A(L,L-1))+S.EQ.S)GO TO 3
13    CONTINUE
      L=1
3     X=A(NN,NN)
      IF(L.EQ.NN)THEN
      WR(NN)=X+T
      WI(NN)=0.
      NN=NN-1
      ELSE
      Y=A(NN-1,NN-1)
      W=A(NN,NN-1)*A(NN-1,NN)
      IF(L.EQ.NN-1)THEN
      P=0.5*(Y-X)
      Q=P**2+W
      Z=SQRT(ABS(Q))
      X=X+T
      IF(Q.GE.0.)THEN
      Z=P+SIGN(Z,P)
      WR(NN)=X+Z
      WR(NN-1)=WR(NN)
      IF(Z.NE.0.)WR(NN)=X-W/Z
      WI(NN)=0.
      WI(NN-1)=0.
      ELSE
      WR(NN)=X+P
      WR(NN-1)=WR(NN)
      WI(NN)=Z
      WI(NN-1)=-Z
      ENDIF
      NN=NN-2
      ELSE
      IF(ITS.EQ.30)PAUSE 'TOO MANY ITERATIONS'
      IF(ITS.EQ.10.OR.ITS.EQ.20)THEN
```

```
      T=T+X
      DO 14 I=1,NN
      A(I,I)=A(I,I)-X
14    CONTINUE
      S=ABS(A(NN,NN-1))+ABS(A(NN-1,NN-2))
      X=0.75*S
      Y=X
      W=-0.4375*S**2
      ENDIF
      ITS=ITS+1
      DO 15 M=NN-2,L,-1
      Z=A(M,M)
      R=X-Z
      S=Y-Z
      P=(R*S-W)/A(M+1,M)+A(M,M+1)
      Q=A(M+1,M+1)-Z-R-S
      R=A(M+2,M+1)
      S=ABS(P)+ABS(Q)+ABS(R)
      P=P/S
      Q=Q/S
      R=R/S
      IF(M.EQ.L)GO TO 4
      U=ABS(A(M,M-1))*(ABS(Q)+ABS(R))
      V=ABS(P)*(ABS(A(M-1,M-1))+ABS(Z)+ABS(A(M+1,M+1)))
      IF(U+V.EQ.V)GO TO 4
15    CONTINUE
4     DO 16 I=M+2,NN
      A(I,I-2)=0.
      IF (I.NE.M+2) A(I,I-3)=0.
16    CONTINUE
      DO 19 K=M,NN-1
      IF(K.NE.M)THEN
      P=A(K,K-1)
      Q=A(K+1,K-1)
      R=0.
      IF(K.NE.NN-1)R=A(K+2,K-1)
      X=ABS(P)+ABS(Q)+ABS(R)
      IF(X.NE.0.)THEN
      P=P/X
      Q=Q/X
      R=R/X
      ENDIF
      ENDIF
      S=SIGN(SQRT(P**2+Q**2+R**2),P)
      IF(S.NE.0.)THEN
      IF(K.EQ.M)THEN
      IF(L.NE.M)A(K,K-1)=-A(K,K-1)
      ELSE
      A(K,K-1)=-S*X
      ENDIF
      P=P+S
      X=P/S
```

```
      Y=Q/S
      Z=R/S
      Q=Q/P
      R=R/P
      DO 17 J=K,NN
      P=A(K,J)+Q*A(K+1,J)
      IF(K.NE.NN-1)THEN
      P=P+R*A(K+2,J)
      A(K+2,J)=A(K+2,J)-P*Z
      ENDIF
      A(K+1,J)=A(K+1,J)-P*Y
      A(K,J)=A(K,J)-P*X
17    CONTINUE
      DO 18 I=L,MIN(NN,K+3)
      P=X*A(I,K)+Y*A(I,K+1)
      IF(K.NE.NN-1)THEN
      P=P+Z*A(I,K+2)
      A(I,K+2)=A(I,K+2)-P*R
      ENDIF
      A(I,K+1)=A(I,K+1)-P*Q
      A(I,K)=A(I,K)-P
18    CONTINUE
      ENDIF
19    CONTINUE
      GO TO 2
      ENDIF
      ENDIF
      GO TO 1
      ENDIF
      RETURN
      END
```

Appendix C

A Fortran Program for L-Shaped Waveguide Analysis by Multi-domain DQ Method

This program is based on the algorithm shown in the Section 9.2.4. The governing equation for the waveguide analysis is Helmholtz equation which was described in Chapters 5 and 8. In the program, 3 subdomains are used to compute the cutoff wavenumbers of a *L*-shaped waveguide. In each subdomain, the same mesh size is used, and the PDQ method is applied to discretize the derivatives. The program can be used to compute the cutoff wavenumbers of *TM* and *TE* waves. For the *TM* wave, the boundary condition is

$$\phi = 0\,,\ \text{on the boundary} \tag{C.1}$$

while for the *TE* wave, the boundary condition is

$$\frac{\partial \phi}{\partial n} = 0\,,\ \text{on the boundary} \tag{C.2}$$

Here n is normal direction to the boundary. The derivative in Equation C.2 is also discretized by the PDQ method. Similar to the technique shown in the Section 7.3.3.3, the discrete boundary conditions are directly substituted into the discrete governing equations in each subdomain. Globally, the eigenvalue equation system is formed through the connection at the interface of neighbouring subdomains. In the program, the neighbouring subdomains are overlapped by one mesh spacing. Like the vibration analysis of plates, the standard QR algorithm is used to compute the eigenvalues of the resultant eigenvalue equation system.

Description of Variables

NX1 — Number of grid points in the x direction in subdomains 1 and 2 as shown in Figure 9.8

NX2 — Number of grid points in the x direction in subdomain 3 as shown in Figure 9.8

NY1 — Number of grid points in the y direction in subdomains 1 and 3 as shown in Figure 9.8

NY2 — Number of grid points in the y direction in subdomain 2 as shown in Figure 9.8

XL — Total length of waveguide in the x direction, which is a as shown in Figure 9.8

YL — Total length of waveguide in the y direction, which is b as shown in Figure 9.8

XI1 — Length of waveguide in the x direction in subdomains 1 and 2, which is c as shown in Figure 9.8

YI1 — Length of waveguide in the y direction in subdomains 1 and 3, which is d as shown in Figure 9.8

IST — Type of waves, which can be taken as 1 for *TM* mode, or 2 for *TE* mode

XX1(I) — x-coordinate of a mesh point in subdomains 1 and 2

XX2(I) — x-coordinate of a mesh point in subdomain 3

YY1(J) — y-coordinate of a mesh point in subdomains 1 and 3

YY2(J) — y-coordinate of a mesh point in subdomain 2

AX1(I,J) — Weighting coefficients of the first order derivative in the x direction, which are used in subdomains 1 and 2

AX2(I,J) — Weighting coefficients of the first order derivative in the x direction, which are used in subdomain 3

BX1(I,J) — Weighting coefficients of the second order derivative in the x direction, which are used in subdomains 1 and 2

BX2(I,J) — Weighting coefficients of the second order derivative in the x direction, which are used in subdomain 3

AY1(I,J) — Weighting coefficients of the first order derivative in the y direction, which are used in subdomains 1 and 3

AY2(I,J) — Weighting coefficients of the first order derivative in the y direction, which are used in subdomain 2

BY1(I,J) — Weighting coefficients of the second order derivative in the y direction, which are used in subdomains 1 and 3

BY2(I,J) — Weighting coefficients of the second order derivative in the y direction, which are used in subdomain 2

PDD(I,J) — Matrix of the Helmholtz eigenvalue equation system

U(I) — Real part of obtained eigenvalues, which is the wavenumber for the problem

V(I) — Imaginary part of obtained eigenvalues, which is zero for the problem

EIGOT(I) — Distribute the computed eigenvalues in sequence from low to high

```
C ANALYSIS OF L-SHAPED RECTANGULAR WAVEGUIDES BY MULTI-DOMAIN PDQ
C METHOD. THREE SUBDOMAINS ARE USED. THE NEIGHBOURING SUBDOMAINS
C ARE OVERLAPPED BY ONE MESH SPACING. THE PROGRAM CAN BE USED FOR
C ANALYSIS OF TM (DIRICHLET CONDITION) AND TE (NEUMANN CONDITION)
```

```
C WAVES.
C
      PROGRAM MAIN
      IMPLICIT REAL*8(A-H,O-Z)
      DIMENSION AX1(50,50),BX1(50,50),AX2(50,50),BX2(50,50)
      DIMENSION AY1(50,50),BY1(50,50),AY2(50,50),BY2(50,50)
      DIMENSION XX1(50),YY1(50),XX2(50),YY2(50)
      DIMENSION PDD(1401,1401),U(1401),V(1401),EIGOT(1401),IEIG(1401)
C
C INPUT NUMBER OF MESH POINTS IN EACH SUBDOMAIN AND GEOMETRIC LENGTHS
C OF THE WAVEGUIDE
C
      WRITE(6,*) 'READ THE VALUES OF NX1,NX2,NY1,NY2,XL,XI1,YL,YI1'
      READ(5,*) NX1,NX2,NY1,NY2,XL,XI1,YL,YI1
C
C INPUT A PARAMETER TO INDICATE THE PROBLEM TO BE CONSIDERED (TM OR
TE)
C
      WRITE(6,*) 'READ THE VALUES OF IST,1 FOR TM MODE,2 FOR TE MODE'
      READ(5,*) IST
C
      PI=4.0*ATAN(1.0)
C
C MESH GENERATION FOR THREE SUBDOMAINS
C
      CALL GRID(NX1,NX2,NY1,NY2,N,M,XL,XI1,YL,YI1,XX1,XX2,YY1,YY2)
C
C COMPUTATION OF WEIGHTING COEFFICIENTS OF THE FIRST AND SECOND ORDER
C DERIVATIVES IN THE X DIRECTION FOR THREE SUBDOMAINS. THE SUBROUTINE
C WCOEF HAS BEEN SHOWN IN APPENDIX A
C
      CALL WCOEF(1,NX1,XX1,AX1,BX1)
      CALL WCOEF(1,NX2,XX2,AX2,BX2)
C
C COMPUTATION OF WEIGHTING COEFFICIENTS OF THE FIRST AND SECOND ORDER
C DERIVATIVES IN THE Y DIRECTION FOR THREE SUBDOMAINS
C
      CALL WCOEF(1,NY1,YY1,AY1,BY1)
      CALL WCOEF(1,NY2,YY2,AY2,BY2)
      RAMBD=XL/YL
C
C NM1 IS THE TOTAL NUMBER OF INTERIOR POINTS IN THE SUBDOMAINS 1 AND 2
C
      NM1=(M-2)*(NX1-2)
C
C NM2 IS THE NUMBER OF INTERIOR POINTS IN THE SUBDOMAIN 3
C
      NM2=(NY1-2)*(NX2-2)
      NMD=NM1+NM2
      IF(IST.EQ.1) THEN
C
C FORM THE MATRIX [PDD] OF HELMHOLTZ EIGENVALUE SYSTEM FOR TM WAVES
```

```
C
      CALL PDCO(N,NX1,NX2,M,NY1,NY2,RAMBD,AX1,BX1,AX2,
     1          BX2,AY1,BY1,AY2,BY2,PDD)
      ELSE
C
C FORM THE MATRIX [PDD] OF HELMHOLTZ EIGENVALUE SYSTEM FOR TE WAVES
C
      CALL PDCOE(N,NX1,NX2,M,NY1,NY2,RAMBD,AX1,BX1,AX2,
     1           BX2,AY1,BY1,AY2,BY2,PDD)
      END IF
C
C COMPUTE EIGENVALUES OF THE MATRIX [PDD] BY USING QR ALGORITHM.
C THE SUBROUTINES HBG(PDD,NMD), HQR(PDD,NMD,U,V) HAVE BEEN SHOWN
C IN APPENDIX B
C
      CALL HBG(PDD,NMD)
      CALL HQR(PDD,NMD,U,V)
C
C NORMALIZE THE OBTAINED EIGENVALUES
C
      RAU=XL
      DO 909 I=1,NMD
      U(I)=SQRT(ABS(U(I)))/RAU
 909  CONTINUE
C
C RE-DISTRIBUTE THE EIGENVALUES IN SEQUENCE. THE SUBROUTINE
C EIGVMX(NMD,U,EIGOT(I),IEIG(I)) HAS BEEN SHOWN IN APPENDIX B
C
      DO 905 I=1,NMD
      CALL EIGVMX(NMD,U,EIGOT(I),IEIG(I))
 905  CONTINUE
C
C OUTPUT EIGENVALUES IN SEQUENCE
C
      WRITE(6,*) 'READ THE NUMBER FOR OUTPUT EIGENVALUES,NEIG'
      READ(5,*) NEIG
      WRITE(6,902) (EIGOT(I),I=1,NEIG)
 902  FORMAT(1X,4F15.6)
      STOP
      END
C
C THIS SUBROUTINE IS TO GENERATE THE COORDINATES OF MESH POINTS
C
      SUBROUTINE GRID(NX1,NX2,NY1,NY2,N,M,XL,XI1,YL,YI1,
     1                XX1,XX2,YY1,YY2)
      IMPLICIT REAL*8(A-H,O-Z)
      DIMENSION XX1(50),YY1(50),XX2(50),YY2(50)
      DIMENSION XXI(50),YYI(50)
      PI=4.*ATAN(1.)
      XND1=XI1/XL
      XND2=(XL-XI1)/XL
```

```
      YND1=YI1/YL
      YND2=(YL-YI1)/YL
C
C GENERATE X-COORDINATES FOR SUBDOMAINS 1 AND 2
C
      DO 10 I=1,NX1
      ALFA=PI*FLOAT(I-1)/FLOAT(NX1-1)
      ALBA=0.5*(1.-COS(ALFA))
  10  XX1(I)=ALBA*XND1
C
C GENERATE X-COORDINATES FOR SUBDOMAIN 3. NOTE THAT ONE MESH SPACING
C IS OVERLAPPED BETWEEN XX1 AND XX2
C
      DO 12 I=1,NX2
      ALFA=PI*FLOAT(I-1)/FLOAT(NX2-1)
      ALBA=0.5*(1.-COS(ALFA))
  12  XXI(I)=XX1(NX1)+ALBA*XND2
      XX2(1)=XX1(NX1-1)
      DO 15 I=2,NX2+1
  15  XX2(I)=XXI(I-1)
C
C GENERATE Y-COORDINATES FOR SUBDOMAINS 1 AND 3
C
      DO 20 J=1,NY1
      ALFA=PI*FLOAT(J-1)/FLOAT(NY1-1)
      ALBA=0.5*(1.-COS(ALFA))
  20  YY1(J)=ALBA*YND1
C
C GENERATE Y-COORDINATES FOR SUBDOMAIN 2. NOTE THAT ONE MESH
C SPACING IS OVERLAPPED BETWEEN YY1 AND YY2
C
      DO 22 J=1,NY2
      ALFA=PI*FLOAT(J-1)/FLOAT(NY2-1)
      ALBA=0.5*(1.-COS(ALFA))
  22  YYI(J)=YY1(NY1)+ALBA*YND2
      YY2(1)=YY1(NY1-1)
      DO 25 J=2,NY2+1
  25  YY2(J)=YYI(J-1)
C
      N=NX1+NX2-1
      M=NY1+NY2-1
      NX2=NX2+1
      NY2=NY2+1
C
      WRITE(6,102)
 102  FORMAT(15X,'          HORIZONTAL COORDINATE                ')
      WRITE(6,100) (XX1(I),I=1,NX1)
      WRITE(6,100) (XX2(I),I=1,NX2)
 100  FORMAT(1X,6F12.6)
      WRITE(6,103)
 103  FORMAT(15X,'          LONGITUDINAL COORDINATE               ')
      WRITE(6,100) (YY1(J),J=1,NY1)
```

```
      WRITE(6,100) (YY2(J),J=1,NY2)
      RETURN
      END
C
C THIS SUBROUTINE IS TO FORM THE MATRIX OF HELMHOLTZ EIGENVALUE
C EQUATION SYSTEM FOR TM WAVES. THE CONTRIBUTION FROM THREE
C SUBDOMAINS IS CONSIDERED
C
      SUBROUTINE PDCO(N,NX1,NX2,M,NY1,NY2,RAMBD,AX1,BX1,AX2,
     1                BX2,AY1,BY1,AY2,BY2,PDD)
      IMPLICIT REAL*8(A-H,O-Z)
      DIMENSION AX1(50,50),BX1(50,50),AX2(50,50),BX2(50,50),
     1          AY1(50,50),BY1(50,50),AY2(50,50),BY2(50,50),
     2          PDD(1401,1401)

      NM1=(M-2)*(NX1-2)
      NM2=(NY1-2)*(NX2-2)
      NM=NM1+NM2

      DO 604 I=1,NM
      DO 604 J=1,NM
      PDD(I,J)=0.0
 604  CONTINUE
C
C CONTRIBUTION FROM SUBDOMAIN 1
C
      DO 10 I=2,NX1-1
      DO 10 J=2,NY1-1
      IJ=(M-2)*(I-2)+J-1
      DO 11 K2=2,NY1
      IK=(M-2)*(I-2)+K2-1
      PDD(IJ,IK)=RAMBD**2*BY1(J,K2)
   11 CONTINUE
      DO 14 K1=2,NX1-1
      KJ=(M-2)*(K1-2)+J-1
      PDD(IJ,KJ)=PDD(IJ,KJ)+BX1(I,K1)
   14 CONTINUE
C
C KJ IS IN THE SUBDOMAIN 3
C
      KJ=NM1+J-1
      PDD(IJ,KJ)=PDD(IJ,KJ)+BX1(I,NX1)
  10  CONTINUE
C
C CONTRIBUTION FROM SUBDOMAIN 2
C
      DO 110 I=2,NX1-1
      DO 110 J=2,NY2-1
      IJ=(M-2)*(I-2)+NY1+J-3
      DO 16 K2=1,NY2-1
      IK=(M-2)*(I-2)+NY1+K2-3
```

```
      PDD(IJ,IK)=RAMBD**2*BY2(J,K2)
   16 CONTINUE
      DO 12 K1=2,NX1-1
      KJ=(M-2)*(K1-2)+NY1+J-3
      PDD(IJ,KJ)=PDD(IJ,KJ)+BX1(I,K1)
 12   CONTINUE
 110  CONTINUE
C
C CONTRIBUTION FROM SUBDOMAIN 3
C
      DO 20 I=2,NX2-1
      DO 20 J=2,NY1-1
      IJ=NM1+(NY1-2)*(I-2)+J-1
      DO 21 K2=2,NY1-1
      IK=NM1+(NY1-2)*(I-2)+K2-1
      PDD(IJ,IK)=RAMBD**2*BY1(J,K2)
   21 CONTINUE
      DO 24 K1=2,NX2-1
      KJ=NM1+(NY1-2)*(K1-2)+J-1
      PDD(IJ,KJ)=PDD(IJ,KJ)+BX2(I,K1)
   24 CONTINUE
C
C KJ1 IS IN THE SUBDOMAIN 1
C
      KJ1=(M-2)*(NX1-3)+J-1
      PDD(IJ,KJ1)=PDD(IJ,KJ1)+BX2(I,1)
   20 CONTINUE
      RETURN
      END
C
C THIS SUBROUTINE IS TO FORM THE MATRIX OF HELMHOLTZ EIGENVALUE
C EQUATION SYSTEM FOR TE WAVES. THE CONTRIBUTION FROM THREE
C SUBDOMAINS IS CONSIDERED
C
      SUBROUTINE PDCOE(N,NX1,NX2,M,NY1,NY2,RAMBD,AX1,BX1,AX2,
     1                 BX2,AY1,BY1,AY2,BY2,PDD)
      IMPLICIT REAL*8(A-H,O-Z)
      DIMENSION AX1(50,50),BX1(50,50),AX2(50,50),BX2(50,50),
     1          AY1(50,50),BY1(50,50),AY2(50,50),BY2(50,50),
     2          PDD(1401,1401)
      NM1=(M-2)*(NX1-2)
      NM2=(NY1-2)*(NX2-2)
      NM=NM1+NM2
      DO 604 I=1,NM
      DO 604 J=1,NM
      PDD(I,J)=0.0
 604  CONTINUE
C
C CONTRIBUTION FROM SUBDOMAIN 1
C
      AXN=AX1(1,1)
      AYM=AY1(1,1)
```

```
      DO 10 I=2,NX1-1
      DO 10 J=2,NY1-1
      IJ=(M-2)*(I-2)+J-1
      DO 11 K2=2,NY1
      IK=(M-2)*(I-2)+K2-1
C
C THE DERIVATIVE CONDITION AT BOTTOM BOUNDARY IS DIRECTLY SUBSTITUTED
C INTO THE HELMHOLTZ EIGENVALUE EQUATION SYSTEM
C
      PDD(IJ,IK)=RAMBD**2*(BY1(J,K2)-BY1(J,1)*AY1(1,K2)/AYM)
   11 CONTINUE
      DO 14 K1=2,NX1-1
      KJ=(M-2)*(K1-2)+J-1
C
C THE DERIVATIVE CONDITION AT LEFT BOUNDARY IS DIRECTLY SUBSTITUTED
C INTO THE HELMHOLTZ EIGENVALUE EQUATION SYSTEM
C
      PDD(IJ,KJ)=PDD(IJ,KJ)+BX1(I,K1)-BX1(I,1)*AX1(1,K1)/AXN
   14 CONTINUE
C
C KJ IS IN THE SUBDOMAIN 3
C
      KJ=NM1+J-1
      PDD(IJ,KJ)=PDD(IJ,KJ)+BX1(I,NX1)-BX1(I,1)*AX1(1,NX1)/AXN
  10  CONTINUE
C
C CONTRIBUTION FROM SUBDOMAIN 2
C
      AXN=AX1(1,1)*AX1(NX1,NX1)-AX1(NX1,1)*AX1(1,NX1)
      AYM=AY2(NY2,NY2)
      DO 110 I=2,NX1-1
      DO 110 J=2,NY2-1
      IJ=(M-2)*(I-2)+NY1+J-3
      DO 16 K2=1,NY2-1
      IK=(M-2)*(I-2)+NY1+K2-3
C
C THE DERIVATIVE CONDITION AT TOP BOUNDARY IS DIRECTLY SUBSTITUTED
C INTO THE HELMHOLTZ EIGENVALUE EQUATION SYSTEM
C
      PDD(IJ,IK)=RAMBD**2*(BY2(J,K2)-BY2(J,NY2)*AY2(NY2,K2)/AYM)
   16 CONTINUE
      DO 12 K1=2,NX1-1
      KJ=(M-2)*(K1-2)+NY1+J-3
C
C THE DERIVATIVE CONDITION AT LEFT AND RIGHT BOUNDARIES IS DIRECTLY
C SUBSTITUTED INTO THE HELMHOLTZ EIGENVALUE EQUATION SYSTEM
C
      AXK1=AX1(1,K1)*AX1(NX1,NX1)-AX1(NX1,K1)*AX1(1,NX1)
      AXKN=AX1(NX1,K1)*AX1(1,1)-AX1(1,K1)*AX1(NX1,1)
      PDD(IJ,KJ)=PDD(IJ,KJ)+BX1(I,K1)-(BX1(I,1)*AXK1+
     1           BX1(I,NX1)*AXKN)/AXN
```

```
 12    CONTINUE
 110   CONTINUE
C
C CONTRIBUTION FROM SUBDOMAIN 3
C
       AXN=AX2(NX2,NX2)
       AYM=AY1(1,1)*AY1(NY1,NY1)-AY1(NY1,1)*AY1(1,NY1)
       DO 20 I=2,NX2-1
       DO 20 J=2,NY1-1
       IJ=NM1+(NY1-2)*(I-2)+J-1
       DO 21 K2=2,NY1-1
       IK=NM1+(NY1-2)*(I-2)+K2-1
C
C THE DERIVATIVE CONDITION AT BOTTOM AND TOP BOUNDARIES IS DIRECTLY
C SUBSTITUTED INTO THE HELMHOLTZ EIGENVALUE EQUATION SYSTEM
C
       AYK1=AY1(1,K2)*AY1(NY1,NY1)-AY1(NY1,K2)*AY1(1,NY1)
       AYKM=AY1(NY1,K2)*AY1(1,1)-AY1(1,K2)*AY1(NY1,1)
       PDD(IJ,IK)=RAMBD**2*(BY1(J,K2)-(BY1(J,1)*AYK1+
     1             BY1(J,NY1)*AYKM)/AYM)
   21 CONTINUE
       DO 24 K1=2,NX2-1
       KJ=NM1+(NY1-2)*(K1-2)+J-1
C
C THE DERIVATIVE CONDITION AT RIGHT BOUNDARY IS DIRECTLY
C SUBSTITUTED INTO THE HELMHOLTZ EIGENVALUE EQUATION SYSTEM
C
       PDD(IJ,KJ)=PDD(IJ,KJ)+BX2(I,K1)-BX2(I,NX2)*AX2(NX2,K1)/AXN
   24 CONTINUE
C
C KJ1 IS IN THE SUBDOMAIN 1
C
       KJ1=(M-2)*(NX1-3)+J-1
       PDD(IJ,KJ1)=PDD(IJ,KJ1)+BX2(I,1)-BX2(I,NX2)*AX2(NX2,1)/AXN
   20 CONTINUE
       RETURN
       END
```

References

Anderson JD (1995) Computational fluid dynamics: the basic with applications. McGraw-Hill, Inc

Bar-Lev M, and Yang HT (1975) Initial flow field over an impulsively started circular cylinder. J Fluid Mech, Vol 72, pp 625-647

Bartels RH, Stewart GW (1972) Solution of the matrix equation AX+XB=C. Comm ACM, Vol 15, pp 820-826

Bellman RE, Casti J (1971) Differential quadrature and long-term integration. J Math Anal Appl, Vol 34, pp 235-238

Bellman RE, Kashef BG, Casti J (1972) Differential quadrature: a technique for the rapid solution of nonlinear partial differential equations. J Comput Phys, Vol 10, pp 40-52

Bellman RE, Kashef BG, Lee ES, Vasudevan R (1975*a*) Solving hard problems by easy methods: differential and integral quadrature. Comp & Math with Appl, Vol 1, pp 133-143

Bellman RE, Kashef BG, Lee ES, Vasudevan R (1975*b*) Differential quadrature and splines. Comp & Math with Appl, Vol 1, pp 371-376

Bellman RE, Kashef BG, Vasudevan R (1974) The inverse problem of estimating heart parameters from cardiograms. Math Biosci, Vol 19, pp 221-230

Bellman RE, Roth RS (1986) Methods in approximation: techniques for mathematical modeling. D Reidel Publishing Company, Dordrecht, Holland

Bellomo N (1997) Nonlinear models and problems in applied sciences from differential quadrature to generalized collocation methods. Math Comput Modelling, Vol 26, Iss 4, pp 13-34

Bert CW, Jang SK, Striz AG (1988) Two new approximate methods for analyzing free vibration of structural components. AIAA J, Vol 26, pp 612-618

Bert CW, Jang SK, Striz AG (1989) Nonlinear bending analysis of orthotropic rectangular plates by the method of differential quadrature. Comput Mech, Vol 5, pp 217-226

Bert CW, Kang KJ (1996) Stress analysis of closely-coiled helical springs using differential quadrature. Mech Res Commun, Vol 23, Iss 6, pp 589-598

Bert CW, Malik M (1996*a*) The differential quadrature method for irregular domains and application to plate vibration. Int J Mech Sci, Vol 38, Iss 6, pp 589-606

Bert CW, Malik M (1996*b*) Semianalytical differential quadrature solution for free vibration analysis of rectangular plates. AIAA J, Vol 34, Iss 3, pp 601-606

Bert CW, Malik M (1996*c*) Free vibration analysis of tapered rectangular plates by differential quadrature method: a semi-analytical approach. J Sound Vib, Vol 190, Iss 1, pp 41-63

Bert CW, Malik M (1996*d*) Free vibration analysis of thin cylindrical shells by the differential quadrature method. J Pressure Vessel Technol, Vol 118, Iss 1, pp 1-12

Bert CW, Malik M (1996*e*) Differential quadrature method in computational mechanics. Appl Mech Rev, Vol 49, No 1, pp 1-28

Bert CW, Malik M (1997*a*) Differential quadrature: a powerful new technique for analysis of composite structures. Compos Struct, Vol 39, Iss 3-4, pp 179-189

Bert CW, Malik M (1997*b*) On the buckling characteristics of symmetrically laminated cross-ply plates. Mech Compos Mater Struct, Vol 4, Iss 1, pp 39-67

Bert CW, Malik M (1997*c*) Transient analysis of gas-lubricated journal bearing systems by differential quadrature. J Tribol-Trans ASME, Vol 119, Iss 1, pp 91-99

Bert CW, Wang X, Striz AG (1993) Differential quadrature for static and free-vibration analyses of anisotropic plates. Int J Solids Struct, Vol 30, Iss 13, pp 1737-1744

Bert CW, Wang X, Striz AG (1994*a*) Convergence of the DQ method in the analysis of anisotropic plates. J Sound Vib, Vol 170, Iss 1, pp 140-144

Bert CW, Wang X, Striz AG (1994*b*) Static and free vibrational analysis of beams and plates by differential quadrature method. ACTA Mech, Vol 102, Iss 1-4, pp 11-24

Björck AA and Pereyra V (1970) Solution of Vandermonde system of equations. Math Comput, Vol 24, pp 893-903

Blevins RD (1984) Formulas for natural frequency and mode shape. Robert E Krieger Publishing Co, Malabar, Florida

Bodonyi RJ, Stewartson K (1977) The unsteady laminar boundary layer on a rotating disk in a counter-rotating fluid. J Fluid Mech, Vol 79, pp 669-688

Bredif M (1984) Calculation of laminar flow over a step by a finite element method based on the stream function-vorticity formulation. Notes on Numerical Fluid Mechanics, Vol 9, Vieweg, Braunschweig

Canuto C, Hussaini MY, Quarteroni A, Zang TA (1988) Spectral Methods in Fluid Dynamics. Springer, Berlin.

Cebeci T (1975) Calculation of three-dimensional boundary layers, II. three-dimensional flows in Cartesian coordinates. AIAA J, Vol 13, pp 1056-1064

Cebeci T (1979) The laminar boundary layer on a circular cylinder started impulsively from rest. J Comput Phys, Vol 31, pp 153-172

Cebeci T (1986) Unsteady boundary layers with an intelligent numerical scheme. J Fluid Mech, Vol 163, pp 129-140

Chang CT, Tsai CS, Lin TT (1993) The modified differential quadratures and their applications. Chem Eng Commun, Vol 123, pp 135-164

Chen CN (1997) The two-dimensional frame model of the differential quadrature element method. Comput Struct, Vol 62, Iss 3, pp 555-571

Chen CN (1998*a*) Solution of beam on elastic foundation by DQEM. J Eng Mech-Asce, Vol 124, Iss 12, pp 1381-1384

Chen CN (1998*b*) The warping torsion bar model of the differential quadrature element method. Comput Struct, Vol 66, Iss 2-3, pp 249-257

Chen CN (1999*a*) The development of irregular elements for differential quadrature element method: steady-state heat conduction analysis. Comput Method Appl Mech Eng, Vol 170, Iss 1-2, pp 1-14

Chen CN (1999*b*) The differential quadrature element method: irregular element torsion analysis model. Appl Math Mode, Vol 23, Iss 4, pp 309-328

Chen W (1996) Differential quadrature method and its applications in Engineering. PhD dissertation, Shanghai Jiao Tong University, China

Chen W, Wang X, Zhong T (1996) The structure of weighting coefficient matrices of harmonic differential quadrature and its applications. Commun Numer Methods Eng, Vol 12, Iss 8, pp 455-459

Chen W, Yu Y (1993*a*) Calculation and analysis of weighting coefficient matrices in differential quadrature. Proc of the 1st Pan-Pacific Conference on Computational Engineering, Seoul, Korea, 1-5 Nov., 1993, Elsevier Science Publ B V, Netherlands, pp 157-162

Chen W, Yu Y (1993*b*) Differential quadrature method for high order boundary value problems. Proc of the 1st Pan-Pacific Conference on Computational Engineering, Seoul, Korea, 1-5 Nov., 1993, Elsevier Science Publ B V, Netherlands, pp163-168

Chen W, Zhong TX, He WX (1997) A note on the DQ analysis of anisotropic plates. J Sound Vib, Vol 204, Iss 1, pp 180-182

Chen W, Zhong TX, Shu C (1998) A Lyapunov formulation for efficient solution of the Poisson and convection-diffusion equations by the differential quadrature method. J Comput Phys, Vol 141, Iss 1, pp 78-84

Chen WL, Striz AG, Bert CW (1997) A new approach to the differential quadrature method for fourth-order equations. Int J Numer Method Eng, Vol 40, Iss 11, pp 1941-1956

Civan F (1989) Differential cubature for multidimensional problems. Proc of the 20th Annual Pittsburgh Conference on Modeling and Simulation, Vogt WG and Mickle MH (eds), Instrument Society of American, North Carolina, pp 1843-1847.

Civan F (1991) Quadrature and cubature methods for numerical solution of integro-differential equations. Haji-Sheikh A, Corduneanu C, Fry JL, Huang T and Payne FR (eds), Integral Methods in Science and Engineering, Vol 90, pp 282-297, Hemisphere Publishing, NY

Civan F (1994) Rapid and accurate solution of reactor models by the quadrature method. Comput in Chemical Eng, Vol 18, pp 1005-1009

Civan F, Sliepcevich CM (1983*a*) Application of differential quadrature to transport processes. J Math Anal Appl, Vol 93, pp 206-221

Civan F, Sliepcevich CM (1983*b*) Solution of the Poisson equation by differential quadrature. Int J Numer Methods Eng, Vol 19, pp 711-724

Civan F, Sliepcevich CM (1984*a*) On the solution of the Thomas-Fermi equation by differential quadrature. J Comput Phys, Vol 56, pp 343-348

Civan F, Sliepcevich CM (1984*b*) Differential quadrature for multi-dimensional problems. J Math Anal Appl, Vol 101, pp 423-443

Coutanceau M, Bouard R (1977) Experimental determination of the main features of the viscous flow in the wake of a circular cylinder in uniform translation, Part I: steady flow. J Fluid Mech, Vol 79, pp 231-256

Datta L, Morgera SD (1989) On the reducibility of centrosymmetric matrices applications in engineering problems. CSSP, Vol 8, pp 71-96

Dennis SCR, Chang G (1970) Numerical solution for steady flow past a circular cylinder at Reynolds numbers up to 100. J Fluid Mech, Vol 42, pp 471-489

Dong DC, Leong MS, Kooi PS, Lam KY, Shu C (1997) Computation of the propagation characteristics of TE and TM modes in waveguides with the use of the generalized differential quadrature method. Microwave Opt Technol Lett, Vol 14, Iss 1, pp 39-44

Du H, Liew KM, Lim MK (1996) Generalized differential quadrature method for buckling analysis. J Eng Mech-Asce, Vol 122, Iss 2, pp 95-100

Du H, Lim MK, Lin RM (1994) Application of generalized differential quadrature method to structural problems. Int J Numer Method Eng, Vol 37, Iss 11, pp 1881-1896

Du H, Lim MK, Lin RM (1995) Application of generalized differential quadrature to vibration analysis. J Sound Vib, Vol 181, Iss 2, pp 279-293

Farsa J, Kukreti AR, Bert CW (1993*a*) Fundamental-frequency analysis of laminated rectangular-plates by differential quadrature method. Int J Numer Method Eng, Vol 36, Iss 14, pp 2341-2356

Farsa J, Kukreti AR, Bert CW (1993*b*) Fundamental-frequency analysis of single specially orthotropic, generally orthotropic and anisotropic rectangular layered plates by the differential quadrature method. Comput Struct, Vol 46, Iss 3, pp 465-477

Fornberg B (1996) A practical guide to pseudospectral methods. Cambridge University Press

Ghia U, Ghia KN, Shin CT (1982) High-Re solutions for incompressible flow using the Navier-Stokes equations and a multi-grid method. J Comput Phys, Vol 48, 387-411

Good IJ (1970) The inverse of a centrosymmetric matrix. Technometrics, Vol 12, pp 925-928.

Gottlieb D, Orszag SA (1977) Numerical analysis of spectral methods: theory and applications. SIAM-CBMS, Philadelphia

Gresho PM, Chan ST, Lee RL, Upson CD (1984) A modified finite element method for solving the time-dependent, incompressible Navier-Stokes equations, Part II: applications. Int J Numer Methods Fluids, Vol 4, pp 619-640

Grotch SL (1969) The solution of a boundary value problem in reactor design using Galerkin's method. AIChE Jl, Vol 15, pp 463-465

Gu HZ, Wang XW (1997) On the free vibration analysis of circular plates with stepped thickness over a concentric region by the differential quadrature element method. J Sound Vib, Vol 202, Iss 3, pp 452-459

Guan JM, Su CC (1995) Analysis of metallic waveguides with rectangular boundaries by using the finite-difference method and the simultaneous iteration with the chebyshev acceleration. IEEE Trans Microwave Theory Tech, Vol 43, pp 374-382

Gutierrez RH, Laura PAA (1994) Solution of the Helmholtz-equation in a parallelogrammic domain with mixed boundary-conditions using the differential quadrature method. J Sound Vib, Vol 178, Iss 2, pp 269-271

Gutierrez RH, Laura PAA (1995*a*) Vibrations of nonuniform rings studied by means of the differential quadrature method. J Sound Vib, Vol 185, Iss 3, pp 507-513

Gutierrez RH, Laura PAA (1995*b*) Analysis of vibrating, thin, rectangular-plates with point supports by the method of differential quadrature. Ocean Eng, Vol 22, Iss 1, pp 101-103

Gutierrez RH, Laura PAA (1999) Use of the differential quadrature method when dealing with transverse vibrations of a rectangular plate subjected to a non-uniform stress distribution field. J Sound Vib, Vol 220, Iss 4, pp 765-769

Gutierrez RH, Laura PAA, Bambill DV, Jederlinic VA, Hodges DH (1998) Axisymmetric vibrations of solid circular and annular membranes with continuously varying density. J Sound Vib, Vol 212, Iss 4, pp 611-622

Gutierrez RH, Laura PAA, Rossi RE (1994) The method of differential quadrature and its application to the approximate solution of ocean engineering problems. Ocean Eng, Vol 21, Iss 1, pp 57-66

Gutierrez RH, Laura PAA, Sanzi HC, Elvira G (1995) Vibrations of a rectangular plate of nonuniform thickness partially embedded in a winkler medium. J Sound Vib, Vol 185, Iss 5, pp 910-914

Gutierrez RH, Romanelli E, Laura PAA (1996) Vibrations and elastic stability of thin circular plates with variable profile. J Sound Vib, Vol 195, Iss 3, pp 391-399

Han JB, Liew KM (1997*a*) Analysis of moderately thick circular plates using differential quadrature method. J Eng Mech-Asce, Vol 123, Iss 12, pp 1247-1252

Han JB, Liew KM (1997*b*) Numerical differential quadrature method for Reissner/Mindlin plates on two-parameter foundations. Int J Mech Sci, Vol 39, Iss 9, pp 977-989

Han JB, Liew KM (1997*c*) An eight-node curvilinear differential quadrature formulation for Reissner/Mindlin plates. Comput Method Appl Mech Eng, Vol 141, Iss 3-4, pp 265-280

Han JB, Liew KM (1998) Analysis of annular Reissner/Mindlin plates using differential quadrature method. Int J Mech Sci, Vol 40, Iss 7, pp 651-661

Han JB, Liew KM (1999) Axisymmetric free vibration of thick annular plates. Int J Mech Sci, Vol 41, Iss 9, pp 1089-1109

Hirsch C (1988) Numerical computation of internal and external flows, Vol I: fundamentals of numerical discretization. John Wiley & Sons, New York

Hu LC, Hu CR (1974) Identification of rate constenls by differential quadrature in partly measurable compartmental models. Math Biosci, Vol 21, pp 71-76

Hyman CJ (1956) Theory of approximation. Frederick Ungar Publishing Co, New York

Jang SK, Bert CW, Striz AG (1989) Application of differential quadrature to static analysis of structural components. Int J Numer Method Eng, Vol 28, Iss 3, pp 561-577

Kang KJ, Bert CW (1997) Flexural-torsional buckling analysis of arches with warping using DQM. Eng Struct, Vol 19, Iss 3, pp 247-254

Kang KJ, Bert CW, Striz AG (1995) Vibration analysis of shear-deformable circular arches by the differential quadrature method. J Sound Vib, Vol 183, Iss 2, pp 353-360

Kang KJ, Bert CW, Striz AG (1996*a*) Vibration and buckling analysis of circular arches using DQM. Comput Struct, Vol 60, Iss 1, pp 49-57

Kang KJ, Bert CW, Striz AG (1996*b*) Static analysis of a curved shaft subjected to end torques. Int J Solids Struct, Vol 33, Iss 11, pp 1587-1596

Kashef BG, Bellman RE (1974) Solution of the partial differential equation of the Hodgkin-Huxley model using differential quadrature. Math Biosci, Vol 19, pp 1-8

Keller HK, Cebeci T (1970) Accurate numerical methods for boundary layer flows, I. two-dimensional laminar flows. Lecture Notes in Physics, Vol 8, 92-100

Ku HC, Hirsh RS, Taylor TD (1987) A pseudospectral method for solution of the three-dimensional incompressible Navier-Stokes equations. J Comput Phys, Vol 70, pp 439-462

Kuehn TH, Glodstein RJ (1976) An experimental and theoretical study of natural convection in the annulus between horizontal concentric cylinders. J Fluid Mech, Vol 74, pp 695-719

Kueny JL, Binder G (1984) Viscous flow over backward facing steps: an experiment investigation. Notes on Numerical Fluid Mechanics, Vol 9, Vieweg, Braunschweig

Kukreti AR, Farsa J, Bert CW (1992) Fundamental-frequency of tapered plates by differential quadrature. J Eng Mech, Vol 118, Iss 6, pp 1221-1238

Kukreti AR, Farsa J, Bert CW (1993) Fundamental-frequency of tapered plates by differential quadrature – closure. J Eng Mech-Asce, Vol 119, Iss 12, pp 2534-2534

Kukreti AR, Farsa J, Bert CW (1996) Differential quadrature and Rayleigh-Ritz methods to determine the fundamental frequencies of simply supported rectangular plates with linearly varying thickness. J Sound Vib, Vol 189, Iss 1, pp 103-122

Kukreti AR, Siddiqi ZA (1997) Analysis of fluid storage tanks including foundation-superstructure interaction using differential quadrature method. Appl Math Model, Vol 21, Iss 4, pp 193-205

Lam SSE (1993) Application of the differential quadrature method to 2-dimensional problems with arbitrary geometry. Comput Struct, Vol 47, Iss 3, pp 459-464

Laura PAA, Gutierrez RH (1993) Analysis of vibrating Timoshenko beams using the method of differential quadrature. Shock Vib, Vol 1, pp 89-93

Laura PAA, Gutierrez RH (1994) Analysis of vibrating rectangular-plates with nonuniform boundary-conditions by using the differential quadrature method. J Sound Vib, Vol 173, Iss 5, pp 702-706

Laura PAA, Gutierrez RH (1995) Analysis of vibrating circular plates of nonuniform thickness by the method of differential quadrature. Ocean Eng, Vol 22, Iss 1, pp 97-100

Lee ES (1968) Quasilinearization and invariant imbedding. Academic Press, New York

Leissa AW (1973) The free vibration of rectangular plates. J Sound Vib, Vol 31, pp 257-293

Leissa AW, Narita Y (1984). Vibrations of completely free shallow shells of rectangular planform. J Sound Vib, Vol 96, pp 207-218

Li H, Lam KY (1998) Frequency characteristics of a thin rotating cylindrical shell using the generalized differential quadrature method. Int J Mech Sci, Vol 40, Iss 5, pp 443-459

Liakopoulos A (1988) Pseudospectral solutions of separated flows. AIAA Paper 88-3643-CP

Lienhard JH (1987) A heat transfer textbook. Prentice-Hall, Englewood Cliffs, NJ

Liew KM, Han JB (1997*a*) A four-node differential quadrature method for straight-sided quadrilateral Reissner/Mindlin plates. Commun Numer Methods Eng, Vol 13, Iss 2, pp 73-81

Liew KM, Han JB (1997*b*) Bending analysis of simply supported shear deformable skew plates. J Eng Mech-Asce, Vol 123, Iss 3, pp 214-221

Liew KM, Han JB (1998) Bending solution for thick plates with quadrangular boundary. J Eng Mech-Asce, Vol 124, Iss 1, pp 9-17

Liew KM, Han JB, Xiao ZM (1996) Differential quadrature method for thick symmetric cross-ply laminates with first-order shear flexibility. Int J Solids Struct, Vol 33, Iss 18, pp 2647-2658

Liew KM, Han JB, Xiao ZM (1997) Vibration analysis of circular mindlin plates using the differential quadrature method. J Sound Vib, Vol 205, Iss 5, pp 617-630

Liew KM, Han JB, Xiao ZM, Du H (1996) Differential quadrature method for Mindlin plates on Winkler foundations. Int J Mech Sci, Vol 38, Iss 4, pp 405-421

Liew KM, Liu FL (1997) Differential cubature method: a solution technique for Kirchhoff plates of arbitrary shape. Comput Method Appl M, Vol 145, pp 1-10

Liew KM, Teo TM (1998) Modeling via differential quadrature method: three-dimensional solutions for rectangular plates. Comput Method Appl Mech Eng, Vol 159, Iss 3-4, pp 369-381

Liew KM, Teo TM (1999) Three-dimensional vibration analysis of rectangular plates based on differential quadrature method. J Sound Vib, Vol 220, Iss 4, pp 577-599

Lin RM, Lim MK, Du H (1994*a*) Deflection of plates with nonlinear boundary supports using generalized differential quadrature. Comput Struct, Vol 53, Iss 4, pp 993-999

Lin RM, Lim MK, Du H (1994*b*) Large deflection analysis of plates under thermal loading. Comput Method Appl Mech Eng, Vol 117, Iss 3-4, pp 381-390

Liu FL, Liew KM (1998*a*) Static analysis of Reissner-Mindlin plates by differential quadrature element method. J Appl Mech, Vol 65, Iss 3, pp 705-710

Liu FL, Liew KM (1998*b*) Differential cubature method for static solutions of arbitrarily shaped thick plates. Int J Solids Struct, Vol 35, Iss 28-29, pp 3655-3674

Liu FL, Liew KM (1999) Vibration analysis of discontinuous Mindlin plates by differential quadrature element method. J Vib Acoust, Vol 121, Iss 2, pp 204-208

Loy CT, Lam KY, Li H, Liu GR (1999) Vibration of antisymmetric angle-ply laminated cylindrical panels with different boundary conditions. Quart J Mech Appl Math, Vol 52, pp 55-71

Loy CT, Lam KY, Shu C (1997) Analysis of cylindrical shells using generalized differential quadrature. Shock Vibration, Vol 4, Iss 3, pp 193-198

Malik M, Bert CW (1994) Differential quadrature solutions for steady-state incompressible and compressible lubrication problems. J Tribol-Trans ASME, Vol 116, Iss 2, pp 296-302

Malik M, Bert CW (1995) Differential quadrature analysis of free-vibration of symmetrical cross-ply laminates with shear deformation and rotatory inertia. Shock Vibration, Vol 2, Iss 4, pp 321-338

Malik M, Bert CW (1996) Implementing multiple boundary conditions in the DQ solution of higher-order PDE's: application to free vibration of plates. Int J Numer Method Eng, Vol 39, Iss 7, pp 1237-1258

Malik M, Civan F (1995) A comparative-study of differential quadrature and cubature methods vis-a-vis some conventional techniques in context of convection-diffusion-reaction problems. Chem Eng Sci, Vol 50, Iss 3, pp 531-547

Mansell G, Merryfield W, Shizgal B, Weinert U (1993) A comparison of differential quadrature methods for the solution of partial-differential equations. Comput Method Appl Mech Eng, Vol 104, Iss 3, pp 295-316

Mingle JO (1977) The method of differential quadrature for transient nonlinear diffusion. J Math Anal Appl, Vol 60, 559-569.

Mirfakhraei P, Redekop D (1998) Buckling of circular cylindrical shells by the differential quadrature method. Int J Pressure Vessels Piping, Vol 75, Iss 4, pp 347-353

Moradi S, Taheri F (1998) Differential quadrature approach for delamination buckling analysis of composites with shear deformation. AIAA J, Vol 36, Iss 10, pp 1869-1873

Moradi S, Taheri F (1999) Application of differential quadrature method to the delamination buckling of composite plates. Comput Struct, Vol 70, Iss 6, pp 615-623

Naadimuthu G, Bellman RE, Wang KM, Lee ES (1984) Differential quadrature and partial differential equations: some numerical results. J Math Anal Appl, Vol 98, pp 220-233

Napolitano M, Orlandi P (1985) Laminar flow in a complex geometry: a Comparison. Int J Numer Methods Fluids, Vol 5, pp 667-683

Ng TY, Li H, Lam KY, Loy CT (1999) Parametric instability of conical shells by the generalized differential quadrature method. Int J Numer Method Eng, Vol 44, Iss 6, pp 819-837

Pandey MD, Sherbourne AN (1991) Buckling of anisotropic composite plates under stress gradient. J Eng Mech-Asce, Vol 117, Iss 2, pp 260-275

Patankar SV, Spalding DB (1972) A calculation procedure for heat, mass and momentum transfer in three-dimensional parabolic flows. Int J Heat Transfer, Vol 15, pp 1787-1806

Pike J, Roe PL (1985) Accelerated convergence of Jameson's finite volume Euler scheme using Van Der Houwen integrators. Comput Fluids, Vol 13, pp 223-236

Quan JR, Chang CT (1989*a*) New insights in solving distributed system equations by the quadrature methods – I. Comput Chem Engrg, Vol 13, pp 779-788.

Quan JR, Chang CT (1989*b*) New insights in solving distributed system equations by the quadrature methods – II. Comput Chem Engrg, Vol 13, pp 1017-1024.

Raimondi AA, Boyd J (1958) A solution for the finite journal bearing and its application to analysis and design. ASLE Trans, Vol 1, pp 159-209

Roache P (1981) Scaling of high Reynolds number weakly separated channel flows. Symposium on Numerical and Physical Aspects of Aerodynamic Flows

Schroeder W (1994) The origin of spurious modes in numerical solutions of electromagnetic field eigenvalue problems. IEEE Trans Microwave Theory Tech, Vol 42, pp 644-653

Sherbourne AN, Pandey MD (1991) Differential quadrature method in the buckling analysis of beams and composite plates. Comput Struct, Vol 40, Iss 4, pp 903-913

Shu C (1991) Generalized differential-integral quadrature and application to the simulation of incompressible viscous flows including parallel computation. PhD thesis, Univ of Glasgow, UK

Shu C (1996*a*) Free vibration analysis of composite laminated conical shells by generalized differential quadrature. J Sound Vib, Vol 194, Iss 4, pp 587-604

Shu C (1996*b*) An efficient approach for free vibration analysis of conical shells. Int J Mech Sci, Vol 38, Iss 8-9, pp 935-949

Shu C (1999) Application of differential quadrature method to simulate natural convection in a concentric annulus. Int J Numer Methods Fluids, Vol 30, pp 977-993

Shu C, Chen W (1999) On optimal selection of interior points for applying discretized boundary conditions in DQ vibration analysis of beams and plates. J Sound Vib, Vol 222, Iss 2, pp 239-257

Shu C, Chew YT (1997) Fourier expansion-based differential quadrature and its application to Helmholtz eigenvalue problems. Commun Numer Methods Eng, Vol 13, Iss 8, pp 643-653

Shu C, Chew YT (1998) On the equivalence of generalized differential quadrature and highest order finite difference scheme. Comput Method Appl Mech Eng, Vol 155, Iss 3-4, pp 249-260

Shu C, Chew YT (1999) Application of multi-domain GDQ method to analysis of waveguides with rectangular boundaries. In: Kong JA (ed) Electromagnetic waves: PIER 21. EMW Publishing, Cambridge, Massachusetts, USA, pp 1-18

Shu C, Chew YT, Khoo BC, Yeo KS (1995) Application of GDQ scheme to simulate incompressible viscous flows around complex geometries. Mech Res Commun, Vol 22, pp 319-325

Shu C, Chew YT, Khoo BC, Yeo KS (1996) Solutions of three-dimensional boundary layer equations by global methods of generalized differential-integral quadrature. Int J Numer Method Heat Fl F, Vol 6, Iss 2, pp 61-75

Shu C, Chew YT, Liu Y (1997) An efficient approach for numerical simulation of flows in Czochralski crystal growth. J Cryst Growth, Vol 181, Iss 4, pp 427-436

Shu C, Chew YT, Liu Y (1998) Different interface approximations in multi-domain GDQ simulation of Czochralski bulk flows. Int J Numer Method Heat Fl F, Vol 8, Iss 4, pp 424-444

Shu C, Chew YT, Richards BE (1995) Generalized differential and integral quadrature and their application to solve boundary-layer equations. Int J Numer Method Fluid, Vol 21, Iss 9, pp 723-733

Shu C, Du H (1997*a*) Implementation of clamped and simply supported boundary conditions in the GDQ free vibration analysis of beams and plates. Int J Solids Struct, Vol 34, Iss 7, pp 819-835

Shu C, Du H (1997*b*) A generalized approach for implementing general boundary conditions in the GDQ free vibration analysis of plates. Int J Solids Struct, Vol 34, Iss 7, pp 837-846

Shu C, Du H (1997*c*) Free vibration analysis of laminated composite cylindrical shells by DQM. Compos Part B-Eng, Vol 28, Iss 3, pp 267-274

Shu C, Khoo BC, Chew YT, Yeo KS (1996) Numerical studies of unsteady boundary layer flows past an impulsively started circular cylinder by GDQ and GIQ approaches. Comput Method Appl Mech Eng, Vol 135, Iss 3-4, pp 229-241

Shu C, Khoo BC, Yeo KS (1994) Numerical solutions of incompressible Navier-Stokes equations by generalized differential quadrature. Finite Elem Anal Design, Vol 18, Iss 1-3, pp 83-97

Shu C, Khoo BC, Yeo KS, Chew YT (1994) Application of GDQ scheme to simulate natural convection in a square cavity. Int Commun Heat Mass Trans, Vol 21, Iss 6, pp 809-817

Shu C, Richards BE (1990) High resolution of natural convection in a square cavity by generalized differential quadrature. Proc of 3rd Conf on Adv in Numer Methods in Eng: Theory and Appl, Swansea, UK, Vol 2, pp 978-985

Shu C, Richards BE (1992*a*) Application of generalized differential quadrature to solve 2-dimensional incompressible Navier-Stokes equations. Int J Numer Method Fluid, Vol 15, Iss 7, pp 791-798

Shu C, Richards BE (1992*b*) Parallel simulation of incompressible viscous flows by generalized differential quadrature. Comput Syst Eng, Vol 3, pp 271-281

Shu C, Wang CM (1999) Treatment of mixed and nonuniform boundary conditions in GDQ vibration analysis of rectangular plates. Eng Struct, Vol 21, Iss 2, pp 125-134

Shu C, Xue H (1997) Explicit computation of weighting coefficients in the harmonic differential quadrature. J Sound Vib, Vol 204, Iss 3, pp 549-555

Shu C, Xue H (1998) Comparison of two approaches for implementing stream function boundary conditions in DQ simulation of natural convection in a square cavity. Int J Heat Fluid Flow, Vol 19, Iss 1, pp 59-68

Shu C, Xue H (1999) Solution of Helmholtz equation by differential quadrature method. Comput Method Appl M, Vol 175, Iss 1-2, pp 203-212

Shu C, Yeo KS, Yao Q (1998) On the performance of three iterative methods for solution of GDQ algebraic equations. Comput Method Appl Mech Eng, Vol 167, Iss 1-2, pp 1-15

Siddiqi ZA, Kukreti AR (1998) Analysis of eccentrically stiffened plates with mixed boundary conditions using differential quadrature method. Appl Math Model, Vol 22, Iss 4-5, pp 251-275

Streeter VL, Wylie EB, Bedford KW (1998) Fluid Mechanics. WCB/McGraw-Hill

Striz AG, Chen WL (1994) Application of the differential quadrature method to the driven cavity problem. Int J Nonlinear Mech, Vol 29, Iss 5, pp 665-670

Striz AG, Chen WL, Bert CW (1994) Static analysis of structures by the quadrature element method (QEM). Int J Solids Struct, Vol 31, Iss 20, pp 2807-2818

Striz AG, Chen WL, Bert CW (1997) Free vibration of plates by the high accuracy quadrature element method. J Sound Vib, Vol 202, Iss 5, pp 689-702

Striz AG, Jang SK, Bert CW (1988) Nonlinear bending analysis of thin circular plates by differential quadrature. Thin Wall Struct, Vol 6, Iss 1, pp 51-62

Striz AG, Wang X, Bert CW (1995) Harmonic differential quadrature method and applications to analysis of structural components. ACTA Mech, Vol 111, Iss 1-2, pp 85-94

Szeri AZ (1980) Tribology: friction lubrication and wear. McGraw-Hill, New York

Teo TM, Liew KM (1999) A differential quadrature procedure for three-dimensional buckling analysis of rectangular plates. Int J Solids Struct, Vol 36, Iss 8, pp 1149-1168

Tomasiello S (1998) Differential quadrature method: application to initial-boundary-value problems. J Sound Vib, Vol 218, Iss 4, pp 573-585

Tong L (1994) Free vibration of laminated conical shells including transverse shear deformation. Int J Solids Structures, Vol 31, pp 443

Tritton DJ (1959) Experiments on the flow past a circular cylinder at low Reynolds numbers. J Fluid Mech, Vol 16, No 4

Van Dommelen LL, Shen SF (1980) The spontaneous generation of the singularity in a separating laminar boundary layer. J Comput Phys, Vol 38, pp 125-140

Wang KM (1982) Solving the model of isothermal reactors with axial mixing by the differential quadrature method. Int J Numer Methods Eng, Vol 18, pp 111-118

Wang X, Bert CW (1993) A new approach in applying differential quadrature to static and free vibrational analyses of beams and plates. J Sound Vib, Vol 162, Iss 3, pp 566-572

Wang X, Bert CW, Striz AG (1993) Differential quadrature analysis of deflection, buckling, and free-vibration of beams and rectangular-plates. Comput Struct, Vol 48, Iss 3, pp 473-479

Wang X, Bert CW, Striz AG (1994) Buckling and vibration analysis of skew plates by the differential quadrature method. AIAA J, Vol 32, pp 886-889

Wang X, Striz AG, Bert CW (1993) Free-vibration analysis of annular plates by the DQ-method. J Sound Vib, Vol 164, Iss 1, pp 173-175

Wang X, Striz Ag, Bert CW (1994) Buckling and vibration analysis of skew plates by the differential quadrature method. AIAA J, Vol 32, Iss 4, pp 886-889

Wang X, Yang J, Xiao J (1995) On free-vibration analysis of circular annular plates with nonuniform thickness by the differential quadrature method. J Sound Vib, Vol 184, Iss 3, pp 547-551

Wang XW (1995) Differential quadrature for buckling analysis of laminated plates. Comput Struct, Vol 57, Iss 4, pp 715-719

Wang XW, Gu HZ (1997) Static analysis of frame structures by the differential quadrature element method. Int J Numer Meth Eng, Vol 40, Iss 4, pp 759-772

Wang XW, Wang YL, Chen RB (1998) Static and free vibrational analysis of rectangular plates by the differential quadrature element method. Commun Numer Meth En, Vol 14, Iss 12, pp 1133-1141

White FM (1999) Fluid mechanics. WCB/McGraw-Hill

Williamson JH (1980) Low-storage Runge-Kutta schemes. J Comput Phys, Vol 35, pp 48-56

Wu CP, Hung YC (1999) Asymptotic theory of laminated circular conical shells. Int J Eng Sci, Vol 37, Iss 8, pp 977-1005

Zhang QD, Khoo BC, Yeo KS (1997) A numerical study of the effect of free surface and water depth on the stability of wakes: use of GDQ formulation. Int J Numer Meth Fl, Vol 24, Iss 11, pp 1079-1090

Zhang YM, Mirfakhraei P, Xu B, Redekop D (1998) A computer program for the elastostatics of a toroidal shell using the differential quadrature method. Int J Pres Ves Pip, Vol 75, Iss 13, pp 919-929

Zhang SJ, Shen YC (1995) Eigenmode sequence for an elliptical waveguide with arbitrary ellipticity. IEEE Trans Microwave Theory Tech, Vol 43, pp 227

Zhong HZ (1998) Elastic torsional analysis of prismatic shafts by differential quadrature method. Commun Numer Meth En, Vol 14, Iss 3, pp 195-208

Zhong HZ, He YH (1998) Solution of Poisson and Laplace equations by quadrilateral quadrature element. Int J Solids Struct, Vol 35, Iss 21, pp 2805-2819

Index

Adams-Bashforth schemes, 126
anisotropic plates, 187, 198
antisymmetric cross-ply conical shell, 222
Approach 1, 178, 181-182, 184-185
Approach 2, 180-181, 184
average equivalent heat conductivity, 173
axial compressive load, 188
backward facing step, 245, 249-250
base polynomials, 29-34, 36, 55, 57, 268-269
base vector, 8, 10-11, 16, 19, 22-25, 29, 49, 56-57, 61, 70-71, 73, 75-77, 81-83, 86-87, 94, 99
beams, 110, 186-188, 200-201, 242-243
Bellman's approaches, 2-3, 26-28
bending analysis, 188, 194, 266
Bernoulli-Euler beam, 188
Blasius, 12, 279-280, 283
boundary layer, 11, 168, 174, 275, 277, 279-286
Boussinesq approximation, 156
Buckling, 186-188, 190, 197-198
buoyancy force, 156
Burgers equation, 143, 145
Cartesian coordinate system, 148, 156-157, 245
CDQ, 70, 83, 85, 87, 99-100, 107-110, 113, 119, 149-151
CDQ weighting coefficients
-first order derivative, 70-74, 78-80
-second order derivative, 74-75, 78-80
-third order derivative, 75-76, 78-80
centrosymmetric matrix, 101, 103
Chebyshev collocation method, 26, 52, 55
Chebyshev-Gauss-Lobatto grid, 62, 113, 168
Chebyshev grid, 113
Chebyshev polynomial, 17
chemical reactor, 228
circular cylinder, 170, 172, 277, 285-287
clamped, 110, 194-196, 201-202, 207, 262-263
collocation method, 44, 52, 251
column buckling analysis, 188
complete transformation, 261, 265
complex geometries, 2, 187, 254, 258
complex-valued function, 18
composite laminated, 187
computational sequence, 169, 181
concentric annulus, 170
conical shell, 187, 209-210, 219, 221, 223
conjugate gradient iteration, 139
continuity equation, 154-155, 159, 175-183, 185, 256, 275
continuous function, 13, 19
convection-diffusion operator, 119-120
convergence criterion, 146-147, 169, 181-182
convergence rate, 91, 147
coordinate transformation, 60, 63, 85, 90, 92, 187, 236, 245, 254, 256, 259-260, 264
cosine expansion-based differential quadrature, 83, 85, 91-93, 99, 108
cosine functions, 19, 93
curvilinear coordinate system, 187, 245, 254, 256, 258, 260-261
cutoff wavelength, 235, 238-239
cutoff wavenumber, 236, 238, 252
cylindrical shell, 187, 209-210
derivative approximation, 26, 38, 40-43, 46, 55, 91
determinant, 96-98, 100, 102-106

diagonal weighting coefficients, 62, 67-68, 76, 85, 87
differential cubature, 60, 266
differential-integral equation, 267
differential operator, 36, 111-112, 119, 123
differential quadrature, 2, 3, 5, 25, 93, 186
diffusion operator, 117-119
dimension of a linear vector space, 8
direct method, 124, 129, 134-135, 168, 190, 226, 233, 240, 274
direct substitution of boundary conditions into discrete governing equations, 194-196, 202-205
Dirichlet condition, 118-119, 128, 141, 150-151, 163-164, 166, 191, 194, 202
divergence formulation, 256
domain decomposition, 187, 251
DQ algebraic equations, 96, 128-129, 143, 145-147, 168
DQ discretization matrix, 96, 111, 125, 134
DQ eigenvalue problems, 143
DQ element method, 187
DQ-GIQ, 275-276, 279-281, 284-287
DQ-GIQ-FD, 285-287
DQ weighting coefficient matrices, 96-98, 100, 107, 109-110, 164-166, 191
eccentric sectorial plate, 264-265
eigenvalues, 2, 96, 103-104, 106, 111-120, 124-125, 133, 148-150, 152, 168, 186, 190
elliptical waveguide, 235-236, 238
energy equation, 154-156, 275
error analysis, 38, 41
error distribution, 38, 41-44
Euler backward scheme, 124
Euler forward scheme, 125-126
even function, 21, 70, 80
external distributed load, 188, 198
FDQ, 3, 8, 69, 71, 86, 91-93, 99-100, 107-110, 113-115, 118-120, 124, 143, 149-152, 169, 173, 189, 199, 234, 237, 239-241
FDQ weighting coefficients
-first order derivative, 86-91
-second order derivative, 86-91
-third order derivative, 86-91
field, 6-8, 10, 15-16, 154, 170-171, 176, 179-181, 184, 239, 286
finite difference, 1, 26, 44-45, 47-51, 65-66, 123, 147, 154, 184, 186, 228, 231-232, 235, 245, 247-248, 252, 275, 277-278, 280-281, 285-286
finite element, 1, 66, 152, 166, 186, 194, 232, 250-251, 259, 264
finite volume, 1, 154, 178
flat plate, 11, 281, 283
flexural rigidity, 188, 198
flexural vibration, 110, 265
fluid dynamics, 154
Fourier expansion-based differential quadrature, 3, 69, 93, 99, 108
Fourier series expansion, 3, 13, 18, 21, 23, 69-70, 81, 86, 91
free condition, 196
free corner condition, 207
free vibration, 187, 190, 196-198, 200, 204, 207, 222, 251, 264
free vibration analysis of
beams, 196-197
plates, 197-209, 251
shells, 209-223
function approximation, 3, 6, 11, 13, 25, 38, 41, 91
Gauss-Seidel iteration, 136
Gaussian elimination method, 130
Gaussian quadrature, 267
general approach, 29, 33-34, 188, 205, 207
general function, 18, 86
generalized integral quadrature, 268
generation of low order difference schemes, 65-68
Gibbs phenomenon, 92

GIQ, 267-268, 271, 273-276, 278, 280, 283-285
GIQ weighting coefficients, 274, 276
global method, 1, 2, 114, 286
harmonic function, 13, 241
heat conduction, 144, 224, 254
heat transfer, 224-225
Helmholtz equation, 1, 12, 148, 224, 235-236, 239, 252
high order derivative, 76, 260
Howarth, 275, 280
implicit scheme, 124
incompressible flow, 153-157, 159, 245, 249
incompressible Navier-Stokes equation, 134, 153, 159, 175, 181, 256
initial condition, 126, 143, 228, 230, 278, 283
instability, 66, 120, 286
integral domain, 4, 268, 270-271, 275
integral quadrature, 2-5, 26
interface topology, 246
irregular domain, 55, 60, 254
isoparametric shape functions, 264
isotropic conical shell, 221
iterative method, 124, 129, 134-135, 138, 147, 168-169, 190, 226, 233
J-OE iteration, 136, 139
Jacobi iteration, 135
Keller-box difference scheme, 285
Kronecker operator, 30, 71, 87
Lagrange interpolation polynomials, 25, 28-32, 34, 36, 59, 92, 268
Laplacian operator, 148, 155, 173, 256
Legendre polynomial, 2, 17-18, 25, 27-29
Levy-Lees coordinates, 276
lid-driven cavity flow, 170, 175, 182
linear independence, 8-9
linear operator, 10-11, 26, 30-32, 34, 71
linear vector space, 3, 6-11, 14, 19, 22-23, 25, 29-31, 33, 36, 56, 61, 69-70, 73, 75, 77, 81, 86, 91, 94, 268
low order methods, 1-2, 186, 232
low order Simpson's rule, 267
lower triangular matrix, 131, 136
L-shaped waveguide, 252-253
LU decomposition, 130-131, 134, 168, 170, 190, 240, 274
lubrication problem, 232
Lyapunov matrix, 129
matrix-diagonalization, 133-134
matrix multiplication method, 76, 81, 86, 110
mixed condition, 142
model equation, 119, 228
modification of weighting coefficient matrix, 191-194, 201-202
momentum equation, 153-157, 159, 178, 180-181
multi-dimensional, 26, 55, 62, 69, 92, 129, 266
multi-domain DQ method, 187, 245, 249-250, 252
natural convection, 156-161, 170, 224
natural wave number, 241
Navier-Stokes equation, 1, 119, 143, 153-154, 175, 256, 287
Neumann condition, 118-119, 141-142, 146, 149-150, 163-164, 175, 237
Newton interpolation polynomials, 29
Newtonian fluid, 1, 153
no-slip boundary condition, 171, 258
non-diagonal weighting coefficients, 62, 87, 90
non-dimensional, 12, 157, 188, 197, 224, 232, 277
non-homogeneous boundary conditions, 193
nonlinear differential operator, 124
non-orthogonal grid, 259
non-periodic problems, 90, 93
non-staggered grid, 154, 176, 178
non-uniform grid, 110, 120, 227
odd function, 23, 81

one-dimensional, 5, 26, 46, 48, 55, 60, 62, 69, 92, 95, 119, 140, 143, 151-152, 224, 234, 272-273, 279
ordinary differential equation, 96, 100, 111-112, 123, 125, 143-144, 168, 188, 231, 277-278, 280, 285
orthogonal transformation, 132-134
overlapped, 66, 246, 248-249, 251-252
parallel simulation, 245
partial differential equations, 1, 11, 123, 224
partial transformation, 262, 265
patched, 66, 246, 248-249, 251
PDQ, 3, 8, 25, 44, 48-53, 55, 59, 65-66, 69, 71, 73, 75, 77-78, 91-92, 98-100, 107-110, 113-115, 118-121, 124, 144-146, 149-152, 160, 168-170, 172-173, 182, 189, 196-197, 199, 207-208, 222-223, 234, 237, 239-242, 244, 250, 252, 259, 268-274, 280, 284-286
PDQ weighting coefficients
-first order derivative, 26-32
-second order derivative, 32-34
-high order derivative, 34-36
periodic problems, 90, 93, 151-152
physical domain, 2, 60, 63-64, 90, 92, 143, 170, 230, 245-246, 254, 264
plain journal bearing, 232, 234-235
plate deflection, 197
plates, 1, 110, 186-187, 197-198, 200-201, 254, 266
Poisson equation, 128, 145, 153
Poisson's ratio, 198, 211
polynomial approximation, 1, 3, 13, 25, 29, 33, 44-45, 49, 55, 59, 69, 244, 248
polynomial-based differential quadrature, 3, 25, 69, 91
polynomial series expansion, 12
pressure, 153, 159, 176, 178-182, 184, 228-230, 233-234, 284
pressure correction, 154, 176, 178-182
primitive variable form, 153, 159-160, 175, 181
principal material coordinates, 211
properties of a linear vector space, 8, 25, 30, 34, 36, 57, 71, 73, 75
pseudospectral method, 1, 184
Quan and Chang's approach, 29, 32
rank, 96-100
real-valued function, 18
rectangular domain, 12, 55, 61, 92, 128, 149, 170, 239-240, 246
rectangular plate, 187, 197, 199, 201, 203, 205
recurrence relationship, 3, 25, 32, 38, 58, 67, 75, 88, 109-110
regular domain, 55-56, 60, 93, 236, 245, 254, 262
relaxation factor, 134, 146-147
Reynolds number, 157, 168-169, 250, 258-259
Richardson extrapolation, 285
Richardson iteration, 135-136
Rolle's theorem, 40-42
round-off error, 122, 151
Runge-Kutta schemes, 126-127
scalar function, 96, 255
scalar multiplication, 7-8, 14-15, 19, 23, 29, 56
Schur-decomposition, 132-134
SDQ, 81, 85, 99-100, 107-110, 113, 119, 149-151
SDQ weighting coefficients
-first order derivative, 81-85
-second order derivative, 81-85
-third order derivative, 81-85
second order derivative, 3, 17, 25, 31-32, 37, 43, 68, 69-70, 73, 91, 98, 100, 128, 144, 146, 148, 165-166, 176, 193-194, 225, 233, 255
self-starting schemes, 126
shells, 186-187, 209
shell's coordinates, 211
shifted Legendre polynomial, 2, 25, 27
Shu's general approach, 29, 33-34
simply supported, 110, 191, 194-197, 201-202, 207, 219, 222, 251, 262

sine expansion-based differential quadrature, 85, 92-93, 99, 108
sine functions, 20, 93
skew centrosymmetric matrix, 101
skew symmetric vector, 102
solution procedures, 187
SOR iteration, 136, 138-139
spatial derivatives, 123, 176, 178, 182, 186, 231, 254, 261, 275, 278, 284-286
spurious results, 152
square matrix, 96-97
square plate, 207-208
stability condition, 111-113, 120, 125, 135
staggered grid, 154, 178
steady boundary layer, 279, 285
stream function, 12, 134, 153, 159, 162, 164, 169-170, 249, 257, 275-276, 279
structured grid, 26, 69
structural and vibration analysis, 186-188, 197, 245, 251, 254
structural mechanics problems, 186-187
subdomain, 245-249, 252
trial-and-error process, 147
trigonometric polynomials, 21
truncated Fourier series expansion, 19, 22-23, 91
truncation error, 2
two-dimensional, 7, 12-13, 55, 60, 62, 128, 145, 148, 153, 159-160, 175, 185, 236, 239, 256, 272, 275, 277, 279, 285
uniaxial load, 198
uniform grid, 62, 114, 120, 121
unsteady boundary layer, 277, 285
unstructured grid, 26
upper triangular matrix, 131, 136
Vandermonde form, 14, 27
vector addition, 7-8, 14-15, 19, 23, 29, 56
vibration analysis, 2, 186-189, 209, 260
submatrix, 97-100
summation form, 193-194, 201
symmetric vector, 102
Taylor series expansion, 1, 44-46, 48
temperature, 154-156, 160, 168-170, 224-228
temporal discretization, 111, 125
test function, 3, 25-28, 30, 32, 60-61, 90, 266
thin beam and plate, 187
thin conical shell, 214
thin triangular fin, 224
third order derivative, 12, 34, 75-76, 81-83, 85-87
three-dimensional, 55, 153, 175, 185, 187, 273, 275, 279, 281
time-dependent problem, 110, 123-124
time-independent problem, 124
time step size, 113, 116, 118, 124-125, 144, 169, 280
transformation matrix, 211

virtual storage, 2, 129-130, 134, 148, 159, 186, 207, 278
vorticity, 153, 159, 162, 168-170, 184, 249, 257, 259
vorticity-stream function formulation, 153, 159-160, 168, 175, 181, 249, 257
wave propagation, 235
waveguide analysis224, 235, 245, 252
Weierstrass' first theorem, 13
Weierstrass' second theorem, 19
weighted residuals, 1
weighting function, 1, 52
weighting coefficient matrix, 37, 99-100, 108-110, 165, 188, 192, 194, 201, 273-274
Young's modulus, 198, 211
δ-technique, 190-191, 200

9 781852 332099

MIX
Papier aus verantwortungsvollen Quellen
Paper from responsible sources
FSC® C105338

FSC
www.fsc.org

If you have any concerns about our products,
you can contact us on
ProductSafety@springernature.com

In case Publisher is established outside the EU,
the EU authorized representative is:
Springer Nature Customer Service Center GmbH
Europaplatz 3, 69115 Heidelberg, Germany

Printed by Libri Plureos GmbH
in Hamburg, Germany